Physical Foundations of Quantum Electronics

by *David Klyshko*

Physical Foundations of Quantum Electronics

by David Klyshko

Editors

Maria Chekhova

Lomonosov Moscow State University, Russia

Sergey Kulik

Lomonosov Moscow State University, Russia

NEW JERSEY • LONDON • SINGAPORE • BEIJING • SHANGHAI • HONG KONG • TAIPEI • CHENNAI

Published by

World Scientific Publishing Co. Pte. Ltd.

5 Toh Tuck Link, Singapore 596224

USA office: 27 Warren Street, Suite 401-402, Hackensack, NJ 07601

UK office: 57 Shelton Street, Covent Garden, London WC2H 9HE

British Library Cataloguing-in-Publication Data
A catalogue record for this book is available from the British Library.

PHYSICAL FOUNDATIONS OF QUANTUM ELECTRONICS BY DAVID KLYSHKO

ISBN-13 978-981-4324-50-2
ISBN-10 981-4324-50-7

Printed in Singapore.

Preface

This book belongs to the series of textbooks in electronics and radiophysics written at the Physics Department of Lomonosov Moscow State University. Similarly to the other books of this series [Migulin (1978); Vinogradova (1979)], it is written for undergraduate Physics students and aims at introducing the readers to the most general concepts, rules, and theoretical methods. The main focus is on the three directions in physical optics that appeared after the advent of lasers: *nonstationary interactions between light and matter* (Chapter 5), *optical anharmonicity of matter* (Chapter 6) and *quantum properties of light* (Chapter 7). The first four chapters describe the theoretical base of more traditional parts of quantum electronics. The book starts with a short review of the history of quantum electronics with its main concepts, ideas, and terms. Further, basic methods for describing the interaction of optical radiation with matter are considered, based on quantum transition probabilities (Chapter 2), the density matrix formalism (Chapter 3), and the linear dielectric susceptibility of matter (Chapter 4).

The author tried to combine a systematic approach with a more detailed insight into several interesting ideas and effects, such as, for instance, *superradiance* (Sec. 5.3), *phase conjugation* (Sec. 6.5), and *photon antibunching* (Sec. 7.6).

The reader is expected to know the basics of quantum mechanics and statistical physics; however, much attention is paid to explaining the notations used in the book. The author tried to gradually increase the presentation complexity within each section as well as within the whole book. Each section or chapter starts with a simplified qualitative picture of the phenomenon considered. More complicated sections providing additional information are marked by circles.

The book uses the Gaussian system of units, which is most common in quantum electronics; however, in the numerical estimates, energy and power are given in Joules and Watts.

A large number of general guides in quantum electronics have already been published [Klimontovich (1966); Zhabotinsky (1969); Bertin (1971); Fain (1972); Pantell (1969); Yariv (1989); Piekara (1973); Khanin (1975); Tarasov (1976); Loudon (2000); Apanasevich (1977); Maitland (1969); Svelto (2010); Strakhovskii (1979); Kaczmarek (1981); Tarasov (1981); Elyutin (1982)] at all levels of presentation, from popular books [Klimontovich (1966); Zhabotinsky (1969); Piekara (1973)] to fundamental monographs [Fain (1972); Khanin (1975); Apanasevich (1977)], and in many cases the reader will be referred to them. For instance, the present book does not consider the design and parameters of lasers and masers as well as their various applications. The theory of optical resonators and waveguides is presented, in particular, in the University course of wave theory [Vinogradova (1979)] (see also [Maitland (1969); Yariv (1976)]), while the self-oscillation theory, dynamics, and classical statistics of laser systems can be found in the textbooks on the oscillation theory [Migulin (1978)] and statistical radiophysics [Akhmanov (1981)] (see also [Khanin (1975); Rabinovich (1989)]).

The book is based on the lecture course in quantum electronics taught by the author to undergraduate students for 20 years. This course was started in 1960, after a suggestion by S. D. Gvozdover, even before the appearance of lasers. At first, the course was completely devoted to masers (paramagnetic amplifiers and molecular generators) and radio-spectroscopy. The advent of lasers and the 'laser revolution' in optics, spectroscopy and other fields of science made the author move the 'center of gravity' of the course from the microwave range to the optical one and supply the course with new sections. However, one should keep in mind that lasers and masers are based on common principles and that quantum electronics originated from radio spectroscopy and radiophysics. The latter provided quantum electronics with one of its basic notions, the feedback, and it is not by chance that the founders of quantum electronics and nonlinear optics, such as Basov, Bloembergen, Khokhlov, Prokhorov, Townes, and many others, worked in radiophysics. Sometimes quantum electronics is called 'quantum radiophysics'.

Both the 'Quantum Electronics' lecture course and this book were hugely influenced by Rem Viktorovich Khokhlov whose advice and friendship are unforgettable. The author is indebted to P. V. Elyutin, A. M. Fedorchenko and A. S. Chirkin, who have read the manuscript and helped to eliminate many flaws. The author is also grateful to V. B. Braginsky who stimulated the writing of this book.

D. N. Klyshko

Foreword

Below, we present the translation of a book by David Klyshko (1929–2000), which was originally published in 1986. This is a remarkable book by a remarkable person whose insight into physics in general and quantum electronics in particular was so deep that even now, after nearly 25 years, a lot of new ideas can still be found in this book. The main advantage of the book is that it generalizes seemingly unique effects and joins together seemingly different approaches. Because it is mainly at the boundaries of the explored that one should look for new ideas and discoveries, this book will be helpful for both a researcher and an ambitious student aiming at research in nonlinear optics, laser physics, quantum or atom optics.

Although some parts of the book look very new even now, others are definitely outdated. This statement relates not to the sections or even subsections of the book; rather, it is about numerous references to the technology or parameters of the equipment that were available when the book was written. This requires additional comments and explanations, which we have endeavored to make throughout the whole text, mostly as footnotes but sometimes as additional sections (Secs. 1.3, 7.2.10 and 7.5.7).

At the same time, we by no means think that the additional parts provide a complete view at the modern state of quantum electronics. For this reason, we have also included an additional list of references, containing books or review articles that appeared after the original book had been published.

Maria Chekhova
Sergey Kulik
The Editors

List of Notation and Acronyms

a, transverse size, cm; photon annihilation operator
A, area, cm^2; probability of spontaneous transition, s^{-1}; vector potential, $(erg/cm)^{1/2}$
B, scaling coefficient between the stimulated transition probability and the energy spectral density, $cm^3/(erg{\cdot}s^2)$
c, state amplitude
d, dipole moment, $(erg{\cdot}cm^3)^{1/2}$
D, electric induction, $(erg/cm^3)^{1/2}$
e, unit polarization vector
E, electric field, $(erg/cm^3)^{1/2}$
$\mathcal{E}$, energy, erg
f, frequency, s^{-1}, oscillator strength
F, photon flux density, $cm^{-2}{\cdot}s^{-1}$; free energy, erg
g, degeneracy; form factor, s
G, transfer coefficient, Green's function; field correlation function, erg/cm^3
H, magnetic field, $(erg/cm^3)^{1/2}$
$\mathcal{H}$, Hamiltonian, erg
I, intensity of radiation, $erg/(cm^2{\cdot}s)$; identity operator
j, current density, $erg/(cm^3{\cdot}s^2)^{1/2}$
k, wave vector, cm^{-1}
l, length, cm
n, refractive index
N, density of molecules or photons, cm^{-3}; number of photons per mode
N_i, population of a level, cm^{-3}
$\mathcal{N}$, mean number of photons per mode in equilibrium radiation
p, momentum, $g{\cdot}cm/s$; pressure, erg/cm^3

P, polarization, $(\text{erg/cm}^3)^{1/2}$; probability
$\mathcal{P}$, power, erg/s
q, generalized coordinate
Q, quality factor; generating function
r, radius vector, cm
R, Bloch vector; reflectivity coefficient
s, angular momentum, erg·s
S, Poynting vector, erg/(cm^2·s)
T, time interval, s; temperature, K
u, group velocity, cm/s
U, internal energy, erg; evolution operator
v, phase velocity, cm/s
V, volume, cm^3
$\mathcal{V}$, interaction energy, erg
w, relaxation transition probability per unit time, s^{-1}
W, transition probability per unit time, s^{-1}
Z, statistical sum
α, linear polarisability, cm^3; absorption or amplification coefficient, cm^{-1}
β, quadratic polarisability, $(\text{cm}^9/\text{erg})^{1/2}$
γ, cubic polarisability, cm^6/erg; dissipation constant, s^{-1}
Δ, relative population difference
ϵ, dielectric permittivity
η, quantum efficiency
ϑ, angle or angle of precession, rad
θ, Heaviside step function
κ, Boltzmann's constant, erg/K
λ, wavelength, cm; $\lambdabar = \lambda/2\pi$
μ, magnetic dipole moment, $(\text{erg}\cdot\text{cm}^3)^{1/2}$; Fermi level, erg
ν, polarization index; wavenumber, cm^{-1}
Π, operator of projection or summation over permutations
ρ, density operator or matrix; mass density, g/cm^3; charge density, $(\text{erg/cm}^5)^{1/2}$
σ, interaction cross-section, cm^2; Pauli matrix
τ, relaxation or correlation time, s
φ, phase or azimuthal angle, rad; eigenfunctions of the energy operator
$\chi^{(n)}$, n-th order susceptibility of the medium, $(\text{erg/cm}^3)^{(1-n)/2}=(\text{Hs})^{1-n}$
ψ, Ψ, wave function

ω, circular frequency, rad/s
Ω, Rabi frequency, rad/s; solid angle, sr
CARS, coherent anti-Stokes Raman scattering
CF, correlation function
EPR, electronic paramagnetic resonance
FDT, fluctuation-dissipation theorem
IR, infrared
MBS, Mandelshtam-Brillouin scattering
MW, microwave
NMR, nuclear magnetic resonance
OPO, optical parametric oscillator
PC, phase conjugation
PDC, parametric down-conversion
PMT, photomultiplier tube
SHG, second harmonic generation
SIT, self-induced transparency
SPDC, spontaneous parametric down-conversion
SRS, spontaneous Raman scattering
StRS, stimulated Raman scattering
StPDC, stimulated parametric down-conversion
StTS, stimulated temperature scattering
SVA, slowly varying amplitude
UV, ultraviolet

Contents

Chapter 1

Introduction

Quantum electronics studies the interaction of electromagnetic field with matter in various wavelength ranges, from radio to X-rays and gamma rays. Investigation of the basic laws of this interaction led to the creation of lasers, sources of coherent (i.e., monochromatic and directed) intense light. Optimization of the existing lasers and the development of new laser types, as well as advances in experimental technology, in their turn, stimulated further development of quantum electronics. This avalanche process, typical for modern science, led to new directions in optics (nonlinear and quantum optics, holography, optoelectronics) and spectroscopy (nonlinear and coherent spectroscopy), to numerous applications of lasers in technology, communications, medicine. We are probably close to solving the problem of laser thermonuclear fusion and laser isotope separation on an industrial scale.[a]

Not so diverse but also important applications were found by the 'elder brothers' of lasers, masers, which operate in the radio range, at wavelengths on the order of 0.1 – 10 cm, and are used as super-stable frequency etalons and super-sensitive paramagnetic amplifiers.

The term 'quantum electronics' appeared as a counterpart of classical electronics, mainly dealing with free electrons, which have continuous energy spectrum and, as a rule, are well described by classical mechanics. However, some essentially quantum devices, such as, for instance, the ones based on the Josephson junction, are traditionally not considered as part of quantum electronics. The other name, 'quantum radiophysics', is not quite appropriate either, since it does not relate to the optical frequency range.

[a]Editors' note: This opinion was quite common in the laser physics community at the time when the book was written. However, further investigations reduced the optimism in this field, and we are now still witnessing new attempts towards laser thermonuclear fusion (*inertial confinement fusion*).

1.1 Basic notions of quantum electronics

The operation of lasers and masers rests on 'the three whales', basic notions of quantum electronics — namely, *stimulated emission, population inversion*, and *feedback*.

1.1.1 *Stimulated emission*

Stimulated emission leads to the 'multiplication' of photons: a photon hitting an excited atom or molecule causes, with a probability W_{12}, the transition of the atom to one of its lower levels (Fig. 1.1). The released energy, $\mathcal{E}_2 - \mathcal{E}_1$, is transferred to the electromagnetic field in the form of the second photon. This other photon has the same parameters as the incident photon, i.e., energy $\hbar\omega = \mathcal{E}_2 - \mathcal{E}_1$, momentum $\mathbf{p} = \hbar\mathbf{k}$ and the same polarization type. Then, there are two indistinguishable photons, which can turn into four photons through the interaction with other excited atoms. In the classical language, this picture corresponds to the exponential amplification of the amplitude of a classical plane electromagnetic wave with frequency ω and wavevector $\mathbf{k}$.[b]

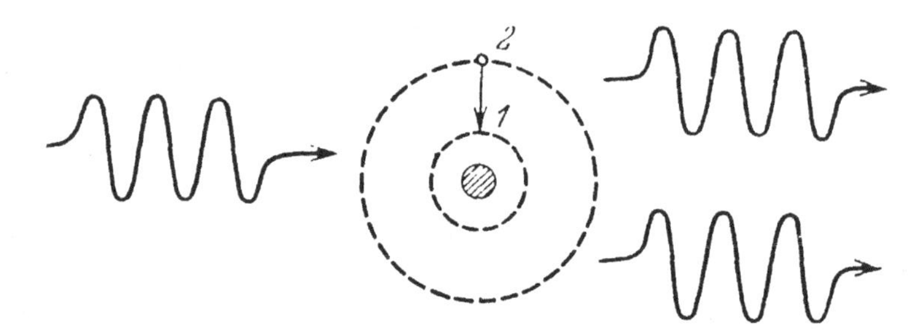

Fig. 1.1 Amplification of light under stimulated transitions. A resonant photon hits an excited atom, which then gives its stored energy to the field. As a result, the field contains two indistinguishable photons.

1.1.2 *Population inversion*

Interaction with atoms that are at the lower level, with the energy $\mathcal{E}_1$, occurs through the absorption of photons, i.e., attenuation of the electromagnetic wave. It is important that the probability W_{21} of this process (per one atom) is exactly

[b]See Editors' note in Sec. 2.5.3.

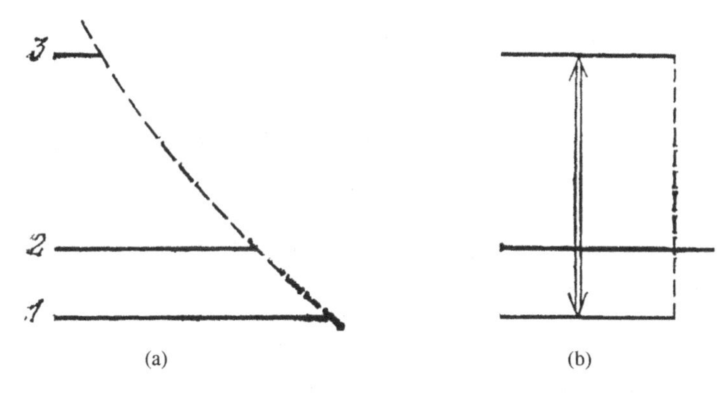

Fig. 1.2 Obtaining population inversion through optical pumping: (a) initial Boltzmann's population distribution; (b) strong resonant radiation balances the populations of levels 1 and 3, so that $N_2 > N_1$.

equal to the probability of stimulated emission, $W_{21} = W_{12}$, and therefore the overall effect depends on the difference of numbers of atoms at the levels 1 and 2, $\Delta N \equiv N_1 - N_2$. Usually, *populations* N_m of the levels are defined per unit volume.

If the matter is at thermodynamic equilibrium with a temperature T, then, according to Boltzmann's distribution, $N_m \propto \exp(-\mathcal{E}_m/\kappa T)$, with κ being the Boltzmann constant. Therefore, if $\mathcal{E}_2 > \mathcal{E}_1$, then $N_2 < N_1$ (Fig. 1.2(a)). As a result, stimulated 'up' transitions occur more frequently than stimulated 'down' transitions, and external electromagnetic radiation in equilibrium medium is attenuated. *Thus, in order to amplify field, the medium should be in a non-equilibrium state, with* $N_2 > N_1$. *One says that such a state has population inversion, or negative temperature.*

A lot of methods have been developed for achieving population inversion. The most important ones are pumping the medium (Fig. 1.2(b)) with auxiliary radiation (used for solid and liquid doped dielectrics), electric discharge in gases and injection in semiconductors.

1.1.3 *Feedback and the lasing condition*

In order to turn an amplifier into an oscillator, one should provide *positive feedback*, which can be realized using a pair of plane or concave spherical mirrors. (In masers, the active medium is placed into a microwave cavity.)

Amplification (or attenuation) can be quantitatively described as follows. Let F [s^{-1}·cm^{-2}] be the flux density of photons propagating along the z axis. The increment of F scales as the product of the stimulated transition probability per unit time, W, and the number of *active particles*, ΔN:

$$dF/dz = -W\Delta N. \tag{1.1}$$

In its turn, the stimulated transition probability scales as F,

$$W = \sigma F, \tag{1.2}$$

where σ [cm^2] is the probability of a transition per unit time for a photon flux with unit density. It is called the *interaction cross-section*. As a result,

$$dF/dz = -\sigma\Delta NF \equiv -\alpha F, \tag{1.3}$$

which leads to exponential intensity variation for a plane wave in matter (for $\alpha > 0$, it is called the *Bouguer law*):

$$F(z) = F(0)e^{-\alpha z}, \alpha \equiv \sigma\Delta N. \tag{1.4}$$

The parameter α is called *absorption* (at $\alpha > 0$) or *amplification* (at $\alpha < 0$) coefficient. Its inverse, α^{-1}, has the meaning of the mean free walk of a photon. The interaction cross section σ, in principle, can be as large as $3\lambda^2/2\pi$ ($\lambda = 2\pi c/\omega$ is the wavelength), so that in the optical range, where $\lambda \propto 10^{-4}$ cm, it is sometimes sufficient to have $\Delta N \propto 10^9$ cm^{-3} for noticeable amplification at a length of 1 cm.

Let the active medium of length l be placed between two mirrors (a Fabry–Perot interferometer) with reflection coefficients R_1, R_2. Then, from (1.4), the threshold condition of lasing is

$$R_1R_2e^{-2\alpha l} = 1. \tag{1.5}$$

For mirrors with dielectric coating, one can easily have $R \gtrsim 0.99$, and for lasing with $l = 10$ cm it is sufficient to have $\alpha = (\ln R)/l = -0.001$ cm^{-1}. Usually, the radiation is fed out from the laser by making one of the mirrors have lower reflection coefficient.

1.1.4 *Saturation and relaxation*

Let us consider some other important notions of quantum electronics. *Saturation* occurs when the populations of some pair of levels become equal ($N_1 = N_2$) due to stimulated transitions in a sufficiently intense external radiation. This effect restricts and stabilizes the intensity of quantum oscillators and the gain coefficient of quantum amplifiers. *Relaxation* processes counteract saturation and tend to restore the equilibrium Boltzmann distribution of populations, which is determined by the temperature of the thermostat. Relaxation processes determine the lifetimes of particles at different levels and the spectral linewidths.

Even in the absence of incident radiation or other external influence, an excited molecule can make a transition into one of the lower-energy states by emitting a photon. This kind of emission is called *spontaneous*. Spontaneous emission plays the role of a 'seed' for self-oscillations in quantum oscillators, restricts their stability, and creates noise in quantum amplifiers. Spontaneous and stimulated transitions in equilibrium matter lead to thermal radiation, which is described by Planck's formula and Kirchhoff's law.

It is important that while stimulated effects can be rather well calculated in the framework of classical electrodynamics with deterministic field amplitudes E, H, spontaneous effects are consistently described only by the laws of quantum statistical optics, where E and H are random values or operators.

The above-mentioned terms and notions relate to different fields of theoretical physics: quantum mechanics (energy levels, transition probabilities), statistical physics (relaxation, populations, fluctuations), oscillation theory (feedback, self-oscillations). Quantum electronics, as a field of physics, is remarkable and attractive because it uses theoretical and experimental tools from a diversity of fields, and also because it poses new problems for these fields and provides them with new experimental methods.

1.2 History of quantum electronics

Quantum electronics can be considered as a new chapter in the theory of light and, more generally, in the theory of the interaction between electromagnetic field and matter. The earliest chapters of this theory were devoted to the empirical description of normal dispersion of light in the transparency ranges of the matter, which was studied by Newton and his contemporaries more than 300 years ago. The next steps, made in the 19th century, were the study of anomalous dispersion within the absorption bands and the classical dispersion theory by Lorentz. The quantum era in optics and generally in physics started at the beginning of the 20th century from Planck's theory of equilibrium radiation, which led Einstein to the notion of photon, and from Bohr's postulates. Quantum theory of dispersion was formulated in the 1920s by Kramers and Heisenberg. Meanwhile, Dirac, Heisenberg, and Pauli developed quantum electrodynamics.

The history of quantum electronics, in its turn, is quite interesting and instructive [Dunskaya (1974)]. In principle, at the beginning of the 20th century the level of laboratory technique was high enough for building, for instance, a gas laser. However, this could not happen before the discovery of certain concepts and laws, which form the base of a quantum generator.

1.2.1 *First steps*

The first step along this way, which took several decades, was made in 1916 by A. Einstein who introduced the notions of stimulated emission and absorption. A quantitative theory of these effects was developed about ten years later by P. Dirac. From the theory, it followed that the photons generated via stimulated emission have all their parameters (energy, propagation direction, and polarization) the same as the ones of the incident photons. This property is called the *coherence* of stimulated emission.

The first experiments demonstrating stimulated emission were reported in 1928 by Ladenburg and Kopfermann. These experiments studied the refractive index dispersion for neon excited by electric discharge. (Note that in the first gas laser, which was built only 33 years later, neon was used as well.) In their paper, Ladenburg and Kopfermann have accurately formulated the condition of population inversion and the resulting necessity to selectively excite the atomic levels. In 1940, V. A. Fabrikant has pointed out, for the first time, that the intensity of light in a medium with population inversion should increase. (He considered this effect only as a proof for the existence of stimulated emission but not as a phenomenon that can have useful applications.) Unfortunately, this paper, as well as an application for an invention filed by V. A. Fabrikant and his colleagues in 1951, was not properly published in time and therefore did not influence further development of quantum electronics.

1.2.2 *Radio spectroscopy*

The first devices of quantum electronics, masers, which were later used in applications such as generation and amplification of waves in the centimeter range, were developed only in the middle of the 1950s. Remarkably, quantum electronics has first conquered the radio range; lasers appeared at the beginning of the 1960s. This is partly because in usual optics experiments, $N_1 \gg N_2$, and therefore stimulated emission, as a rule, plays no role. At the same time, in radio spectroscopy, $N_1 \approx N_2 \gg |N_1 - N_2|$, and the observed absorption of radio waves is caused by the stimulated absorption slightly exceeding the stimulated emission.

An important role was also played by the advanced development of radio spectroscopy in the 1940s, in both theory and experiment. (Experimental base for microwave radio spectroscopy was provided by the development of radar technique.) By that time, the theory of radio waves interaction with gas molecules was developed, the structure of rotational spectra was calculated in detail, the role of relaxation and saturation was understood. Of considerable importance were investigations with beam radio spectroscopes, which had been started as early as in the 1930s. Probably, it was also important that radio spectroscopists, in contrast to opticians, understood very well the operating principles of MW generators and amplifiers based on free-electron beams (klystrons, magnetrons, traveling-wave and backward-wave tubes), they were familiar with the notions of negative resistance and positive feedback, and had practical experience with high-quality MW cavities.

Among the works directly preceding the advent of masers, one should mention the ones by Kastler (France), who developed in 1950 the optical pumping method for increasing the population inversion of close levels in gases. Besides gas and beam radio spectroscopy, an important role was also played by magnetic radio spectroscopy, a direction that was started in the 1940s and studied the interaction of radio waves with ferromagnetics and nuclear or electronic paramagnetics (E. K. Zavoisky, 1944). These are namely the achievements in the theory and technique of magnetic resonance that led to the development of paramagnetic amplifiers, which have an extremely small level of noise. Population inversion has been first obtained in a system of nuclear spins placed into magnetic field (Parcell and Pound, 1951).

1.2.3 *Masers*

The idea of using stimulated emission in a medium with population inversion for the amplification and generation of MW electromagnetic waves was suggested at several different conferences at the beginning of the 1950s by N. G. Basov and A. M. Prokhorov (Lebedev Physics Institute, Academy of Sciences, USSR), C. H. Townes (Columbia University, USA), and J. Weber (University of Maryland, USA). The first quantitative theory of a quantum generator was published by Basov and Prokhorov in 1954. They have found the threshold population difference necessary for self-excitation and suggested a method for obtaining population inversion in a molecular beam using inhomogeneous electrostatic field. Later, Basov, Prokhorov, and Townes were awarded a Nobel Prize for their contributions to the development of quantum electronics.

In 1954, description of the first operating maser was published by Gordon, Zeiger, and Townes. The active medium was ammonium molecular beam, focused with the help of electric field. Nowadays, beam masers are used in the national standards of frequency and time.

The second basic maser type, paramagnetic amplifier, was created in 1957 by Scovill, Feher, and Seidel who followed a suggestion by Bloembergen. In paramagnetic amplifiers, population inversion is created with the help of auxiliary radiation, the *pump*, which saturates the populations of levels 1 and 3 (Fig. 1.2). As a result, levels 1 and 2 (or 2 and 3) get population inversion. The idea of pumping a three-level system, which was later widely used in solid-state and liquid lasers, belongs to Basov and Prokhorov (1955). The active medium of paramagnetic amplifiers, which is a diamagnetic crystal doped with a small amount (on the order of 10^{-3}) of paramagnetic atoms, i.e., atoms with odd electron numbers, is cooled down to helium temperatures. Cooling is necessary for reducing the noise and slowing down the relaxation processes, which counteract the population inversion. (In paramagnetics, relaxation of populations is caused by the interaction between crystal lattice vibrations and the magnetic moments of non-compensated electrons.)

1.2.4 *Lasers*

Transition from radio to the optical frequencies took about five years: the first operating laser emitting coherent red light was described by Maiman in 1960. As the active medium, the laser used a pink ruby crystal (aluminium oxide doped with chrome) and population inversion was achieved using blue and green light from a pulsed flash lamp. An important step was realizing that a Fabry-Perot interferometer, i.e., two parallel plane mirrors, is a high-quality resonator, i.e., an oscillation system for light waves (Prokhorov, Dicke, 1958).

The laser era of physics started. Soon after the appearance of solid-state lasers with optical pumping, a number of other laser types was developed: gas discharge lasers (1961), semiconductor lasers based on $p-n$ transitions (1962), liquid lasers based on the solutions of organic dyes (1966). Rather quickly, the wavelength range from far infrared (IR) to far ultraviolet (UV) was covered. The parameters of the lasers (power, monochromaticity, directivity, stability, tunability) were continuously improving; their field of application rapidly broadened. An important role was played by the invention of methods to shorten the duration of laser light pulses (q-switching and mode locking).

First experiments on light frequency doubling (Franken et al., 1961) started the explosive development of nonlinear optics, which studies and uses the nonlinearity of the matter at optical frequencies. Holography and optical spectroscopy had their second birth; new fields appeared, such as optoelectronics, coherent spectroscopy, and quantum optics. X-ray and gamma-ray lasers are to arrive soon.[c]

It should be stressed once again that the rapid development of quantum electronics was provided by a large amount of ideas and information stored by the

[c]Editors' note: While X-ray lasers have been indeed constructed in the end of the 20th century [Svelto (2010)], making a gamma-ray laser is still a challenge.

beginning of the 1950s in the fields of radio and optical spectroscopy. Such directions of physics as magnetic resonance or molecular-beam spectroscopy, seemingly far from practical applications, led to a 'laser revolution' in many fields of science and technology.

1.3 Recent progress in quantum electronics (added by the Editors)

This textbook was published in 1987, almost a quarter century ago. At that time, it was a very modern book; it reflected the latest events in quantum electronics and provided a complete picture of its directions and tendencies. Since then, many changes took place in this field. New technologies appeared, new laser sources were developed, and new effects were discovered. In this section, we will try to briefly review the advances in quantum electronics that happened after the book had been published.

1.3.1 *Physics of lasers*

During the last two decades, important progress has been achieved in laser technology, and all parameters of lasers have been considerably improved. Mean powers of laser radiation achieved at present amount to hundreds of kW, while peak powers reach the *petawatt* (10^{15} W) range. Such radiation provides the values of electric and magnetic fields comparable to atomic ones and threfore opens a perspective for observing principally new effects in optics and particle physics. The ultra-fast laser technology is now capable of producing pulses as short as tens of *attoseconds*, containing only *few optical cycles*. The spectral range covered by modern commercial laser systems, in particular, achieved by continuous frequency tuning, is from vacuum UV (about 100 nm) to mid-IR (tens of microns).

These achievements became possible due to both the development of existing methods, such as frequency conversion, generation of higher optical harmonics, mode locking etc., and the discovery of new technologies. In particular, dye-laser systems were gradually replaced by solid-state ones. The most famous among them are *titanium-sapphire lasers* and similar systems, providing ultra-short pulse generation, as well as *optical combs*, via mode locking. Huge progress has been achieved in the development of semiconductor lasers. A totally novel step in laser technology, with respect to the 1980s, was the invention of *fibre laser systems*, which can have extremely high efficiency and therefore provide record output powers.

Apparently, lasers became widely used devices which penetrate into all fields of human activity starting from toys up to the high technologies and medicine.

1.3.2 *Laser physics*

Laser physics, or research in physics essentially based on the use of lasers, underwent considerable progress as well. Modern laser physics covers several branches of science and various applications like nonlinear and quantum optics, fiber optics, optical pulse shaping, optoelectronics (including integrated optics), optical communications, different aspects of general optics etc. New directions appeared, such as, for instance, *high resolution spectroscopy* or *atom optics*. Some of the new directions will be discussed in more detail below; the rest will be briefly mentioned here. Application of laser methods to *metrology* resulted in the development of *caesium atomic clock* to a high-technology level; recently, this device has been made on a chip and is now available as a consumer product. Laser methods became extremely helpful in the manipulation with microscopic and nanoscopic objects; in particular, the technique of *laser tweezers* enables trapping and displacing small particles, including biological objects. Laser cooling of atoms and molecules is another example of progress in laser physics. Finally, lasers are now widely used in the technique of *scanning near-field optical microscopy (SNOM)*, which successfully complemented the existing methods of scanning tunnel microscopy and atomic-force microscopy.

1.3.3 *New trends in nonlinear optics*

Huge progress in nonlinear optics is due to the development of the material science, which led to the production of new nonlinear optical materials. Among them, there were newly synthesized crystals with high nonlinear susceptibilities and broad transparency range, such as BBO, LBO, KTP, and many others. Further opportunities in realizing various types of phase matching were provided by the use of spatially inhomogeneous structures such as *periodically and aperiodically poled crystals*, *photonic crystals* and *microstructured fibres* (*photonic-crystal fibres*). The opportunities offered by such structures are: making use of new components of nonlinear susceptibility tensors, *non-critical phase matching* and simultaneous phase matching for different nonlinear processes, as well as processes in different frequency ranges.

One of the novel trends is development of *integrated nonlinear optics*. Due to the miniaturization of optical elements, involving fibre optics and *waveguide structures*, it became possible to realize most of nonlinear optical processes on a chip. Optical fibres are now used not only for light transmission, but also for beam splitting, polarization transformations, as nonlinear elements and as active elements [Agrawal (2007)]. *Nonlinear waveguides*, based on KTP and lithium niobate crystals, and sometimes on semiconductor layers, are used as extremely efficient and compact elements for frequency conversion, requiring very low pump powers and allowing for relatively easy control. Integrated optics also uses plasmonic structures, which form convenient interfaces between free space or dielectrics and metal surfaces.

We now witness a certain shift of interest to novel frequency ranges. Among them, attention is drawn to the terahertz (10^{12} Hz) range of frequencies, which is important for spectroscopic studies in biology, for astronomy, and for the security applications (detection of explosive materials and weapons). For more details on the recent developments in nonlinear optics, one can see, for instance, [Boyd (2008)].

1.3.4 *Atom optics*

A completely novel direction that appeared in the end of the 20th century is atom optics, i.e., manipulation of individual atoms by means of laser beams. It is worth noticing that manipulating single quantum objects characterizes the modern development of quantum electronics and, probably, physics in general compared with the last century when the ensemble approach dominated.

Forces acting on atoms due to the gradients of light intensity turn a standing wave into a scatterer for atomic beams, causing diffraction, interference, and trapping. Trapping of ions and atoms enables one to address these quantum objects, single ones or in an array, and control their quantum state. In particular, it is possible to organize the interaction between single material quantum objects and single photons. This is extremely important both in fundamental research and for various applications like quantum information.

Furthermore, the effect of Bose-Einstein condensation, predicted as early as in 1925, has been observed in 1995. A *Bose-Einstein condensate (BEC)*, a large group of atoms described by a single wave function, is one of the few examples of a macroscopic object manifesting quantum behavior. Similarly to single atoms and ions, a BEC can be manipulated by means of laser beams.

1.3.5 *Optics of nonclassical light*

Quantum optics, started by the famous Hanbury Brown–Twiss experiment (Sec. 7.2) in 1956, had 'explosive' development in the end of the last century. New types of nonclassical light have been generated. In addition to single-photon and two-photon Fock states in superposition with the vacuum (Sec. 7.5), higher-order Fock states can be conditionally prepared now by using spontaneous parametric down-conversion [Bouwmeester (2000); Mandel (2004)]. The spectral and spatial structure of such states has been studied in detail, as well as their polarization properties. The concept of squeezed states (Sec. 7.5), which were first observed about the same time as the book was published, and the idea of shot-noise suppression [Yamamoto (1999)], were since then considerably developed. Squeezed states became one of the main instruments of experimental quantum optics [Bachor (2004); Walls (1994)], together with the two-photon states (photon pairs). The phenomenon of *polarization squeezing* was observed and studied. Finally, various types of *entangled states* [Scully (1997); Mandel (2004); Bouwmeester (2000)], both faint (few-photon) and bright ones, based on quadrature squeezing, were generated, and numerous experiments on testing Bell's inequalities [Grynberg (2010); Scully (1997); Mandel (2004); Klyshko (1998)] were carried out.

New sources of nonclassical light were discovered. Since the beginning of the 21st century, optical fibres have been used as a very reliable and efficient source of both squeezed states and photon pairs. This source is based on the *cubic susceptibility* (*Kerr nonlinearity*), and the corresponding nonlinear optical effect is *spontaneous four-wave mixing* (originally called *hyper-parametric scattering*, Sec. 6.5). By applying fibres with specially tailored dispersion dependence, which can be achieved by modifying the structure, by doping, or by tapering, one can fully control the phase matching and provide its new types [Agrawal (2007)]. Photon pairs and squeezed light are also generated in waveguide structures having high efficiency, compact sizes, and controllable properties. In addition, modern sources of nonclassical light include nano- and micro-emitters such as *quantum dots*, *vacancies and color centers in diamond*, and others. These sources are in a sense similar to single atoms, which were used for generating nonclassical light in the 1960s and the 1970s; however, an important advantage of solid-state emitters is much easier handling, including preparation and control.

Huge progress has been made in the development of the detection techniques [Leonhardt (1997)]. The only type of photon-counting detector mentioned in the book is a *photomultiplier tube (PMT)*; nowadays, much more common for single-photon counting are *avalanche photodiodes (APDs)* operating in the Geiger mode. Such detectors provide quantum efficiencies of up to 60% and time resolution of about 50 ps in the visible (Si-based APDs) and near-IR (InGaAs- or Ge-based APDs) ranges while having relatively low dark noise (up to tens of pA). Other types of single-photon detectors appeared quite recently, namely, *superconducting photodetectors*, which can operate in the IR and even terahertz range, and *transition-edge sensors (TES)*, capable of photon-number resolution. The latter possibility, nearly impossible at the time when this book was written, is also achieved by combining single-photon counting with time or space multiplexing. Finally, the technique of homodyne detection, which is hardly mentioned in the book, has been hugely developed during the last two decades. Using this technique, it is possible not only to measure the distributions of coordinate and momentum for various quantum states (Sec. 7.5), but also to reconstruct the quasi-probability distributions, such as Wigner or Husimi functions [Schleich (2001); Bachor (2004)].

Probably the most important event in the development of quantum optics is its application to *quantum information*, a field that emerged in the end of the 20th century at the boundary of quantum mechanics, mathematics, and information science [Nielsen (2000)]. Along with the quantum metrology, which is briefly mentioned in the book, quantum information and quantum communication technologies became a real practical output of quantum optics, which at first looked like nothing but a collection of beautiful fundamental experiments. In quantum metrology, in addition to the *absolute calibration* methods (Sec. 7.6), which were developed in the 1980s, there appeared the techniques of *super-resolution* and *precise positioning* [Bachor (2004)] based on squeezed light or high-order Fock states. A lot of experimental techniques, developed earlier in quantum optics for nonclassical state generation, transformation and measurement, were simply transferred to quantum communication. In quantum communication, various states of light are used as information carriers, from *qubits* (quantum information bits), qutrits, ququarts, and high-dimensional *qudits* to entangled states formed by these elementary carriers [Bouwmeester (2000); Nielsen (2000)]. Transformations of these states by linear optical elements, as well as interactions between these states, can form the basis for *quantum gates*, which, in their turn, may in the nearest future become the key elements of a *quantum computer* [Nielsen (2000)]. Different approaches to the measurement of quantum states serve as a powerful tool for quantum state tomography and quantum process tomography. Finally, the most advanced branch of quantum information is *quantum key distribution*, in which a secret encryption key is distributed between several communicating parties in such a way that eavesdropping is not possible due to the fundamental laws of quantum physics.[d]

[d]This is true provided that the unavoidably introduced error rate exceeds some critical level, depending on the specific type of *protocol* used.

Chapter 2

Stimulated Quantum Transitions

The most important notion in quantum electronics is the probability for an electron in an atom or a molecule to make a quantum transition from one level to another. In this chapter, we will first give the general expression for the probability of a quantum transition in the first order of the perturbation theory (Sec. 2.1), then calculate the probability of a transition due to monochromatic radiation (Sec. 2.2) and find the interaction cross-section and the absorption coefficient (Sec. 2.3). Further, we will consider stimulated transitions under fluctuating (noise) radiation with a broad spectrum (Sec. 2.4). Noise radiation surrounding an atom can play the role of a thermostat and cause relaxation (Sec. 2.5).

A consistent theory of electromagnetic processes should describe both the matter and the field based on the principles of quantum mechanics. However, most part of quantum electronics effects are sufficiently well described by the so-called *semiclassical theory of radiation*, in which only the motion of particles is quantized while the electromagnetic field is considered in terms of classical Maxwell's equations. By avoiding quantum electrodynamics, one gets the theory considerably simplified but, at the same time, loses the chance to consistently describe fluctuations of the electromagnetic field and, in particular, spontaneous emission and the noise of quantum amplifiers. The present book mainly considers stimulated effects in a classical deterministic field and therefore uses the semi-classical theory of radiation. Quantization of the field and spontaneous effects are considered in Chapter 7.

2.1 Amplitude and probability of a transition

In the simplest model of quantum electronics, matter is assumed to consist of separate non-interacting motionless atoms or molecules in external electromagnetic field. Our first task is to find out what happens with a given atom in a given alternating field $\mathbf{E}(t)$. (Usually the effect of the magnetic field is much weaker than the one of the electric field.) At the second stage, we will find the back action of the atoms on the field. The self-consistent solution to the two systems of equations describing the response of the matter to the field and the response of the field to a given motion of charges, under certain simplifying conditions, is the main problem in the theory of interaction between radiation and matter.

The behavior of material particles in given external fields is described by the Schrödinger equation,

$$(i\hbar\partial/\partial t - \mathcal{H})\Psi(\mathbf{r}, t) = 0. \tag{2.1}$$

Here, Ψ is the wave function, whose arguments are the set of coordinates $\mathbf{r} = \{\mathbf{r}_1, \mathbf{r}_2, ...\}$ and the time; $\mathcal{H}$ is the energy operator consisting of the non-perturbed part, $\mathcal{H}_0$, and the alternating energy of the particles in the external field, $\mathcal{V}(t)$,

$$\mathcal{H}(t) = \mathcal{H}_0 + \mathcal{V}(t). \tag{2.2}$$

The non-perturbed energy, in its turn, includes the kinetic energy of the particles and the energy of their interaction $\mathcal{V}_0$. (The latter also includes the energy of the particles in external static fields).

2.1.1 *Unperturbed atom*

In the absence of the alternating field, the wave function can be represented as

$$\Psi^{(0)}(\mathbf{r}, t) = \sum_n c_n^{(0)} \Phi_n(\mathbf{r}, t), \tag{2.3}$$

$$\Phi_n(\mathbf{r}, t) = \varphi_n(\mathbf{r})\exp(-i\mathcal{E}_n t/\hbar), \tag{2.4}$$

where $\mathcal{E}_n$ and $\varphi_n(\mathbf{r})$ are the eigenvalues and the eigenfunctions of $\mathcal{H}_0$, satisfying the stationary Schrödinger equation,

$$(\mathcal{H}_0 - \mathcal{E}_n)\varphi_n = 0. \tag{2.5}$$

The index n numerates the energy levels. (We assume that the particles move within a bounded space domain and therefore the levels are discrete; we also assume the levels to be non-degenerate.) The set of functions $\{\varphi_n\}$ is assumed to be orthogonal and normalized,

$$\int d\mathbf{r}\varphi_n^*\varphi_m = \delta_{nm}, \tag{2.6}$$

so that

$$\int d\mathbf{r}|\Psi|^2 = \sum_n |c_n|^2 = 1. \tag{2.7}$$

The c_n coefficient in the expansion (2.3) gives the relative population $|c_n|^2$ of the level n, i.e., the probability to measure the energy $\mathcal{E}_n$ or, as one says, the probability to find the system 'at the level' n. Indeed, according to the rule of calculating mean values in quantum mechanics, the mean energy of the system, with an account for Eqs. (2.3)–(2.6), is

$$\mathcal{E} \equiv \langle \mathcal{H}_0 \rangle \equiv \int d\mathbf{r} \Psi^* \mathcal{H}_0 \Psi = \sum_n |c_n|^2 \mathcal{E}_n. \tag{2.8}$$

Note that, according to (2.3), in the general case the atom is not necessarily in a stationary state with a definite energy $\mathcal{E}_n$ (even in the absence of the alternating force, $\mathcal{V}(t) = 0$). For instance, let only two coefficients c_n of the superposition (2.3) be nonzero: $c_1 = c_2 = 1/\sqrt{2}$; then the mean ensemble energy of the atom is $(\mathcal{E}_1 + \mathcal{E}_2)/2$ but single energy measurements will give either $\mathcal{E}_1$ or $\mathcal{E}_2$. Then the electron 'cloud', i.e., the probability density to find the electron at point $(\mathbf{r}, t)$, will oscillate with *the Bohr frequency*, $\omega_{21} \equiv \omega_2 - \omega_1 \equiv (\mathcal{E}_2 - \mathcal{E}_1)/\hbar$ (Fig. 2.1):

$$\begin{aligned} P(\mathbf{r}, t) &= |\Psi(\mathbf{r}, t)|^2 = |\varphi_1(\mathbf{r}) + \varphi_2(\mathbf{r}) \exp(-i\omega_{21} t)|^2/2 \\ &= \varphi_1^2/2 + \varphi_2^2/2 + \varphi_1 \varphi_2 \cos(\omega_{21} t). \end{aligned} \tag{2.9}$$

(We assume that $\varphi_n = \varphi_n^*$.) Such nonstationary states are called *coherent* ones. This term is often used in the case where many identical atoms are in a non-

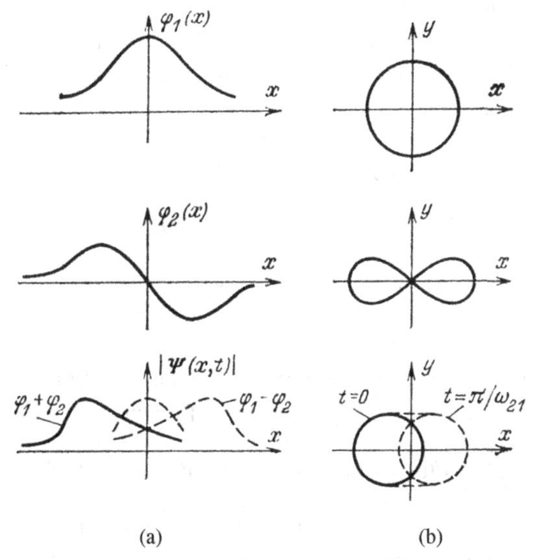

Fig. 2.1 Electron cloud of an atom that is in a coherent (non-stationary) state given by a superposition of two stationary states φ_1 and φ_2 with different symmetries oscillates with the transition frequency ω_{21}: (a) dependencies of the wave functions on one of the space coordinates; (b) corresponding configurations of the electron cloud.

stationary state with the same phase. Then, electrons oscillate synchronously and the system of atoms has a macroscopic dipole moment emitting intense light with the frequency ω_{21}. This effect, called *superradiance*, will be considered in Sec. 5.3.

In the presence of external alternating field $\mathbf{E}(t)$, eigenoscillations of the electron cloud with the frequencies ω_{mn} will be accompanied by stimulated oscillations with the frequency of the field ω.

2.1.2 *Atom in an alternating field*

Consider now the effect of an alternating field on the wave function $\Psi(\mathbf{r}, t)$ of an atom or a molecule. At $\mathcal{V}(t) \neq 0$, the function (2.3) does not satisfy the Schrödinger equation (2.1) any more, but the expansion can be kept in the form (2.3) if the coefficients c_n are considered as time-dependent,

$$\Psi(\mathbf{r}, t) = \sum_n c_n(t)\Phi_n(\mathbf{r}, t). \tag{2.10}$$

(This possibility follows from the *completeness* of the eigenfunctions set $\varphi_n(\mathbf{r})$.)

Thus, due to the effect of the incident light, the relative populations $|c_n(t)|^2$ of the levels are redistributed (with the normalization condition (2.7) maintained). In other words, the atom makes stimulated *transitions* between the levels. Let us find the probability of such transitions.

From the Schrrödinger equation (2.1) for the wave function, we will pass to equations for $c_n(t)$. For this purpose, let us substitute expansion (2.10) in (2.1) and take into account that, according to (2.5), $i\hbar\dot{\Phi}_n = \mathcal{H}_0\Phi_n$:

$$\sum_n (i\hbar\dot{c}_n - c_n\mathcal{V})\Phi_n = 0. \tag{2.11}$$

Left-multiplying this equality by one of the functions Φ_m^* and integrating w.r.t. $\mathbf{r}$, we obtain

$$\sum_n (i\hbar\dot{c}_n \int d\mathbf{r}\Phi_m^*\Phi_n - c_n \int d\mathbf{r}\Phi_m^*\mathcal{V}\Phi_n) = 0. \tag{2.12}$$

Let us take into account the orthogonality (2.6) of the eigenfunctions and introduce the following notation for the matrix elements of the perturbation operator:

$$\mathcal{V}'_{mn} \equiv \int d\mathbf{r}\Phi_m^*\mathcal{V}\Phi_n = \mathcal{V}_{mn}\exp(i\omega_{mn}t), \mathcal{V}_{mn} \equiv \int d\mathbf{r}\varphi_m^*\mathcal{V}\varphi_n. \tag{2.13}$$

As a result, we find the system of equations for the coefficients $c_n(t)$, which is equivalent to the initial Schrödinger equation:

$$i\hbar dc_m/dt = \sum_n \mathcal{V}'_{mn}c_n. \tag{2.14}$$

Note that the coefficients c_n form a function of a discrete argument (energy), $c_n \equiv \tilde{\Psi}(\mathcal{E}_n)$, which can be considered as the wavefunction of the system *in the energy representation* (while $\Psi(\mathbf{r})$ is the wave function *in the coordinate representation*). Correspondingly, (2.14) is the Schrödinger equation in the energy representation. The functions $\tilde{\Psi}$ and Ψ are related through a linear transformation and provide the same information. The inverse of transformation (2.10) can be obtained by left-multiplying it by the integral operator $\int d\mathbf{r}\Phi_m^*$:

$$c_m = \int d\mathbf{r}\Phi_m^*\Psi. \tag{2.15}$$

The change of representation, $\Psi \to \tilde{\Psi}$, is similar to the change of the basis in vector algebra, where the components of a vector are also linearly transformed.

The relation between different representations is most clearly manifested in Dirac's notation (Sec. 7.5). In this notation, the Schrödinger equation (2.1) is written in the invariant form (without specifying the representation) as

$$i\hbar d|t\rangle/dt = \mathcal{H}|t\rangle. \tag{2.16}$$

In order to pass to the energy representation, let us left-multiply (2.16) by the m-th eigenvector of $\mathcal{H}_0$,

$$i\hbar\frac{d}{dt}\langle m|t\rangle = \langle m|\mathcal{H}|t\rangle = \sum_n \langle m|\mathcal{H}|n\rangle\langle n|t\rangle. \tag{2.17}$$

The last equation was obtained using the expansion of the unity, $I = \sum |n\rangle\langle n|$. Let us denote $\langle m|\mathcal{V}|n\rangle \equiv \mathcal{V}_{mn}$ and use (2.2) and (2.5), then

$$i\hbar\frac{d}{dt}\langle m|t\rangle = \mathcal{E}_m\langle m|t\rangle + \sum_n \mathcal{V}_{mn}\langle n|t\rangle.$$

Finally, if we separate the slowly varying part of the $\langle m|t\rangle$ factor,

$$\langle m|t\rangle \equiv c_m(t)\exp(-i\mathcal{E}_m t/\hbar),$$

we once again obtain (2.14).

2.1.3 *Perturbation theory*

In the general case, the solution to the system (2.14) can be found using the perturbation theory, as a series expansion in the external force. Alternatively, the system can be solved without the perturbation theory, using the so-called two-level approximation, which will be considered below, in Sec. 4.3.

Thus, we look for the solution to (2.14) in the form of the sum

$$c_n = c_n^{(0)} + c_n^{(1)} + \cdots = \sum_s c_n^{(s)}, \tag{2.18}$$

in which the s-th term is proportional to the s-th power of the external force, $c_n^{(s)} \propto \mathcal{V}^s$. Substituting (2.18) in (2.14) and setting equality between the terms of the same order in $\mathcal{V}$, we find

$$\dot{c}_m^{(0)} = 0, \tag{2.19}$$

$$i\hbar \dot{c}_m^{(1)} = \sum_n \mathcal{V}'_{mn} c_n^{(0)}, \tag{2.20}$$

$$i\hbar \dot{c}_m^{(s)} = \sum_n \mathcal{V}'_{mn} c_n^{(s-1)}. \tag{2.21}$$

The set of zero-order coefficients $c_m^{(0)}$ provides the initial conditions for equations (2.14).

2.1.4 *Linear approximation*

Usually, it is assumed that only one of the coefficients, for instance, $c_1^{(0)}$, is nonzero, so that at time t_0 the system is in a state with a given energy,

$$c_n(t_0) = c_n^{(0)} = \delta_{n1}. \tag{2.22}$$

In this case, the system of equations (2.20) for the first-order coefficients gets 'decoupled',

$$\dot{c}_m^{(1)} = \mathcal{V}'_{m1}/i\hbar. \tag{2.23}$$

Hence, it follows that in the linear approximation, the response of a quantum system that is initially at level 1, to an external perturbation is given by the formula

$$c_m^{(1)}(t) = \frac{1}{i\hbar}\int_{t_0}^{t} dt' \mathcal{V}_{m1}(t') \exp(i\omega_{m1}t'). \tag{2.24}$$

Thus, at large time delays after the perturbation has been switched on ($t - t_0 \to \infty$), the c_2 coefficient, which determines the perturbed population of the level with the energy $\mathcal{E}_2$, scales as the Fourier transform of the external force at the Bohr frequency $\omega_{21} = (\mathcal{E}_2 - \mathcal{E}_1)/\hbar$. In other words, *in the first-order perturbation theory, a quantum system behaves as a set of linear oscillators and responds only to the resonant harmonics of the external force*. If there is no ω_{21} harmonic in the perturbation spectrum, or its amplitude is small, then level 2 will not be populated.

2.1.5 *Probability of a single-quantum transition*

Given the initial condition $c_m^{(0)} = \delta_{mn}$, the dimensionless complex number c_m is called the *amplitude of the transition* from level n to level m. (Often, the second subscript is added to specify the initial condition, $c_m \equiv c_{mn}$.[a]) The squared modulus of the transition amplitude, $|c_{mn}(t)|^2$, is equal to the conditional probability of finding the system at the level m at time t provided that at time t_0 it was at the level n. In the linear (single-photon) approximation, this probability, according to (2.24), is

$$P_{mn}^{(1)} \equiv P^{(1)}(m,t|n,t_0) = \hbar^{-2}|\int_{t_0}^{t} dt' \mathcal{V}_{mn}(t')\exp(i\omega_{mn}t')|^2. \tag{2.25}$$

From (2.25), an important property of quantum systems follows: *the probability of forward and backward transitions are equal:*

$$P_{mn}^{(1)} = P_{nm}^{(1)}. \tag{2.26}$$

Indeed, the $\mathcal{V}$ operator corresponds to an observable quantity, the energy, and therefore its mean values are real, $\langle\mathcal{V}\rangle = \langle\mathcal{V}\rangle^*$. Hence, this operator is Hermitian (*self-conjugate*), $\mathcal{V} = \mathcal{V}^\dagger$. Matrix elements of Hermitian operators have the property $\mathcal{V}_{mn} = \mathcal{V}_{nm}^*$; therefore, (2.25) leads to (2.26).

Note that for the transition $n \to m$ to be enabled, it is necessary, in addition to the resonance condition, that its matrix element is nonzero,

$$\mathcal{V}_{mn} = \int d\mathbf{r}\varphi_m^* \mathcal{V}\varphi_n \neq 0.$$

This requirement, providing the *selection rules*, ‘forbids’ some transitions for high-symmetry systems. For instance, if $\varphi_n(-\mathbf{r}) = \pm\varphi_n(\mathbf{r})$ (central symmetry) and $\mathcal{V} \sim r$ (dipole approximation), then transitions between states with the same parity are forbidden (since the integrand is odd in this case). If $\mathcal{V}_{mn} \neq 0$, one says that the perturbation ‘couples’ or ‘mixes’ the states m and n.

2.2 Transitions in monochromatic field

2.2.1 *Dipole approximation*

Let us apply the general formula (2.25) to the case of a harmonic perturbation. Most problems of quantum electronics allow the dipole approximation for the energy of the interaction between charges and field (Sec. 7.3),

$$\mathcal{V} = -\mathbf{d}\cdot\mathbf{E} = -\mathbf{d}\cdot(\mathbf{E}_0 e^{-i\omega t} + \mathbf{E}_0^* e^{i\omega t})/2. \tag{2.27}$$

[a]Recall that in quantum mechanics, the subscripts are often read from right to left; therefore, c_{mn} is the amplitude of the transition $n \to m$.

Here, the amplitude $\mathbf{E}_0$ of the electromagnetic wave is assumed constant within the system considered, an atom or a molecule, since in the optical frequency range $\lambda \sim 10^{-4}$cm$\gg a_0 \sim 10^{-8}$cm, where a_0 is the Bohr radius.

In the case of a single-electron atom, the dipole moment operator $\mathbf{d}$ is equal to the product of the electron charge and its radius vector, $\mathbf{d} = -e\mathbf{r}$, so that the matrix elements are

$$\mathcal{V}_{mn} = -\mathbf{d}_{mn} \cdot (\mathbf{E}_0 e^{-i\omega t} + \text{c.c.})/2, \tag{2.28}$$

$$\mathbf{d}_{mn} = -e \int d^3 r \mathbf{r} \varphi_m^* \varphi_n, \tag{2.29}$$

where c.c. means the complex conjugated expression. For allowed transitions, the integral in (2.29) is on the order of the atom size a_0, so that

$$d_{mn} \sim e a_0 \sim 10^{-18}\text{CGS} \equiv 1\text{D}. \tag{2.30}$$

In the case of magneto-dipole transitions, which are used, in particular, in paramagnetic amplifiers, the electric field $\mathbf{E}_0$ in (2.28) should be replaced by the magnetic field $\mathbf{H}_0$ and $\mathbf{d}$ should be replaced by the magnetic dipole moment μ whose absolute value is on the order of *Bohr's magneton*,

$$\mu_{mn} \sim \mu_0 \equiv e\hbar/2mc \approx 0.9 \cdot 10^{-20}\text{CGS}. \tag{2.31}$$

2.2.2 *Transition probability*

Substituting (2.28) in (2.25), for $t_0 = 0$ we find

$$P_{mn} = \frac{1}{4\hbar^2} \left| \mathbf{d}_{mn} \cdot \mathbf{E}_0 \frac{\exp[i(\omega_{mn} - \omega)t] - 1}{\omega_{mn} - \omega} + \mathbf{d}_{mn} \cdot \mathbf{E}_0^* \frac{\exp[i(\omega_{mn} + \omega)t] - 1}{\omega_{mn} + \omega} \right|^2. \tag{2.32}$$

Further, let us consider stimulated transitions up ($\omega_{mn} > 0$) or down ($\omega_{mn} < 0$) under the condition that the frequency of the field ($\omega > 0$) is resonant to the transition:

$$\omega \sim |\omega_{mn}| \gg |\omega - |\omega_{mn}||. \tag{2.33}$$

Then, one of the terms in (2.32) is much larger than the other one, so that the latter can be neglected (the so-called *rotating-wave approximation*[b]). For instance, stimulated absorption is mainly determined by the first term, which is proportional to the *positive-frequency part* of the field $\mathbf{E}_0 e^{-i\omega t}/2$, while for stimulated emission, it is the second term that matters, proportional to the *negative-frequency part*, $\mathbf{E}_0^* e^{i\omega t}/2$. Note that in quantum electrodynamics, the amplitudes $\mathbf{E}_0, \mathbf{E}_0^*$ are operators scaling as the photon annihilation and creation operators, $a, a^\dagger$, see Sec. 7.4.

[b]The title comes from the fact that in the complex plane, the $e^{-i\omega t}$ vector is rotating, in contrast to the vector $\cos \omega t$, which oscillates.

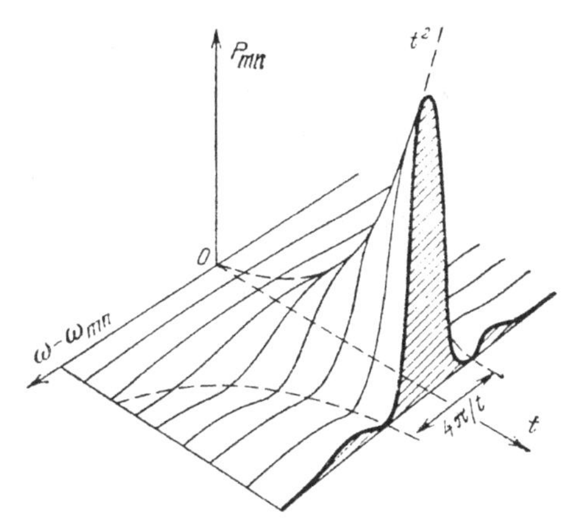

Fig. 2.2 Probability P_{mn} of a stimulated transition as a function of the field frequency ω and the interaction time t.

Thus, in the first-order perturbation theory, the probability to find the atom at level m at time t, under the initial condition $c_n(0) = 1$, is

$$P_{mn} = \left|\frac{\mathbf{d}_{mn} \cdot \mathbf{E}_0}{2\hbar}\right|^2 \left[\frac{\sin(\tilde{\omega}t/2)}{\tilde{\omega}/2}\right]^2, \tag{2.34}$$

where $\tilde{\omega} \equiv \omega - |\omega_{mn}|$. According to this formula, the dependence of the transition probability on the external field frequency is resonant, and *the sharpness of the resonance increases with time* (Fig. 2.2). As a result, in the limit $t \to \infty$, the transition probability is given by Dirac's delta function,

$$\lim_{t\to\infty} \left[\frac{\sin(\tilde{\omega}t/2)}{\tilde{\omega}/2}\right]^2 = 2\pi t\delta(\tilde{\omega}). \tag{2.35}$$

The factor by the delta function is verified by integrating both parts of (2.35) in $\tilde{\omega}$.

Because the transition probability scales as the time duration t of the perturbation, one can introduce the *transition probability per unit time*, also called the *transition rate*,

$$W_{mn} \equiv P_{mn}/t = 2\pi|\mathbf{d}_{mn} \cdot \mathbf{E}_0/2\hbar|^2\delta(\tilde{\omega}). \tag{2.36}$$

Thus, the transition rate scales as the square of the field, i.e., the intensity of the wave. The presence in (2.36) of the delta function, which differs from zero only at exact resonance, can be understood from the photon viewpoint: according to the energy conservation law, the change in the atom energy by the value $\mathcal{E}_m - \mathcal{E}_n \equiv \hbar\omega_{mn}$ should be accompanied by the absorption or emission of a photon with the energy $\hbar\omega$.

2.2.3 *Finite level widths*

In reality, however, there always exist additional perturbations, for instance, collisions in the case of a gas. These perturbations broaden the energy levels and, as a result, even at $t \to \infty$ the resonance has a finite width $\Delta\omega$. (In the case of collisions, $\Delta\omega = 2/\tau$, where τ is the mean time interval between the collisions.) The finite width of the resonance is taken into account by changing the delta function in (2.36) to a *form factor* $g(\omega)$, which describes the true shape of the spectral line and is also normalized to unity:

$$W_{mn} = 2\pi|\mathbf{d}_{mn} \cdot \mathbf{E}_0/2\hbar|^2 g(\tilde{\omega}), \tag{2.37}$$

$$\int_{-\infty}^{\infty} d\omega g(\omega) = 1. \tag{2.38}$$

If the broadening is due to collisions, spectral lines have Lorentzian shapes (Fig. 2.3, curve 1):

$$g_L(\tilde{\omega}) = \frac{2/(\pi\Delta\omega)}{1 + (2\tilde{\omega}/\Delta\omega)^2}. \tag{2.39}$$

The transition probability is maximal at $\omega = |\omega_{mn}|$:

$$W_0 = \Omega^2/\Delta\omega, \tag{2.40}$$

where $\Omega \equiv |\mathbf{d}_{mn} \cdot \mathbf{E}_0|/\hbar$ is *the Rabi frequency*. It has the dimensionality of frequency and characterizes the perturbation of an atom by resonant monochromatic field.

Thus, *the probability of a stimulated transition scales as the intensity of light, squared matrix element of the dipole moment, and the inverse width of the spectral line*.

This dependence on the spectral line width is typical not only for collision broadening: from the normalization condition (2.38) it follows, in the general case, that $g(0) \sim 1/\Delta\omega$. For instance, in gases at low pressure $\Delta\omega$ is often determined by the Doppler effect, which results in a Gaussian line shape. Figure 2.3 shows the comparison of spectral line shapes due to collision broadening (2.39), the Doppler effect,

$$g_D(\omega) = \frac{2}{\Delta\omega}\left(\frac{\ln 2}{\pi}\right)^{1/2} \exp\left[4 - \left(\frac{\omega}{\Delta\omega}\right)^2 \ln 2\right], \tag{2.41}$$

and finite interaction time t (c.w. Eq. (2.34)),

$$g_t(\omega) = \frac{0.886}{\Delta\omega}\mathrm{sinc}^2\left(\frac{2.78\omega}{\Delta\omega}\right), \quad \Delta\omega = \frac{5.56}{t} \tag{2.42}$$

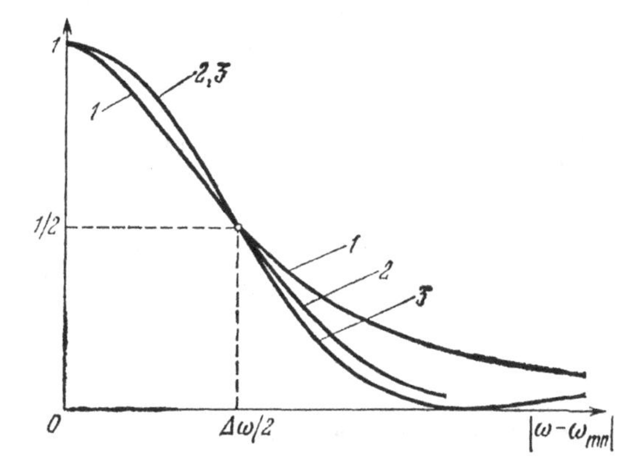

Fig. 2.3 Spectral line shapes. Broadening due to collisions or spontaneous radiation leads to a Lorentzian line shape (1); the Doppler effect results in a Gaussian line shape (2), while limited interaction time t causes a $\text{sinc}^2(\tilde{\omega}t/2)$ shape (3).

where $\text{sinc}\, x \equiv (\sin x/x)$ and $\Delta\omega$ is the full line width at level $1/2$. It is clear from Fig. 2.3 that the particular mechanism of broadening has a noticeable effect only on the tails of the lines. Note that the amplitude ratio of the area-normalized form factors (2.39), (2.41), (2.42) at exact resonance and with the same $\Delta\omega$ is

$$\frac{2}{\pi} : \left(\frac{4\ln 2}{\pi}\right)^{1/2} : \frac{5.56}{2\pi} \approx 1 : 1.47 : 1.39. \tag{2.43}$$

When deriving (2.37), we have replaced the function $(2/\omega)^2 \sin^2(\omega t/2)$, which was obtained from the perturbation theory, by $2\pi t\, g(\omega)$. (Here, ω is the frequency mismatch between the field and the transition.) Let us explain this procedure using the 'strong collision' model. According to this model, every collision instantly brings the atom back to the initial level, after which the interaction between the field and the atom starts anew. Then, t in (2.34) should be replaced by $t - t_0 \equiv \Delta t$, where t_0 is the time instance of the last collision. In gas, the time interval Δt between the last collision of an atom and some fixed time t is a random variable with exponential distribution (see [Rytov (1976)]):

$$P(\Delta t) = \exp(-\Delta t/\tau)/\tau, \tag{2.44}$$

where τ is the mean time interval between the collisions.

The power absorbed (or emitted) by the atom scales as the instant transition rate at a given time t,

$$W(\Delta t) \equiv dP/dt = \Omega^2 \sin(\omega\Delta t)/2\omega. \tag{2.45}$$

Here, we have used expression (2.34) for the transition probability P. The mean transition rate is given by averaging (2.45) using (2.44),

$$W \equiv \int_0^\infty d(\Delta t) P(\Delta t) W(\Delta t) = \frac{\Omega^2 \tau/2}{1 + \omega^2 \tau^2}. \qquad (2.46)$$

The last expression is in agreement with (2.37) and (2.39) if τ is replaced by the inverse half-width of the line, $2/\Delta\omega$. The integral in (2.46) can be easily done by replacing $\sin x$ with $\mathrm{Im}[\exp(ix)]$.

2.3 Absorption cross-section and coefficient

2.3.1 *Relation between intensity and field amplitude*

In order to pass from the transition rate W to the transition cross-section $\sigma = W/F$ and the absorption (or amplification) coefficient $\alpha = \sigma \Delta N$, we have to write the squared field $|\mathbf{E}_0|^2$ in terms of either the photon flux density $F\,[\mathrm{s}^{-1} \cdot \mathrm{cm}^{-2}]$ or intensity $I = \hbar\omega F\,[\mathrm{W/cm}^2]$.

Let us first find the energy $\mathcal{E}$ of the wave. From Maxwell's equations, it follows that the instant energy of the field contained in a volume V of transparent isotropic non-magnetic material with the dielectric constant ϵ is

$$\mathcal{E}(t) = \int_V \mathrm{d}^3 r (\epsilon E^2 + H^2)/8\pi. \qquad (2.47)$$

In the case of a plane monochromatic wave,

$$E = (1/2)\mathbf{e} E_0 e^{i(\mathbf{k}\cdot\mathbf{r} - \omega t)} + c.c., \quad \mathbf{H} = n\hat{\mathbf{k}} \times \mathbf{E}. \qquad (2.48)$$

Here, $\mathbf{e}$ is the unity polarization vector, $\mathbf{k} = \hat{\mathbf{k}} n\omega/c$ is the wave vector, $\hat{\mathbf{k}}$ is the unity vector in the propagation direction, $n = \sqrt{\epsilon}$ is the refractive index. After substituting (2.48) in (2.47) and time averaging, we get the following relationship between the time-averaged energy and the amplitude of the plane wave:

$$\mathcal{E} \equiv \overline{\mathcal{E}(t)} = n^2 V |E_0|^2/8\pi. \qquad (2.49)$$

Intensity of the wave is obviously given by the product of the energy density, $\mathcal{E}/V$, and the velocity of the wave, c/n,

$$I = cn|E_0|^2/8\pi. \qquad (2.50)$$

Note that this formula was derived without considering the frequency dispersion of the medium, $\epsilon(\omega)$. The dispersion can be taken into account by replacing ϵ in

(2.47) by $d(\omega\epsilon)/d\omega$ (see, for instance, Refs. [Landau (1982); Silin (1961)]). As a result, (2.49), instead of n^2, will contain the expression

$$(1/2)[d(\omega\epsilon)/d\omega + \epsilon] = c^2/uv, \tag{2.51}$$

where $v = c/n$ is the phase velocity and $u = d\omega/dk$, the group velocity. Now, the energy density should be multiplied by u, which again leads to relationship (2.50).

2.3.2 *Cross-section of resonance interaction*

After replacing the squared field $|E_0|^2$ in the expression (2.37) for the transition probability W with the photon flux density F, we find the transition cross section, which is, by definition, equal to W/F (we assume $n = 1$):

$$\sigma_{mn} = (4\pi^2/\hbar c)\omega g(\omega)|d_{mn}^{(e)}|^2, \tag{2.52}$$

where $d_{mn}^{(e)} \equiv \mathbf{d}_{mn} \cdot \mathbf{e}$.

The transition cross-section is maximal at exact resonance ($\omega = \omega_{mn} > 0$); in the case of a Lorentzian line shape (2.39),

$$\sigma_{mn0} = 8\pi\omega_{mn}|d_{mn}^{(e)}|^2/\hbar c\,\Delta\omega. \tag{2.53}$$

One can imagine σ as the area of the 'shadow' made by the atom. Let us estimate this area. The line width $\Delta\omega$ cannot be less than the so-called *natural width*, determined by spontaneous transitions. In what follows (Sec. 2.5), it will be shown that

$$\Delta\omega_{nat} = 4\omega_{mn}^3|\mathbf{d}_{mn}|^2/3\hbar\, c^3 = 1/T_{1\,nat} \tag{2.54}$$

where $T_{1\,nat}$ is the lifetime of an atom in the excited state, which is finite due to spontaneous transitions into the ground state. Let $\mathbf{d}_{mn} \parallel \mathbf{e}$, then, after substituting (2.54) in (2.53), we get

$$\sigma_0 = (3/2\pi)(\Delta\omega_{nat}/\Delta\omega)\lambda^2. \tag{2.55}$$

If $\mathbf{d}_{mn}$ has random orientation, we have $|\mathbf{d}_{mn}|^2 = 3|d_{mn}^{(e)}|^2$, hence a factor of $1/3$ appears in (2.55).

Thus, if both collision and Doppler widths are much less than the natural one, *the 'shadow' of the atom with respect to resonance optical transitions has a size on the order of the wavelength*, $\lambda \sim 10^{-4}$cm, and not the atom size, $a_0 \sim 10^{-8}$cm. In rare gases, the main role is played by Doppler broadening, which is on the order of $\Delta f \equiv \Delta\omega/2\pi \sim 1$GHz. The natural width for allowed optical transitions is two orders of magnitude as small; therefore, $\sigma \sim \lambda^2/100$. (In the case of magnetic dipole transitions, $\Delta f_{nat} \sim 10^3$Hz and $\sigma \sim 10^{-6}\lambda^2$.)

2.3.3 *Population kinetics*

Let us consider the evolution of mean populations N_m of the levels, which are defined as

$$N_m \equiv |c_m|^2 N_0, \tag{2.56}$$

where N_0 is the total number of atoms. Thus, from considering a single particle we pass to studying a system of N_0 identical non-interacting particles. The $n \to m$ transition rate is defined as (see Eq. (2.36))

$$W_{mn} \equiv d|c_m|^2/dt. \tag{2.57}$$

Hence, the population increase rate $\dot{N}_m$ of the final state m is $N_0 \sum_n W_{mn}$. However, we have assumed that the initial state n is occupied with a unity probability; here, this probability is equal to N_n/N_0; therefore, $\dot{N}_m = \sum_n W_{mn} N_n$. If we also take into account that particles leave the levels, we get the following system of equations for populations:

$$dN_m/dt = \sum_n (W_{mn} N_n - W_{nm} N_m). \tag{2.58}$$

Here, so far, the relaxation is ignored; in the general case, the rates W_{mn} should also include contributions from chaotic fields created by the surrounding particles. Such equations are studied in non-equilibrium thermodynamics; they are called *kinetic equations or population balance equations*. If only two levels take part in the exchange, then

$$N_1 + N_2 = N_0, \tag{2.59}$$

and only one of equations (2.58) is sufficient. Because the perturbation operator is Hermitian, it follows that $W_{12} = W_{21} \equiv W$ (see (2.26)), hence

$$\dot{N}_1 = -\dot{N}_2 = W(N_2 - N_1), \tag{2.60}$$

2.3.4 *Photon kinetics*

Each transition down is accompanied by the emission of a single photon while each transition up, by the absorption of a single photon; as a result, the rate of photon emission[c] is $\dot{N}_1$, and the *transfer equation* for photons takes the form

$$\partial N/\partial t + \nabla(\mathbf{u}N) = dN_1/dt, \tag{2.61}$$

[c] In the semiclassical theory of radiation, there is no concept of a photon, and it would be more consistent to speak here of the field energy variation by $\hbar\omega$. But 'photon language' is more convenient as it is more visual.

where $N(\mathbf{r}, t)$ is the photon concentration and $\mathbf{u}N = \mathbf{F}$ is the photon flux density vector. Hence, in the stationary one-dimensional case, where $\partial N/\partial t = 0$ and $F = F_z(z)$,

$$dF/dz = W(N_2 - N_1). \tag{2.62}$$

Further, from the definition of the transition cross-section, it follows that

$$dF/dz = -\sigma(N_1 - N_2)F \equiv -\alpha F. \tag{2.63}$$

Let populations be independent of z, then (2.63) leads to the exponential variation of the intensity[d] of light,

$$F(z) = F(0)e^{-\alpha l}, \tag{2.64}$$

where the absorption (or amplification, at $N_1 < N_2$) coefficient, according to (2.52), is

$$\alpha = (4\pi^2/\hbar c)\omega\, g(\omega)|d_{12}^{(e)}|^2(N_1 - N_2). \tag{2.65}$$

2.3.5 *Coefficient of resonance absorption*

Maximal (resonance) value of the absorption coefficient in the case of a Lorentzian line shape (2.39) is

$$\alpha_0 = (8\pi/\hbar c)(\omega_{21}/\Delta\omega)|d_{12}^{(e)}|^2(N_1 - N_2). \tag{2.66}$$

Note that in the stationary case, $\dot{N}_m = 0$, and at $W \neq 0$ it follows from (2.60) that $N_1 = N_2$, so that, seemingly, always $\alpha = 0$. (This population balancing due to the effect of the radiation is called *saturation*.) However, relaxation processes that are ignored in (2.60), such as spontaneous transitions, inelastic collisions of atoms with each other and with electrons in gases, interaction with lattice oscillations in solids, radiation-free transitions etc., tend to restore the initial population difference $N_1 - N_2$. Therefore, in the case of sufficiently weak fields, saturation can be neglected.

In the optical range, for allowed transitions with natural broadening, α can be as high as 1 cm^{-1} at relatively small numbers of active particles $|\Delta N| = N_2 - N_1$. At $\lambda = 0.5\mu$, according to (2.55), $|\Delta N| = 2\pi\alpha_0/3\lambda^2 = 10^9$cm^{-3}. In the X-ray range, λ is 4 orders of magnitude as small, and $|\Delta N| \approx 10^{17}$cm^{-3}.

In microwave paramagnetic amplifiers, the line width is determined by the dipole-dipole interaction of paramagnetic ions. In a ruby crystal ($Al_2O_3 + 10^{-3}Cr$),

---

[d]Recall that intensity scales as F, namely, $I = \hbar\omega F$.

with the concentration of chrome ions $10^{19}\,\mathrm{cm}^{-3}$, the line width is on the order of 50 MHz. Substituting μ for d in (2.53), for $\lambda = 1$ cm we get $\sigma = 8\pi\mu^2/\hbar\lambda\Delta f = 5 \cdot 10^{-20}\,\mathrm{cm}^2$. A realistic number of active particles available in paramagnetics is approximately equal to the equilibrium population difference, $\Delta N^{(0)} = \hbar\omega N_0/\kappa T g \approx N_0/10 = 10^{18}\,\mathrm{cm}^{-3}$, where $g = 2S + 1 = 4$ is the degeneracy of the chrome ion ground level, which is lifted by a constant magnetic field, and S is the spin number. Hence, $\alpha_0 = 0.05\,\mathrm{cm}^{-1}$, and for obtaining amplification $G = 100$ one needs the length of the crystal $l = \ln G/\alpha_0 \approx 1$ m. In order to reduce l, the crystal is placed into a bulk resonator, where radiation can many times pass through the matter, or into a slowing-down system. In the latter case, the above-given formulas for σ and α will be still valid, with the speed of light in the vacuum replaced by the group velocity of waves in the slowing down medium, $u = dK/dz$, where K is the propagation constant.

2.3.6 *Amplification bandwidth*

Due to the exponential relation (2.64) between the *transfer coefficient* of a layer of thickness l, $G \equiv F(l)/F(0)$, and the absorption coefficient α, the shape of the observed frequency dependence $G(\omega)$ at $|\alpha|\,l \gg 1$ (*large optical density*) will differ from the function $\alpha(\omega)$. It is easy to see that this effect will lead to the 'sharpening' of the observed resonance at $\alpha < 0$ and to its 'broadening' at $\alpha > 0$ (Fig. 2.4). Let $\alpha < 0$ and the $\alpha(\omega)$ dependence be Lorentzian. Defining the amplification bandwidth $\Delta\omega'$ by the condition of $G(\omega) - 1$ two-fold reduction with respect to its

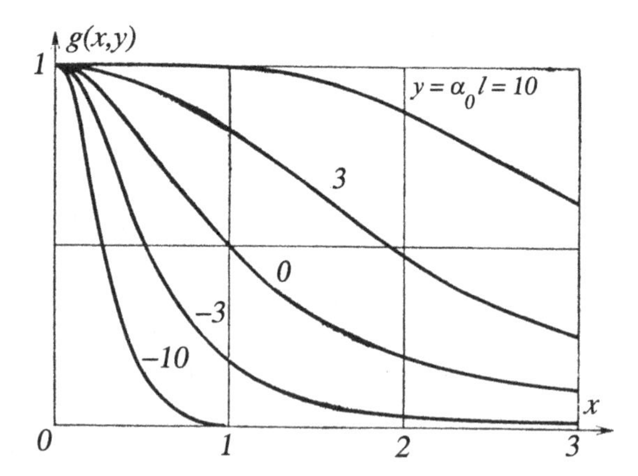

Fig. 2.4 Observed shape of the resonance in the case of a Lorentzian line with width $\Delta\omega$ at different optical densities $y = \alpha_0 l$ at the center of the line ($x \equiv 2(\omega - \omega_0)/\Delta\omega$).

maximal value, we find from (2.64) that

$$\frac{\Delta\omega'}{\Delta\omega} = \left\{\frac{\ln G_0}{\ln[(G_0+1)/2]} - 1\right\}^{1/2}, \tag{2.67}$$

with $G_0 \equiv \exp(-\alpha_0 l)$. Hence, at $G_0 - 1 \ll 1$ it follows that $\Delta\omega' = \Delta\omega$, while at $G_0 - 1 \gg 1$,

$$\frac{\Delta\omega'}{\Delta\omega} \approx \left(\frac{\ln 2}{\ln G_0 - \ln 2}\right)^{1/2} \approx \frac{1}{\sqrt{|\alpha_0|l}}. \tag{2.68}$$

Thus, *the narrowing of the amplification band with the increase of the amplifier length is rather slow*. For instance, at $G_0 = 100$ ($\alpha_0 l = -4.6$) the ratio (2.67) takes the value 0.417 (approximate expressions (2.68) yield 0.42 and 0.46).

2.3.7 °*Degeneracy of the levels*

Expression for the amplification coefficient and the inversion condition $N_2 > N_1$ have been obtained above under the assumption that the atom energy levels were not degenerate. Let now g_1 different (with respect to some parameters) states have the same energy $\mathcal{E}_1$ and g_2 states have energy $\mathcal{E}_2$,

$$\begin{aligned}(\mathcal{H}_0 - \mathcal{E}_1)\varphi_{1i} &= 0 \ \ (i = 1, \ldots, g_1),\\ (\mathcal{H}_0 - \mathcal{E}_2)\varphi_{2j} &= 0 \ \ (j = 1, \ldots, g_2).\end{aligned} \tag{2.69}$$

Note that the conclusions given below will be also valid in the case where the degeneracy is lifted due to sufficiently small perturbations (Fig. 2.5). Now, the

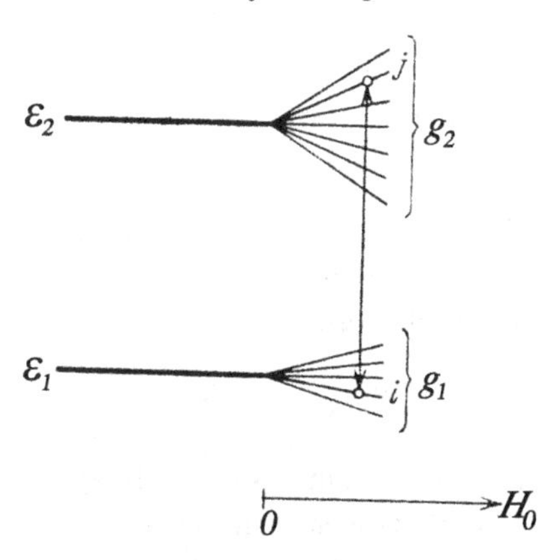

Fig. 2.5 Degeneracy of the levels: g_1 different states have the same energy $\mathcal{E}_1$ while g_2 other states have the energy $\mathcal{E}_2$. In the right-hand side of the figure the degeneracy is lifted due to the external constant field H_0, which breaks the symmetry of the system. Alternating field causes transitions between a certain pair of states i, j.

subscripts 1 or 2 in the perturbation theory should be replaced by double indices, $1i$ or $2j$. The probability of a stimulated transition between the states $1i$ and $2j$, according to (2.37), scales as the corresponding matrix element:

$$W_{1i,2j} = W_{2j,1i} \sim |d^{(e)}_{1i,2j}|^2.$$

The number of transitions up or down is proportional to the population of the initial state N_{1i} or N_{2j}, therefore

$$\dot{N}_{1i} = \sum_{j=1}^{g_2} (W_{1i,2j} N_{2j} - W_{2j,1i} N_{1i}). \tag{2.70}$$

The variation rate of the total population of the level, $N_1 \equiv \sum N_{1i}$, will be equal to the double sum over the degenerate states,

$$\dot{N}_1 = \sum_{ij} W_{1i,2j}(N_{2j} - N_{1i}). \tag{2.71}$$

Now, assume that the saturation effect is absent and relaxation or inversion lead to a uniform distribution of sublevel populations,

$$N_{1i} = N_1/g_1, \; N_{2j} = N_2/g_2. \tag{2.72}$$

As a result, (2.71) takes the form (see (2.60))

$$\dot{N}_1 = -W' \Delta N', \tag{2.73}$$

where

$$W' \equiv \sum_{ij} W_{1i,2j}, \; \Delta N' \equiv N_1/g_1 - N_2/g_2. \tag{2.74}$$

Thus, degeneracy of the levels can be taken into account if W in Eq. (2.66) is understood as W' and ΔN, as the difference of 'state populations' N_m/g_m. The inversion condition then takes the form

$$N_1/g_1 < N_2/g_2. \tag{2.75}$$

Let, for instance, $g_1 = 1$ and $g_2 = 3$, then, one needs $N_2 > 3N_1$ for amplification. Recall that, according to the Boltzmann distribution,

$$N_2^{(0)}/N_1^{(0)} = (g_2/g_1)\exp(-\hbar\omega_{21}/\kappa T) \tag{2.76}$$

and $N_2^{(0)} < 3N_1^{(0)}$.

2.4 Stimulated transitions in a random field

Up to now, the field stimulating a transition was considered as monochromatic. Let now $\mathbf{E}(t)$ have an arbitrary time dependence. According to (2.24), the first-order perturbation theory gives the following expression for the transition amplitude in the dipole approximation, provided that $c_1(t_0) = 1$:

$$c_2(t) = -\frac{1}{i\hbar}\int_{t_0}^{t} dt' \exp(i\omega_{21}t') \sum_{\alpha} d_{21\alpha} E_\alpha(t'), \tag{2.77}$$

where $\alpha = x, y, z$ are indices in the Cartesian frame. Squared module of this expression gives the transition probability,

$$P_{21}(t) = \hbar^{-2} \sum_{\alpha\beta} d_{21\alpha} d^*_{21\beta} \iint_{t_0}^{t} dt' dt'' \exp[i\omega_{21}(t' - t'')] E_\alpha(t') E_\beta(t''). \tag{2.78}$$

2.4.1 *Correlation functions*

Next, consider the case of a chaotic, random field. Then, P should be averaged over a corresponding probability distribution, so that instead of the pair product $E_\alpha E_\beta$, (2.78) will contain the matrix of second-order field *moments*[e]

$$\langle E_\alpha(t') E_\beta(t'')\rangle \equiv G_{\alpha\beta}(t', t'') = G_{\beta\alpha}(t'', t'). \tag{2.79}$$

This equality defines a certain tensor, each of its nine components being a function of two arguments. The matrix of second-order moments (2.79) is also called the field *correlation tensor*. Another equivalent term is the field *coherence function* (of the first order). Statistical properties of a random field are fully described by a set of moments (coherence functions) of all orders and for all possible pairs of 'points' $x \equiv \mathbf{r}, t$. Statistical optics will be described in more detail in Chapter 7; here, we only note that odd field moments, as a rule, are equal to zero, while moments of order $2n$ determine the probabilities of n-photon transitions. Let us also mention that the sum of the diagonal elements of the second-order moment matrix, $\sum G_{\alpha\alpha}$ (the *trace* of the matrix), with coinciding arguments defines the mean energy density $\langle E^2\rangle/8\pi$ of the electric field at the point under consideration.

The most important class of random fields are *stationary fields*, whose statistical characteristics (intensity, spectrum, polarization) do not change with time. The correlation function of a stationary process can only depend on the difference of its two arguments,

$$G_{\alpha\beta}(t', t'') = G_{\alpha\beta}(t' + t_0, t'' + t_0) \equiv G_{\alpha\beta}(t' - t''). \tag{2.80}$$

---

[e]Angular brackets denote averaging over a statistical ensemble of fields (Sec. 7.2).

From (2.80) and the definition (2.79), the symmetry property follows,

$$G_{\alpha\beta}(-t) = G_{\beta\alpha}(t); \tag{2.81}$$

in particular, $G_{\alpha\alpha}(t)$ should be an even function of time.

Thus, according to (2.78) and (2.79), the probability of a transition due to stationary random radiation is determined by the field correlation tensor,

$$P_{21} = \hbar^{-2} \sum_{\alpha\beta} d_{21\alpha} d^*_{21\beta} \iint_{t_0}^{t} dt' dt'' \exp[i\omega_{21}(t' - t'')] G_{\alpha\beta}(t' - t''). \tag{2.82}$$

2.4.2 *Transition rate*

Consider the action of the perturbation at time intervals that are much larger than the field correlation time τ_E (Sec. 7.2). Then the integration limits in (2.82) can be replaced by $\pm\infty$. Let us make a change of variables, $t_1 \equiv t' - t'', t_2 \equiv t' + t''$. Integration in t_1 yields the Fourier transform of $G_{\alpha\beta}(\omega_{21})$, which is called the field *spectral density* tensor,

$$G_{\alpha\beta}(\omega) \equiv (2\pi)^{-1} \int dt e^{i\omega t} G_{\alpha\beta}(t) = G^*_{\alpha\beta}(-\omega) = G^*_{\beta\alpha}(\omega) \equiv G^+_{\alpha\beta}(\omega). \tag{2.83}$$

The inverse transformation has the form

$$G_{\alpha\beta}(t) = \int d\omega e^{-i\omega t} G_{\alpha\beta}(\omega). \tag{2.84}$$

Here, as usual, we omit infinite integration limits and denote the function and its Fourier transform by the same letter. The second integration (in t_2) simply yields the observation time $t - t_0$, so that one can define a time-independent transition rate $W \equiv P/(t - t_0)$. Let the dipole moment of the transition be parallel to the x axis, then we finally find a simple expression for the transition rate,

$$W_{21} = 2\pi\hbar^{-2} |d_{21}|^2 G_{xx}(\omega_{21}). \tag{2.85}$$

Thus, *the rate of a stimulated transition due to a random (noise or incoherent) perturbation scales as the spectral density $G(\omega)$ of the perturbation at the transition frequency.* It is useful to compare (2.85) with formula (2.37), which defines the transition rate in the case of a monochromatic field. The two formulas coincide after the substitution $|E_0|^2 g(\omega) \to G(\omega)$.

In this consideration we did not take into account the broadening $\Delta\omega$ of the levels due to relaxation processes. However, intuitively it is clear that the conclusion should be still valid in the case where $\Delta\omega$ is much less than the width of the perturbation spectrum, $\Delta\omega_E \sim 1/\tau_E$. In this case, the field is called incoherent. In the opposite case, the field can be obviously considered as monochromatic, i.e., coherent.

2.4.3 *Einstein's B coefficient*

Consider now isotropic non-polarized radiation with $G_{\alpha\beta} = G\delta_{\alpha\beta}$. From $G(\omega)$ we will pass to the energy spectral density $\rho(\omega)$. The latter is defined through the energy density as follows (we assume $n = 1$):

$$\mathcal{E}/V = \langle E^2 + H^2 \rangle / 8\pi \equiv \int_0^\infty d\omega \rho(\omega). \tag{2.86}$$

In the radiation, $E = H$, therefore,

$$\mathcal{E}/V = \sum_\alpha G_{\alpha\alpha}(t=0)/4\pi = 3\int_0^\infty d\omega G(\omega)/2\pi, \tag{2.87}$$

where the last equality was obtained using relation (2.84) with $t = 0$ and taking into account that, according to (2.83), $G(\omega)$ is an even function. From the comparison of (2.86) and (2.87) we get the relation between the spectral densities of field amplitude and energy,

$$G(\omega) = 2\pi\rho(\omega)/3. \tag{2.88}$$

By substituting (2.88) in (2.85), we finally find the transition rate in an isotropic non-polarized noise field with a broad spectrum:

$$W_{21} = B_{21}\rho(\omega_{21}), \tag{2.89}$$

$$B_{21} = B_{12} \equiv (2\pi|d_{21}|/\hbar)^2/3. \tag{2.90}$$

The proportionality coefficient B between the transition rate and the energy density is called the *Einstein coefficient for a stimulated transition*. In the next section, using the Planck function $\rho^{(0)}(\omega)$ for equilibrium radiation, we will find the second Einstein coefficient, A, giving the rate of spontaneous transitions.

2.4.4 °*Spectral field density*

Concluding this section, let us clarify the physical meaning of the field spectral density $G(\omega)$. In order to do this, we formally represent $E(t)$ as a Fourier integral (the α subscript is omitted),

$$E(t) = \int d\omega e^{-i\omega t} E(\omega), \tag{2.91}$$

$$E(\omega) = \int dt e^{i\omega t} E(t)/2\pi. \tag{2.92}$$

A rigorous definition for the Fourier representation of a random function can be found in Ref. [Rytov (1976)]. With the help of (2.92) and definitions (2.79), (2.80), we find the correlator of the field Fourier components,

$$\langle E(\omega)E(\omega')\rangle = \iint dtdt' e^{i\omega t + i\omega' t'} G(t-t')/4\pi^2. \tag{2.93}$$

In the time integral, we make a change of variables, $t_1 \equiv t - t'$, then the integral in t_1 yields, according to (2.83), $2\pi G(\omega)$, while the second integral yields $2\pi\delta(\omega+\omega')$, according to one of the delta-function representations,

$$\lim_{T\to\infty}\int_{-T}^{T} dt e^{i\omega t} = 2\pi\delta(\omega). \tag{2.94}$$

As a result,

$$\langle E(\omega)E(\omega')\rangle = G(\omega)\delta(\omega+\omega'). \tag{2.95}$$

Because $E(t)$ is real, it follows from (2.92) that

$$E(\omega) = E^*(-\omega); \tag{2.96}$$

therefore, (2.95) can be also represented in the form

$$\langle E(\omega)E^*(\omega')\rangle = G(\omega)\delta(\omega-\omega'). \tag{2.97}$$

Thus, *in a stationary field, only harmonics of opposite frequencies correlate, and their correlation is determined by the spectral density* $G(\omega)$. It means that the reading of a photodetector measuring the field energy within a frequency band $\Delta\omega$ centered at $\bar{\omega}$ will scale as $G(\bar{\omega})\Delta\omega$.

2.5 Field as a thermostat

Consider population kinetics for atoms in an equilibrium field with the energy spectral density $\rho^{(0)}(\omega)$. From the kinetic equation (2.60) for two nondegenerate levels, it follows that

$$\dot{N}_2 = -\dot{N}_1 = B\rho(N_1 - N_2), \tag{2.98}$$

where $B \equiv B_{12} = B_{21}$ and $\rho \equiv \rho(\omega_{21})$. Thus, a noise broadband field, similarly to a monochromatic one, tends to equalize the populations of the levels, so that in the stationary regime $N_1 = N_2$. However, equilibrium radiation of temperature T should heat or cool the matter to the same temperature, hence the population distribution is given by the Boltzmann formula, according to which $N_1 > N_2$.

2.5.1 *Spontaneous transitions*

This contradiction can be solved by adding to the kinetic equation (2.98) a term describing spontaneous (i.e., field-independent) transitions from the excited level 2 to the ground level 1. Such transitions are accompanied by the emission of light from the heated body, which prevents the equalization of the populations due to the external field. According to Einstein, let us denote the rate of spontaneous transitions as $A_{12} \equiv A$, then (2.98) takes the form

$$\dot{N}_2 = B\rho(N_1 - N_2) - AN_2. \tag{2.99}$$

The A coefficient can be calculated from Boltzmann's and Planck's distributions and the B value found above. If the atoms are in equilibrium with the field, $\dot{N}_2 = 0$, and (2.99) leads to the relation

$$A/B = (N_1/N_2 - 1)\rho. \tag{2.100}$$

Substituting here the Planck distribution,

$$\rho^{(0)}(\omega) = \hbar k^3 \mathcal{N}(\omega)/\pi^2, \tag{2.101}$$

with

$$\mathcal{N}(\omega) = [\exp(\hbar\omega/\kappa T) - 1]^{-1}, \tag{2.102}$$

and the Boltzmann distribution,

$$N_1^{(0)}/N_2^{(0)} = \exp(\hbar\omega_{21}/\kappa T), \tag{2.103}$$

we find the ratio of the coefficients for spontaneous and stimulated transitions:

$$A/B = \hbar k^3/\pi^2, \tag{2.104}$$

where $k = \omega/c = 1/\lambda\!\!\!^-$. Hence, taking into account expression (2.90) for the B coefficient, we find that

$$A = 4k^3|d_{21}|^2/3\hbar. \tag{2.105}$$

For an allowed transition in the visible range ($d_{21} = 1\text{D}$, $\lambda = 0.5\mu$), estimation yields $A = 2 \cdot 10^6\ \text{s}^{-1}$.

The notion of a spontaneous transition plays an important role in the theory of interaction between field and matter and in quantum electronics. Spontaneous transitions determine the minimal linewidths of emission and absorption. They lead to the thermal radiation of heated matter. Similarly to relaxation processes in general, they hinder obtaining population inversion. Also, since spontaneous transitions occur independently of the external field, they are a source of noise and therefore limit the sensitivity of quantum amplifiers and the monochromaticity of quantum generators (Sec. 7.1).

Let us note that the probability of a spontaneous transition has a strong (cubic) dependence on the frequency, which explains why creating UV and X-ray lasers faces a certain difficulty. Although in the opposite spectral range, the microwave one, the probability of spontaneous transitions is very small, the minimal noise temperature of paramagnetic amplifiers is namely determined by spontaneous transitions (see (7.10), (7.11)).

We have found A in an indirect way. Spontaneous transitions can be consistently explained in the framework of quantum electrodynamics, by the interaction between atoms and the vacuum (Sec. 7.7). However, they have simple classical and semiclassical analogues, the radiation of accelerated electrons in an atom (Sec. 5.2). One can also provide a 'semi-quantum model': a classical current, according to Glauber [Glauber (1965)], excites quantized field into a coherent state.

2.5.2 *Natural bandwidth*

Spontaneous transitions limit the lifetime T_1 of an isolated atom in an excited state. One can expect that $T_1 = 1/A$. This simple dependence will be confirmed in what follows, see (5.34). Further, according to the uncertainty relation $\Delta\mathcal{E}\Delta t = \hbar$, where $\Delta\mathcal{E}$ is the accuracy of the energy measurement and Δt is the measurement time, a finite lifetime of an atom leads to a finite width of the energy level. Assuming $\Delta t = T_1$, we obtain $\Delta\mathcal{E}_2 = \hbar A$. This broadening of the level should manifest itself, in stationary experiments, in the variance of the transition frequency, $\Delta\omega_{21} = \Delta\mathcal{E}_2/\hbar$, i.e., in the broadening of spectral lines,

$$\Delta\omega_{nat} = A. \tag{2.106}$$

The width of spectral lines caused by spontaneous transitions is called the *natural bandwidth*. This term stresses that $\Delta\omega_{nat}$ is the minimal possible linewidth, which takes place even in the case of a single isolated atom. Note, however, that natural broadening can be, in principle, eliminated by placing an atom into a bulk resonator that has no oscillations with frequencies in the vicinity of ω_{21}. In practice, observed lines have natural bandwidth only in the rare cases where other perturbations, such as collisions and the Doppler effect in gases, interaction with phonons in crystals, etc., have much smaller contributions and, in addition, the optical thickness of the sample is small (Sec. 7.1). Note that in this case, the width of absorption or amplification lines related to stimulated transitions is also equal to A.

Natural broadening of isolated lines leads to Lorentzian lineshapes. Rigorously, this follows from the Wigner-Weisskopf theory (see, for instance, Ref. [Louisell (1964)]). There is also a simple classical model, according to which an excited atom emits an exponentially decaying quasi-monochromatic oscillation (Sec. 5.2). Fourier transformation of this oscillation provides the Lorentzian (dispersion) lineshape of the emission spectrum.

Let us estimate, with the help of (2.106), the relative value of the natural broadening. For allowed transitions, $d \approx ea_0$ (see (2.30)), so that

$$\Delta\omega_{nat}/\omega \approx (4/3)(1/137)(a_0/\lambda\!\!\!^{-})^2, \tag{2.107}$$

where we assumed the value $1/137$ for the fine structure constant $e^2/\hbar c$ (recall that this number also defines the ratio of the velocity of an electron in a hydrogen atom to the speed of light). Assuming $\lambda = 1/R \equiv 4\pi \cdot 137a_0 \approx 0.1\mu$, with R being the Rydberg constant, we obtain

$$\Delta\omega_{nat}/\omega \approx 0.3/(137)^2 \approx 10^{-7}. \tag{2.108}$$

Displacement of atomic levels due to the interaction with the electromagnetic vacuum (*the Lamb shift*) is on the same order of magnitude or smaller. Thus, relative perturbation of an atom by the vacuum is extremely small.

2.5.3 *Number of photons, spectral brightness, and brightness temperature*

Let us find the ratio of stimulated and spontaneous transition probabilities in the case of incoherent (noise) field. According to (2.104) and (2.89),

$$W_{st}/W_{sp} = B\rho/A \equiv \rho/\hbar\omega g_\omega \equiv N, \tag{2.109}$$

where

$$g_\omega \equiv \omega^2/\pi^2 c^3, \tag{2.110}$$

has the meaning of *the spectral density of field modes* in a unit volume (recall that ρ is the energy spectral density per unit volume). A *mode*, or an *oscillation type*, is, roughly speaking, an oscillation degree of freedom (or a spatial harmonic) of the field (Sec. 7.3). The inverse value, $1/g_\omega$, is equal to the frequency interval between the neighboring modes. According to definition (2.109), N has the meaning of the field energy per one mode, in $\hbar\omega$ units. In other words, N is the *number of photons per mode*. This value is also called the *degeneracy factor* of photon gas. Note that both energy and the number of photons fluctuate; here, ρ and N are mean values.[f]

The value N is the most important parameter of incoherent radiation. Let us show that it scales as the basic photometry characteristic, the *spectral brightness* $I_{\omega\Omega}$. The latter is defined as the radiation intensity within a unit spectral interval and unit solid angle and has dimensionality $[\mathrm{W}/(\mathrm{cm}^2\cdot\mathrm{Hz}\cdot\mathrm{sr})]$. Radiation intensity per unit frequency is equal to half the spectral energy density, $\rho/2$, multiplied by the speed of light. By adding the factor $1/4\pi$, we pass to the spectral brightness $I_{\omega\Omega}$. Hence, with the help of (2.109), we find

$$I_{\omega\Omega} = c\rho/8\pi = \hbar c\lambda^{-3}N. \tag{2.111}$$

Thus, according to (2.109), *stimulated transitions due to incoherent field occur N times as frequently as spontaneous ones*. The total number of transitions up and down can be represented as

$$w_{21} = AN, \quad w_{12} = A(N+1). \tag{2.112}$$

Sometimes spontaneous transitions, which correspond to the second term in the last expression, are interpreted as stimulated transitions due to zero-point (vacuum) fluctuations of the field. However, this interpretation leads to underestimating twice the probability of spontaneous transitions down and does not explain the absence of spontaneous transitions up [Ginzburg (1983)]. The correct result is obtained by distinguishing between *normally and antinormally* ordered fluctuations. In Sec. 7.7, it is shown that *spontaneous transitions are determined by normally ordered fluctuations of the atom dipole moment and by antinormally ordered fluctuations of the vacuum.*

In equilibrium radiation, N only depends on the frequency and temperature and is given by the Planck formula (2.102), $N = \mathcal{N}(\omega)$. In the general case, N depends, in addition to frequency, on the observation direction, polarization type,

[f]Editors' note: Eq. (2.109) has an important consequence in the context of quantum information. As it was first mentioned by Wooters and Zurek and independently by Milonni and Hardies in 1982, if the ratio between the induced and spontaneous transitions tends to unity, then it leads to the impossibility to clone the polarization state of a single photon. Indeed, if $W_{in} = W_{sp}$, then the mean number of photons per field mode equals unity. Then, amplification of a single photon through stimulated transitions in an atom (which was initially prepared in excited state) will be accompanied by the spontaneous emission of a photon that has random polarization with respect to the initial one. If the mean number of photons per mode grows, then the contribution of spontaneous transitions goes down. According to (2.109), cloning becomes possible in the limit of high N, which is often associated with classical field, therefore this fact does not contradict to the non-cloning theorem. However, we would like to stress that this conclusion has to be applied with caution because there is no criteria of non-classicality of light based on the mean photon number! For instance, squeezed states of light or bright squeezed vacuum states (Sec.7.5) are nonclassical despite having large photon numbers [Bachor (2004)].

and observation point: $N = N(\mathbf{k}, \mathbf{r}, \nu)$. Here, $\mathbf{k}$ is the wavevector, which also defines the frequency and the direction, and ν is polarization index taking two values.

In non-equilibrium field, equality (2.102) is used as the definition of the brightness temperature $T_{ef}(N) \equiv \hbar\omega/\kappa \ln(1 + 1/N)$ for radiation with given frequency, direction, and polarization. For instance, solar radiation, taken in the optical range and within an appropriate angular interval, has $T_{ef} \sim 6000$ K, so that, according to (2.102), for green light ($\lambda \sim 0.5\mu$) $N \sim 10^{-2}$. We see that stimulated transitions in solar light are much less frequent than spontaneous ones. Thus, in the visible range the probabilities of stimulated and spontaneous transitions become equal only for radiation that is hundreds of times as bright as the Sun radiation, with $T_{ef} \sim 4 \cdot 10^4$ K. Such brightness can be only achieved with multimode lasers (the notion of brightness cannot be applied to single-mode lasers).

2.5.4 °*Relaxation time*

With the help of (2.99), let us now define the rate of population variation with an account for both spontaneous and stimulated transitions. For this, we replace N_1 by $N_0 - N_2$ and use relation (2.112):

$$\dot{N}_2 = A[NN_0 - (2N + 1)N_2]. \tag{2.113}$$

Here N is the number of photons per mode and $N_0 = N_1 + N_2$ is the total number of atoms at two levels. Hence, in a stabilized regime,

$$\begin{aligned} N_1^{(0)}/N_0 &= (N + 1)/(2N + 1) = \nu/(1 + \nu), \\ N_2^{(0)}/N_0 &= N/(2N + 1) = 1/(1 + \nu), \end{aligned} \tag{2.114}$$

where $\nu \equiv \exp(\hbar\omega_{21}/\kappa T)$. The solution to (2.113) has the form

$$\begin{aligned} N_2(t) &= N_2^{(0)} + [N_2(t_0) - N_2^{(0)}]e^{-t/T_1}, \\ 1/T_1 &\equiv A(2N + 1) = 2B\rho + A = w_{12} + w_{21}. \end{aligned} \tag{2.115}$$

Thus, the time T_1 of heating (or cooling) of the atomic internal degrees of freedom due to the interaction with incoherent radiation at small N equals $1/A$, the life time of an atom due to spontaneous transitions, while at large N it reduces $2N + 1 = \coth(\hbar\omega_{21}/2\kappa T)$ times.

In fact, here we have considered a simple model of relaxation where the thermostat is formed by incoherent electromagnetic radiation surrounding the atom (Sec. 7.7).

Chapter 3

Density Matrix, Populations, and Relaxation

The probability method used above allowed us to describe the energy exchange between radiation and atoms. At the same time, another known manifestation of the interaction between field and matter, namely, the slowing down of the waves propagation, was not considered. Another, and more important, drawback of the probability approach is that it does not provide a sufficiently rigorous account for relaxation processes, whose consistent consideration should be performed in the framework of statistical physics and kinetics. A more complete theory of the interaction of atoms with the external field and the thermostat is based on the density matrix formalism, which combines quantum and statistical considerations.

Below, in Sec. 3.1, we discuss the definition and the general properties of the density matrix. In Sec. 3.2, its diagonal elements are considered, which give the populations of the levels, and the notion of negative temperature is introduced. Section 3.3 describes the time evolution of the density matrix and the relaxation processes.

3.1 Definition and properties of the density matrix

3.1.1 *Observables*

In Chapter 2, we have defined the transition probability in terms of the amplitudes of energy states c_n. Let us now write an arbitrary observable f of a quantum system (further, f will be understood as the dipole moment of an atom, $f \equiv d_\alpha$) in terms of similar coefficients. We will start from the basic 'measurement' postulate of the quantum mechanics: multiple measurements of a value f performed on an ensemble of identical systems, i.e., systems 'prepared' in the same state $\Psi(\mathbf{r}, t)$, will yield, on the average, the value

$$\langle f(t)\rangle = \int d\mathbf{r}\Psi^*(\mathbf{r}, t)\hat{f}\Psi(\mathbf{r}, t) \equiv \langle t|f|t\rangle, \tag{3.1}$$

where $\hat{f}$ is the operator corresponding to f and $\mathbf{r}$ is the set of the system coordinates. (In what follows, we will often omit the hats of the operators.)

It is important that f in (3.1) can be also understood as a product of operators, $f \equiv g^2$ or $f \equiv gh$. This enables one to determine not only the mean values, $\langle f \rangle$, called first-order *moments*, but also higher-order moments, $\langle g^2 \rangle$, $\langle g^n \rangle$, $\langle gh \rangle, \ldots$, which characterize the quantum fluctuations of g and the quantum correlation of g and h. Of course, Eq. (3.1) is written in the Schrödinger picture and is therefore applicable only to the case where f is a single-time operator, for instance, $f(t) = g(t)\,h(t)$. In order to define the correlation functions $\langle g(t_1)\,h(t_2)\rangle$, one has to switch to the Heisenberg picture where the time dependence is attributed to the operators and not to the wave functions.

Knowing the wave function, one can find not only the moments $\langle f^n(t)\rangle$ of an observable f but also its distribution at time t, $P(f,t)$. This function is given by Eq. (3.1) with the operator $\hat{f}$ replaced by the diade operator $|f\rangle\langle f|$ (Sec. 7.5).

3.1.2 *Density matrix of a pure state*

Let us expand the wave function over the set of the eigenstates of some operator (not necessarily the energy operator),

$$\Psi(\mathbf{r},t) = \sum_n b_n(t)\varphi_n(\mathbf{r}). \tag{3.2}$$

In the Dirac notation, simply $|t\rangle = \sum |n\rangle\langle n|t\rangle$. Note that if φ_n are energy functions, then the coefficients b_n and c_n differ in only exponential factors (see (2.13)). Substituting (3.2) in (3.1), we obtain

$$\langle f \rangle = \sum_{nm} b_n^* b_m f_{nm}. \tag{3.3}$$

Here, the matrix

$$f_{nm} \equiv \int d\mathbf{r}\varphi_n^* f \varphi_m \equiv \langle n|f|m\rangle$$

is assumed to be known, and the problem is reduced to the one of finding pairwise products $b_n^* b_m$, which also form a matrix, called the *density matrix* or the *statistical matrix*,

$$\rho_{mn} \equiv b_m b_n^*. \tag{3.4}$$

Thus, the state vector of the system Ψ is put into correspondence with a matrix. One can also define an operator $\hat{\rho}$ corresponding to Ψ:

$$\int d\mathbf{r}\varphi_m^* \hat{\rho} \varphi_n \equiv b_m b_n^*, \tag{3.5}$$

which, in Dirac's notation, reads $\hat{\rho} = |t\rangle\langle t|$ (Sec. 7.5).

In terms of the density matrix and operator, the mean (3.3) can be written in a more compact manner,

$$\langle f \rangle = \sum_{mn} \rho_{mn} f_{nm} = \mathrm{Tr}(\rho f), \tag{3.6}$$

where $\mathrm{Tr}f$ means the sum of the diagonal elements, $\sum f_{nn}$, called the *trace*, or *Spur*, of the matrix. The trace of a matrix is one of its invariants, since it does not change under the matrix transformations of the form $f' = UfU^{-1}$. Such operator transformations describe changes of the representation in quantum mechanics, and the invariance of the trace provides that the observable quantities are independent of the choice of representation. The property $\mathrm{Tr}f' = \mathrm{Tr}f$ immediately follows from another property, $\mathrm{Tr}(gh) = \mathrm{Tr}(hg)$, which can be easily verified from the definitions of the Tr and multiplication operations.

3.1.3 *Mixed states*

In the transition from classical mechanics to statistical physics, the main notion is the Gibbs ensemble, which is composed of identical systems distributed with the probabilities $P(q, p)$ over the possible states of the system. A quantum statistical ensemble is constructed in a similar way: we assume that its P_1N systems are in the state Ψ_1, P_2N in the state Ψ_2, P_iN in the state Ψ_i, and so on. Here, N is the total number of systems in the ensemble and $\sum P_i = 1$.

A *mixed state* is a state of the system for which the exact wave function is not defined but only the set of numbers P_i is known, each of them giving the probability that the system is in the Ψ_i state. The system is then characterized as a weighed *mixture* of states, in contrast to a *pure* state, for which the wave function of the system is known.

We stress that a linear combination $\alpha_1\Psi_1 + \alpha_2\Psi_2$ is still a pure state with a well-defined wave function. The mean value of an operator contains in this case an interference term, which depends on the relative phase of the states,

$$\langle f \rangle = P_1 f_{11} + P_2 f_{22} + 2\mathrm{Re}(\alpha_1^* \alpha_2 f_{12}), \tag{3.7}$$

where

$$P_i \equiv |\alpha_i|^2, \quad f_{ij} \equiv \int d\mathbf{r} \Psi_i^* f \Psi_j.$$

In a similar mixed state, the last term in (3.7) is absent. There is an analogy with the superposition of two light fields: coherent fields interfere, while a non-coherent mixture simply yields double intensity.

Additional uncertainty of mixed states leads to additional 'thermal' fluctuations of observables over an ensemble and, according to the *ergodicity hypothesis*, also over time. However, these fluctuations are not as principal and inevitable as quantum ones.

In real experiments, the 'purity' of the prepared states depends on the skill of the experimentalist. Near the absolute zero, a crystal is in a pure (ground) state with a definite energy, but the coordinates of its atoms still have quantum fluctuations. In a good maser or laser with complete population inversion, the atoms are in the excited state with the energy $\mathcal{E}_2$. In this case, there are no energy fluctuations, $\langle \mathcal{H}^2 \rangle = \mathcal{E}_2^2$, but the coordinate of the electron and the dipole moment still fluctuate.

Thus, depending on its prehistory, a quantum object can be found in one of the three possible state types:

1) in an eigenstate φ_n of a given operator $\hat{f}$, where one knows *a priori* that $\langle f \rangle = f_{nn}$ and f does not fluctuate,

$$\langle f^k \rangle = \langle f \rangle^k;$$

2) in a pure state Ψ formed by a superposition $\sum b_n \varphi_n$, with quantum fluctuations observed and only the probability $|b_n|^2$ of measuring a certain value f_{nn} is known,

$$\langle f^k \rangle = \sum |b_n|^2 (f_{nn})^k;$$

3) in a mixed state, where quantum uncertainty is combined with the lack of information about the wave function.

In the case of mixed states, mean values should be calculated via double averaging: quantum averaging over the wave function Ψ_i, according to Eq. (1), and classical averaging, with the help of the P_i distribution and the usual rules of the probability theory,

$$\overline{\langle f \rangle} = \sum_i P_i f_{ii} = \sum_i P_i \int d\mathbf{r} \Psi_i^* f \Psi_i. \tag{3.8}$$

Now, the amplitudes in expansion (3.2) and the density matrix (3.4) depend on the index i,

$$\Psi_i = \sum b_n^{(i)} \varphi_n,$$

so that (3.8) takes the form

$$\overline{\langle f \rangle} = \sum_{mni} P_i (b_n^{(i)})^* b_m^{(i)} f_{nm} = \mathrm{Tr}(\overline{\rho f}), \tag{3.9}$$

where we have defined the density operator of a mixed state,

$$\hat{\bar{\rho}} \equiv \sum_i P_i \hat{\rho}_i, \; \bar{\rho}_{mn} \equiv \overline{b_m b_n^*}. \tag{3.10}$$

Due to its linearity, the averaging operation could be included into the definition of the density matrix. Equation (3.6) maintains its form in this case. In future, the bar denoting additional averaging will be omitted.

3.1.4 °*More general definition of the density matrix*

Often, one defines a mixed state and the density matrix in a different, more general way. In this case, the term 'mixed state' is applied not to the whole system but to a part of it.

Let a system consist of two parts, A and B. Its state is given by a wave function $\Psi(\mathbf{r}_A, \mathbf{r}_B)$, which, in the general case, is not factorable, $\Psi(\mathbf{r}_A, \mathbf{r}_B) \neq \Psi_A(\mathbf{r}_A)\Psi_B(\mathbf{r}_B)$. Therefore, the wave function Ψ_A of a subsystem does not exist. Indeed, factorability means independence of the subsystems, hence it is impossible if A and B interact or have interacted in the past. There is an analogy with classical statistical physics: for interacting particles, the joint probability distribution $P(\mathbf{r}_A, \mathbf{r}_B)$ cannot be represented as $P(\mathbf{r}_A)P(\mathbf{r}_B)$.

However, in classical statistical physics we can separately define the probability distribution function for subsystem A by summing P over the variables that are of no interest for us,

$$P_A(\mathbf{r}_A) = \int d\mathbf{r}_B P(\mathbf{r}_A, \mathbf{r}_B). \tag{3.11}$$

The question is whether a similar procedure is possible in quantum mechanics.

In order to define the density matrix of a subsystem, let us expand Ψ over some complete set of functions $\psi_{in}(\mathbf{r}_A, \mathbf{r}_B)$,

$$\Psi = \sum_{in} b_{in}\psi_{in}. \tag{3.12}$$

Such a set $\psi_{in} = \chi_i\varphi_n$ is generated by two operators, each of them acting on the variables of only one subsystem. For instance,

$$(\mathcal{H}_A - \mathcal{E}_{An})\varphi_n = 0, \;\; (\mathcal{H}_B - \mathcal{E}_{Bi})\chi_i = 0.$$

Let f_A be the observable of interest,

$$\begin{aligned} \langle f_A \rangle &= \int d\mathbf{r}_A d\mathbf{r}_B \Psi^* f_A \Psi \\ &= \sum_{ii'nn'} b^*_{in} b_{i'n'} f_{Ann'} \delta_{ii'} \equiv \sum_{nn'} \rho_{An'n} f_{Ann'} = \mathrm{Tr}(\rho_A f_A). \end{aligned} \tag{3.13}$$

We have once again obtained Eq. (3.6) by introducing the notation

$$\rho_{An'n} \equiv \sum_i b^*_{in} b_{in'}, \tag{3.14}$$

which is equivalent to the definition (compare with (3.11))

$$\rho_A \equiv \mathrm{Tr}_B(\rho_{AB}). \tag{3.15}$$

In quantum electronics, the A system usually corresponds to a particular atom (or a molecule or an electron in a crystal) while the B system includes all other matter particles and quantized electromagnetic field. (Apparently, a classical field acting on a system does not destroy the 'purity' of the state.) Sometimes, on the contrary, A is understood as a particular field mode while B corresponds to the matter. If system B has a sufficient number of degrees of freedom and a continuous energy spectrum, i.e., has a large heat capacity, then its state can be considered as independent of A and it plays the role of a *thermostat*. The influence of B on A causes the relaxation of A.

If the back-action of A on B can be neglected, then A can be described by means of a wave function or a density matrix of a pure state (3.4), and we come back to the problem of a quantum system in a given noise field, which was considered in Sec. 2.4. Solving this problem in the framework of some model enables one to calculate the relaxation properties and the shapes of spectral lines. In the simplest model, the thermostat for an atom is formed by the equilibrium Planck field, and the probabilities of relaxation transitions are given by the Einstein coefficients A, B (Sec. 2.5).

3.1.5 *Properties of the density matrix*

Using definition (3.10), one can easily show that the density matrix has the following properties:

$$\mathrm{Tr}\rho = 1, \;\; 0 \leq \rho_{nn} \leq 1, \;\; \rho^+ = \rho. \tag{3.16}$$

In most cases, one uses the energy representation, in which the diagonal elements of the density matrix $\rho_{nn} \equiv \rho_n$ have the meaning of relative occupation numbers of the levels. The first property in (3.16) means that the probability to find the system on some level is equal to unity, the second one provides the non-negativity of the probability, and the third one (Hermiticity), that the observable quantities are real,

$$\langle f \rangle^* = \sum \rho^*_{mn} f^*_{nm} = \sum \rho_{nm} f_{mn} = \langle f \rangle.$$

The non-diagonal element of the density matrix $\overline{b^*_m b_n}$ characterizes the degree of correlation between the m and n states in a statistical ensemble. If the state amplitudes of various systems of the ensemble contain random phase factors, $b_n^{(i)} \sim \exp(i\varphi_n^{(i)})$, then, for $m \neq n$,

$$\rho_{mn} \sim \overline{\exp i(\varphi_m - \varphi_n)} = 0, \tag{3.17}$$

and the state of the ensemble is fully characterized by the state populations ρ_n.

For a pure state, definition (3.4) leads to the property $|\rho_{mn}|^2 = \rho_{mm}\rho_{nn}$. In a mixed state, the elements of the density matrix satisfy the Cauchy-Bunyakovsky inequality,

$$|\rho_{mn}|^2 < \rho_{mm}\rho_{nn}. \tag{3.18}$$

3.1.6 °*Density matrix and entropy*

Let a closed system be in a pure energy state, $\Psi = \varphi_1 \exp(-i\mathcal{E}_1 t/\hbar)$. Then, according to definition (3.4), there is only one nonzero element of the density matrix, $\rho_{mn} = \delta_{mn}\delta_{n1}$. Such a trivial matrix satisfies the matrix equation

$$\hat{\rho^2} = \hat{\rho}. \tag{3.19}$$

This is a property of all pure states. It follows from Eq. (3.4), the matrix multiplication rule, and the normalization condition $\mathrm{Tr}\hat{\rho} = 1$.

Violation of equality (3.19), or its corollary $\mathrm{Tr}\rho^2 = 1$, can be a sign of a mixed state. However, there exists a more convenient quantitative measure of statistical indeterminacy of quantum systems, *the entropy* (see, for instance, [Fain (1972); Landau (1964)]). Let us define the entropy operator in terms of the density operator in the following way: $\hat{S} \equiv -\ln\hat{\rho}$. Then the entropy S is equal to $\langle\hat{S}\rangle$, i.e.,

$$S = -\langle\ln\hat{\rho}\rangle = -\mathrm{Tr}(\hat{\rho}\ln\hat{\rho}). \tag{3.20}$$

In the representation where $\hat{\rho}$ is diagonal, (3.20) takes the form

$$S = -\sum_n \rho_n \ln\rho_n. \tag{3.21}$$

(This follows from the fact that in the diagonal representation, $[F(\hat{f})]_{nn} = F(f_{nn})$.) In a pure state, ρ_n equals 0 or 1, therefore $S = 0$, the indeterminacy ('chaoticity') is minimal. The opposite limiting case of a maximal indeterminacy is realized for a uniform mixture of states, $\rho_n = \mathrm{const} = 1/g$, where g is the number of states with a given energy (the Gibbs *microcanonical ensemble*). Then, $\hat{\rho} = \hat{I}/g$, $\hat{\rho^2} = \hat{I}/g^2$ and, according to (3.21)

$$S = -\sum_{n=1}^{g}(1/g)\ln(1/g) = \ln g. \tag{3.22}$$

Thus, $0 \leq S \leq \ln g$.

3.1.7 °*Density matrix of an atom*

In statistical physics, one usually considers macroscopic objects consisting of $N \sim 10^{22}$ identical particles. The terms 'state', 'energy level', 'density matrix' relate in this case to the matter as a whole. In principle, one can speak about the wave function of a 1 cm monocrystal, which depends on about 10^{22} space arguments $\mathbf{r}_i$ and time. Correspondingly, the number of possible states and, hence, the dimensionality of the matrices f_{mn}, ρ_{mn} are also extremely high. Furthermore, in order to realize an ensemble one has to have, say, 10^3 similar crystals.

On the other hand, active media in quantum electronics, such as gases, doped crystals or dye solutions, as a rule, consist of weakly interacting atoms or molecules. Then, it is sufficient to consider the state of a single atom, or, to be precise, of a single external electron. The rest of the particles are then considered as a thermostat, which has a weak influence on the wave function of the atom.

This transition from about 10^{22} degrees of freedom to a few ones leads to a crucial simplification of the theory, i.e., to the ideal gas model. The theory is further simplified by excluding from considerations all states that are not populated and not resonant with respect to the external field. This transition to the *two-level system* is valid in the case of a quasi-monochromatic external field and the absence of degeneracy. Note that the density matrix of an n-level non-degenerate system consists of n^2 elements, $n(n - 1)$ of them being complex. However, the normalization and Hermiticity conditions (3.16) reduce the number of independent elements, so that the state of the system is described by $n^2 - 1 \equiv m$ real numbers. For a two-level system, $m = 3$, and its state can be represented as a point in a three-dimensional phase space, with the coordinates $2\rho'_{21}$, $2\rho''_{21}$, and $\rho_1 - \rho_2 \equiv \Delta$ (Sec. 4.4). In the case of a pure state, condition $\rho^2 = \rho$ reduces the number of independent parameters to two, and the state can be shown by a point on a unit sphere.

Since the atoms are identical and independent, the additive macroscopic parameters of the matter, such as polarization $\mathbf{P}$, are calculated by simply multiplying single-atom mean values by N, $\mathbf{P} = N\langle\mathbf{d}\rangle$. Note that if all gas atoms are under the same conditions, the gas as a whole can be considered as an ensemble (quantum or quantum-statistical) containing approximately 10^{22} systems. Summation over atoms is then equivalent to ensemble averaging, and a diagonal element of the density matrix, ρ_n, defines the average relative population N_n/N of a level $\mathcal{E}_n$ in a real gas rather than in a hypothetical ensemble of 10^3 similar gas volumes.

3.2 Populations of levels

3.2.1 *Equilibrium populations*

In thermodynamic equilibrium, all statistical properties of a system are determined by the Gibbs canonical distribution. This distribution is applicable to both isolated macroscopic systems and systems of any size interacting with a thermostat. If single atoms or molecules of an ideal gas are considered, the Gibbs distribution corresponds to the density matrix of the form

$$\rho_{mn}^{(0)} = \delta_{mn}\rho_m^{(0)} = \delta_{mn}\exp(-\mathcal{E}_m/\kappa T)/Z, \tag{3.23}$$

where the normalization factor Z is called the *statistical sum* and can be found from the normalization condition,

$$Z = \sum_m \exp(-\mathcal{E}_m/\kappa T). \tag{3.24}$$

Here, m numerates various states of an atom, therefore the population of a g_n-fold degenerate level is

$$N_n^{(0)} = g_n N \exp(-\mathcal{E}_n/\kappa T)/Z. \tag{3.25}$$

This equation is called the Boltzmann distribution. Note that the equilibrium density operator (3.23) can be represented in the form

$$\hat{\rho}^{(0)} = \exp(-\hat{\mathcal{H}}_0/\kappa T)/\mathrm{Tr}\{\exp(-\hat{\mathcal{H}}_0/\kappa T)\}. \tag{3.26}$$

As it was shown in Sec. 2.3, interaction of the external field with the matter is determined by the populations of the 'resonant' levels N_1, N_2. In the first order of the perturbation theory, alternate field only creates the non-diagonal elements of the density matrix, $\rho_{12}^{(1)} \sim E$, while the diagonal elements remain unchanged, $\rho_n^{(1)} \approx 0$. Therefore, at sufficiently weak fields one can calculate populations using the Boltzmann distribution (3.25).

According to the Boltzmann distribution, the only populated states are the ones that are apart from the ground state by an energy not much exceeding κT. Hence, field at frequencies much larger than $\kappa T/\hbar \equiv \omega_T$ can only cause transitions up. At room temperature, this boundary frequency is in the far IR range ($\nu_t \equiv \omega_T/2\pi c \approx 200\text{cm}^{-1}$, $\lambda_T = 1/\nu_T \approx 50\mu$), while at helium temperatures, in the microwave range ($\nu_t \approx 1\text{cm}^{-1}$).

In the case of atomic gases and dopant ions in crystals, the lowest excited levels, as a rule, are well above this boundary, and almost all particles are in the ground state, so that they all participate in the absorption of light, $\Delta N \sim N_1 \sim N$. Often, the ground level has a degeneracy g_1, which can be lifted (completely or partly) due to the spin-orbit interaction (the *fine structure*) or due to static fields (the *Stark and Zeeman effects*). In this case, particles are distributed over sublevels, and if the splitting is much less than κT, then the populations of the sublevels are approximately equal to N/g_1, and the population differences are, according to (3.25), on the order of

$$\Delta N \sim (\hbar\omega/g_1\kappa T)N \ll N. \tag{3.27}$$

Transitions between such sublevels in doped crystals are used in paramagnetic amplifiers, and relation (3.27) explains why it is necessary to cool the active media of amplifiers down to helium temperatures.

In the case of molecular gases or solutions of organic dyes, the ground electronic level has a rich rotational-vibrational structure, which covers the microwave and the middle-IR spectral ranges. Therefore, the molecules are distributed over many levels, and the population differences are small as well.

3.2.2 *Two-level system and the negative temperature*

Consider populations of two non-degenerate levels as functions of the temperature. Let the zero energy be placed exactly between the two levels, so that $\mathcal{E}_{1,2} = \pm\hbar\omega/2$, then it follows from (3.23) that $\rho_{1,2} = e^{\pm x}/Z$, with $x \equiv \hbar\omega/2\kappa T$. From the condition $\rho_1 + \rho_2 = 1$ we find that $Z = e^x + e^{-x}$ and, as a result,

$$\rho_1 = N_1/N = (e^{-2x} + 1)^{-1}, \; \rho_2 = N_2/N = (e^{2x} + 1)^{-1}, \tag{3.28}$$

$$\Delta = \frac{N_1 - N_2}{N} = \tanh x. \tag{3.29}$$

The active medium of a laser, in principle, is in a strongly non-equilibrium state, and the Boltzmann distribution (3.25) is not applicable to it, as is, strictly speaking, any notion related to temperature. However, in the case of non-equilibrium systems it is convenient to keep the equations in the form (3.28), (3.29) but to understand T as some effective parameter. *Effective*, or *spin*, temperature for a given pair of non-degenerate levels is defined through the population ratio as follows:

$$N_m/N_n \equiv \exp(\hbar\omega_{nm}/\kappa T_{ef}), \tag{3.30}$$

i.e., the effective temperature is simply a logarithmic measure of the population ratio. It follows from (3.30) that in the case of population inversion, $T_{ef} < 0$.

It is easy to see that Eqs. (3.28), (3.29) maintain their form even for non-equilibrium systems, provided that T is understood as the effective temperature. Figure 3.1 shows the relative population difference as a function of the effective

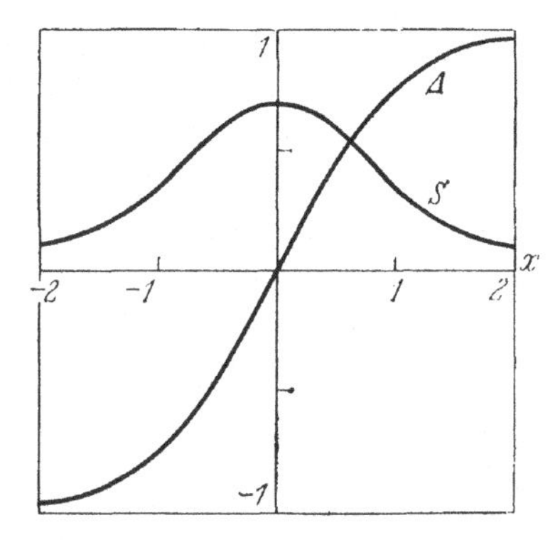

Fig. 3.1 Relative population difference Δ and the entropy S of a two-level system as functions of the parameter $x = \hbar\omega_0/2\kappa T$.

temperature. The dependence is plotted according to Eq. (3.29) for all temperatures, both positive and negative. Full inversion ($\rho_1 - \rho_2 \equiv \Delta = -1$) corresponds to $T_{ef} = -0$, full saturation ($\rho_1 = \rho_2 = 1/2, \Delta = 0$) corresponds to $T_{ef} = \pm\infty$, at $\Delta = 1$ $T_{ef} = +0$. Note that a two-level system with a negative temperature has more energy stored than a system with a positive temperature:

$$\mathcal{E} = \rho_1\mathcal{E}_1 + \rho_2\mathcal{E}_2 = -(\hbar\omega/2)\Delta = -(\hbar\omega/2)\tanh x, \tag{3.31}$$

where $x \equiv \hbar\omega/2\kappa T_{ef}$.

The entropy of a non-equilibrium two-level system can be also defined in terms of T_{ef}. According to definition (3.21) and to (3.28),

$$S = -\rho_1 \ln\rho_1 - \rho_2 \ln\rho_2 = \ln(2\cosh x) - x\tanh x. \tag{3.32}$$

Thus, entropy is an even function of the temperature, with the maximum $S_0 = \ln 2$ at $T_{ef} = \pm\infty$ (Fig. 3.1).

Further, we will show that the intensity of thermal radiation from a two-level system can be also written in terms of the effective temperature (Sec. 7.1). At $T_{ef} < 0$, it is this radiation that causes the noise of quantum amplifiers (the Kirchhoff law for negative temperatures). In particular, at $\hbar\omega \ll \kappa|T_{ef}|$ the noise temperature of an amplifier has the same absolute value as the effective one, $T_n = |T_{ef}|$.

3.2.3 °*Populations in semiconductors*

Boltzmann's distribution (3.25) is not valid for calculating the number of active particles in the case of inter-band transitions in semiconductors. (Such transitions are used in semiconductor lasers.) In contrast to bound electrons in gases or in doped dielectric crystals, electrons in the valence and conduction bands of a semiconductor are not localized and can exchange locations. This possibility allows one to consider the multi-electron problem and to take into account the anti-symmetry of the total wave function with respect to the permutation of two electrons, which leads one to the Pauli principle.

In the first approximation, electrons behave like particles of an ideal quantum gas with high density. By applying to an ideal gas the general Gibbs distribution, which has the form (3.23) provided that m numbers all possible states of a multi-particle system, and taking into account the Pauli principle, we come to the Fermi-Dirac distribution $f^{(0)}(\mathcal{E})$. For comparing it with (3.25), we will represent this distribution as (see Fig. 3.2(b))

$$N_m^{(0)} = 2f^{(0)}(\mathcal{E}_m) = 2\{\exp[(\mathcal{E}_m - \mu)/\kappa T] + 1\}^{-1}, \tag{3.33}$$

where the factor 2 takes into account spin saturation, μ is the Fermi level, defined by the normalization condition $\sum N_m^{(0)} = N$, with N being the total number of electrons, and $\mathcal{E}_m$ are allowed energy values for one electron. The spectrum is discrete due to periodic boundary conditions for the electron wave function. According to (3.33), mean population of any level cannot exceed two electrons, in agreement with the Pauli principle.

The energy levels $\mathcal{E}_m$ of electrons in semiconductors have almost continuous distribution within the allowed bands. As a result, population N_m can be considered as a function of a continuous argument $\mathcal{E}$, and the normalization condition $\sum N_m = N$, which indirectly defines the Fermi level, takes the form

$$\int d\mathcal{E} g(\mathcal{E}) N(\mathcal{E}) = N, \tag{3.34}$$

with the integration running over the valence and conduction bands and $g(\mathcal{E})$ being the energy density of states.

For pure semiconductors, the Fermi level is approximately at the centre of the energy gap. If there were dopant levels, each of them would contain one electron; *the Fermi level can be formally defined as the one that is half-occupied.* At low temperatures, the boundary between full and empty levels is very sharp (Fig. 3.2(b)).

In the case of sufficiently high levels, for which $\mathcal{E} - \mu \gg \kappa T$, one can neglect the unity in the denominator of Eq. (3.33), and the equation takes the form of the Boltzmann distribution (3.25),

$$N_m = 2Z^{-1} \exp(-\mathcal{E}_m/\kappa T) \ll 2, \tag{3.35}$$

where $Z = \exp(-\mu/\kappa T)$.

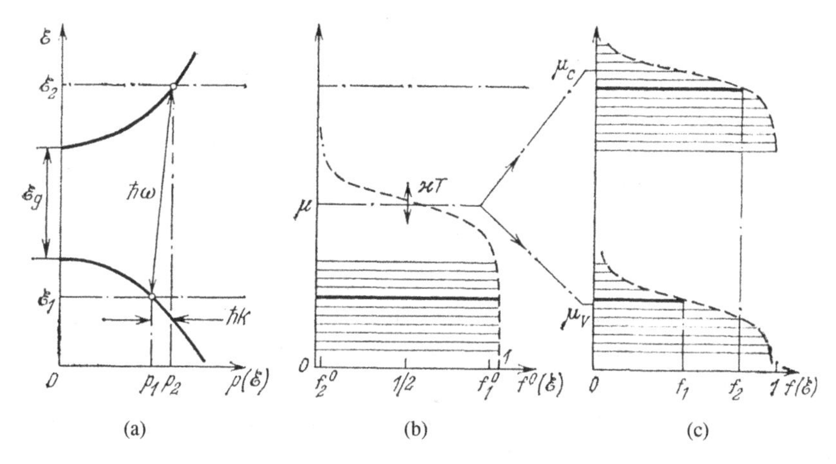

Fig. 3.2 Population inversion in a semiconductor: (a) relation between the momentum p and the energy $\mathcal{E}$, i.e., the dispersion relation, for electrons and holes, $\mathcal{E}_g$ being the energy gap; light with the frequency ω and wavevector k transfers electrons from the $\mathcal{E}_1$ level to the $\mathcal{E}_2$ level (or vice versa); (b) populations of energy levels in an equilibrium semiconductor (the Fermi-Dirac distribution); (c) due to the injection of carriers, the Fermi level μ splits in quasi-levels μ_v, μ_c, and for some pairs of levels, inversion takes place: $f_1 < f_2$

3.2.4 °*Inversion in semiconductors*

Consider the condition for quantum amplification through inter-band transitions in semiconductors. The incident field, with the frequency ω exceeding the gap width $\mathcal{E}_g/\hbar$, leads to almost 'vertical' transitions of electrons from level 1 of the valence band to level 2 of the conduction band (Fig. 3.2(a)). The levels 1,2 within the bands are unambigously defined by the conservation laws of energy, $\hbar\omega = \mathcal{E}_2 - \mathcal{E}_1$, and momentum (or, to be precise, quasi-momentum), $\hbar\mathbf{k} = \mathbf{p}_2 - \mathbf{p}_1$.

The number of stimulated transitions up scales as the probability of filling the ground level, $N_1/2 = f(\mathcal{E}_1) \equiv f_1$, multiplied, in accordance with the Pauli principle, by the probability $1 - f_2$ of a hole being on the excited level. Similarly, the number of transitions down scales as $f_2(1 - f_1)$, with the same proportionality factor (see (2.26)). The overall effect of field energy amplification or absorption scales as the difference,

$$\alpha \sim f_1(1 - f_2) - f_2(1 - f_1) = f_1 - f_2 = (N_1 - N_2)/2. \tag{3.36}$$

Thus, the contribution of a single pair of resonant levels into absorption scales as the difference of their populations, similarly to the case of localized electrons, and the inversion condition has the form

$$f(\mathcal{E}_2) > f(\mathcal{E}_1). \tag{3.37}$$

In an equilibrium semiconductor, $f = f^{(0)}$, and this condition is not satisfied. However, if, for instance, a sufficient number of carriers (electrons and holes) is injected into the bands with the help of an external DC current source, then condition (3.37) can be satisfied, see Fig. 3.2(c). One can easily show that this requires degeneracy of the carriers in the bands,

$$\mu_c - \mu_v > \hbar\omega > \mathcal{E}_g. \tag{3.38}$$

Here, μ_c, μ_v are the Fermi quasi-levels in the conduction and valence bands. In addition to the injection method, semiconductor lasers use optical pumping, either single- or two-photon one, and pumping with an electron beam.

Let us mention here that amplifiers and oscillators using free electrons, such as gyrotrons, free-electron lasers etc., can be also described in terms of population inversion (occupation numbers). For instance, in a quasi-monochromatic beam with the mean energy $\mathcal{E}_0$, only a small group of levels in the vicinity of $\mathcal{E}_0$ is occupied; therefore, inversion takes place with respect to all lower levels, $f(\mathcal{E}_0) > f(\mathcal{E})$.

3.3 Evolution of the density matrix

3.3.1 *Non-equilibrium systems*

The density matrix ρ_{mn} of a system, similarly to the distribution function $P(q, p)$ in classical physics, contains complete statistical information about the properties of the system, i.e., allows one to calculate ensemble means $\langle f\rangle = \mathrm{Tr}(f\rho)$, higher-order moments, correlation coefficients $\langle fg\ldots\rangle = \mathrm{Tr}(fg\ldots\rho)$ etc. Thermodynamics mainly deals with equilibrium systems where the density matrix and the ensemble means are time-independent, $\dot{\rho}^{(0)} = \langle\dot{f}\rangle^{(0)} = 0$. Note, however, that the correlation functions $\langle f(t)g(t')\rangle^{(0)}$ may depend on the time difference $t - t'$.

In quantum electronics, on the contrary, of most interest are systems where, due to the effect of external fields, essentially non-equilibrium state is formed, $\rho \neq \rho^{(0)}$. If the external perturbation is alternating, $\mathcal{V} = \mathcal{V}(t)$, then the density matrix and the ensemble means are naturally time-dependent, $\rho = \rho(t)$, $\langle f\rangle = \langle f(t)\rangle$. On the other hand, after the external field is switched off, the initially non-equilibrium $(\rho(t_0) \neq \rho^{(0)})$ system will relax and tend to equilibrium, and its density matrix and means will be again functions of time. However, the relaxation processes can be also described by alternating perturbation, $\mathcal{V}(t)$, acting on the system from the thermostat.

Non-equilibrium and non-stationary systems are studied by non-equilibrium statistical thermodynamics, also called the kinetic theory. In contrast to dynamics, kinetics studies not the time dependencies of coordinates and momenta of separate particles, $q(t)$, $p(t)$, or of the wave function, $\Psi(q,t)$, but the behavior of the means, $\langle f(p,q,t)\rangle$, distribution functions, $P(q,p,t)$, or the density matrix, $\rho_{mn}(t)$, for systems interacting with the thermostat and (or) with external alternating fields.

3.3.2 *Von Neumann equation*

Let us first consider, in terms of the Schrödinger equation, the dynamic problem about the behavior of the density matrix for a system with a known energy operator $\mathcal{H}$. For this, we substitute expansion (3.2) in the Schrödinger equation and left-multiply the resulting equation by the operator $\int d\mathbf{r}\varphi_m^*$. Because the functions φ_m are orthogonal and normalized, we obtain the following system of equations determining the dynamics of the b_m coefficients:

$$i\hbar\dot{b}_m = \sum_n \mathcal{H}_{mn}b_n. \tag{3.39}$$

Recall that, in contrast to Eq. (2.14), this equation includes the matrix elements of the total Hamiltonian $\mathcal{H}$ rather than the interaction operator $\mathcal{V}$. This is due to a different definition of state amplitudes. In addition, the basic functions φ_m used here are not necessarily eigenfunctions of the energy operator.

We multiply (3.39) by b_k^* and write its complex conjugate,

$$\begin{aligned} i\hbar b_k^*\dot{b}_m &= \sum_n \mathcal{H}_{mn}b_k^*b_n, \\ i\hbar b_k\dot{b}_m^* &= -\sum_n \mathcal{H}_{nm}b_kb_n^*. \end{aligned} \tag{3.40}$$

Here, we used the Hermiticity of the energy operator, $\mathcal{H}^+ = \mathcal{H}$. Let us interchange the m, k indices in the second equation and take the sum of the two equations. As a result, taking into account the definition of the density matrix for a pure state, (3.4), we find the following equation of motion:

$$i\hbar\dot{\rho}_{mk} = \sum_n (\mathcal{H}_{mn}\rho_{nk} - \rho_{mn}\mathcal{H}_{nk}). \tag{3.41}$$

According to definition (3.10), the equation for a mixed-state density matrix has the same form. Using the matrix multiplication rule and the commutator notation, $[f,g] \equiv fg - gf$, one can write Eq. (3.41) in a compact invariant form,

$$i\hbar\dot{\rho} = [\mathcal{H},\rho]. \tag{3.42}$$

This equation, describing the evolution of the density matrix, is called the *von Neumann equation*. It is the starting point for non-equilibrium thermodynamics. Its classical analogue is the Liouville equation for the distribution function $P(q,p,t)$.

3.3.3 *Interaction with the thermostat*

In the most general approach, ρ in Eq. (3.42) is understood as the density matrix of a closed system in a pure state, with the energy containing the following terms:

$$\mathcal{H} = \mathcal{H}_0 + \mathcal{V}, \ \mathcal{H}_0 = \mathcal{H}_A + \mathcal{H}_B, \ \mathcal{V} = \mathcal{V}_1 + \mathcal{V}_2, \tag{3.43}$$

where $\mathcal{H}_A$ and $\mathcal{H}_B$ are unperturbed Hamiltonians of the system and the thermostat while $\mathcal{V}_1$ and $\mathcal{V}_2$ describe, respectively, the interaction of the system with the thermostat, i.e., relaxation, and with the external field. The von Neumann equation is solved in terms of the perturbation theory, and then averaging over the variables of the thermostat is performed, see the second definition of the density matrix (3.15).

In a less rigid approach, ρ only relates to the system under consideration, i.e., an atom, a molecule, etc., $\mathcal{H}_0 \equiv \mathcal{H}_A$, and $\mathcal{V}_1$ is assumed to be a stochastic function of the time with given statistical parameters. Let the indices k, m, n numerate non-perturbed energy functions ($\mathcal{H}_0\varphi_k = \mathcal{E}_k\varphi_k$), then Eq. (3.41) takes the form

$$\left(\frac{d}{dt} + i\omega_{mk}\right)\rho_{mk} = \frac{1}{i\hbar}\sum_n(\mathcal{V}_{mn}\rho_{nk} - \rho_{mn}\mathcal{V}_{nk}), \tag{3.44}$$

where the $\mathcal{V}$ operator includes the action of the thermostat and the external field. Note, however, that this approach does not explain the non-equality of the 'up' and 'down' relaxation transition probabilities, $w_{12} > w_{21}$, see the next section.

Finally, in quantum electronics, as a rule, relaxation is taken into account phenomenologically, using a small number of constants, which are assumed to be known from a more detailed theory or from experiment.

3.3.4 *Evolution of a closed system*

Before introducing relaxation parameters into the density-matrix equation, consider the case of a closed system. Let φ_n be eigenfunctions of the energy operator, then $\mathcal{H}_{mn} = \mathcal{E}_n\delta_{mn}$, and Eq. (3.41) takes the form

$$\dot{\rho}_{mk} = -i\omega_{mk}\rho_{mk}. \tag{3.45}$$

Thus, the density matrix of a closed system has a trivial dependence on time,

$$\rho_{mk}(t) = \rho_{mk}(0)\exp(-i\omega_{mk}t), \tag{3.46}$$

i.e., non-diagonal elements of the density matric oscillate with the corresponding Bohr frequencies while the diagonal elements (relative populations) are constant. Note that this result also follows directly from the exponential time dependence of the state amplitudes, $b_n = c_n\exp(-i\mathcal{E}_nt/\hbar)$, and the definition of ρ (3.4).

The dipole moment of an isolated atom can be calculated from Eq. (3.46),

$$\langle \mathbf{d}(t)\rangle = \mathrm{Tr}\{\mathbf{d}\rho(t)\} = \sum_{mn} \mathbf{d}_{nm}\rho_{mn}(0)\exp(-i\omega_{mn}t). \tag{3.47}$$

At the same time, it follows from Maxwell's equations that an oscillating dipole, similarly to an aerial, emits electromagnetic waves into free space; therefore, within a certain time an atom should lose all its energy and get into the ground state, i.e., $\rho_{mn}(\infty) = \delta_{mn}\delta_{n0}$. Thus, an atom cannot be isolated from electromagnetic vacuum, which plays the role of the thermostat with $T = 0$. This example reminds us that isolated systems do not exist, and hence (3.45) should be completed by relaxation terms describing the evolution into equilibrium, $\rho(\infty) \to \rho^{(0)}$.

3.3.5 *Transverse and longitudinal relaxation*

The most simple models of relaxation, based, in particular, on the Markovian approximation, lead to the following form of kinetic equations for the density matrix (see, for instance, [Fain (1972); Apanasevich (1977)]):

$$\left(\frac{d}{dt} + i\omega_{mk}\right)\rho_{mk} = -\gamma_{mk}\rho_{mk}, \quad m \neq k \tag{3.48}$$

$$\frac{d\rho_m}{dt} = \sum_n (w_{mn}\rho_n - w_{nm}\rho_m), \quad \rho_m \equiv \rho_{mm}. \tag{3.49}$$

According to (3.48), non-diagonal elements of the density matric behave like amplitudes of exponentially decaying oscillators,

$$\rho_{mk}(t) = \rho_{mk}(0)\exp[(-i\omega_{mk} - \gamma_{mk})t]. \tag{3.50}$$

The damping constant for a given pair of levels, $\gamma_{12} \equiv \gamma_{21}$, is often denoted as $1/T_2$. The relaxation time T_2 of the non-diagonal component ρ_{12} is called the time of *spin-spin, or transverse*, relaxation. (The meaning of the second term will be clarified in Sec. 4.4.)

From the experimental viewpoint, transverse relaxation is usually revealed in the broadening of spectral lines. (So far, we ignore nonstationary experiments, which will be considered in Chapter 5.) It will be shown in Sec. 4.2 that (3.48) leads to a Lorentzian line shape with the FWHM

$$\Delta\omega = 2\gamma_{12} = 2/T_2. \tag{3.51}$$

In rarefied gases, relaxation is only caused by the interaction of atoms with electromagnetic vacuum. This interaction leads to the spontaneous emission, with the probability $A_{12} = 2\gamma_{12}$, and the corresponding broadening of the upper level $\Delta\mathcal{E}_2 = \hbar A_{12}$, as well as the spectral broadening, called the *natural* broadening,

$$\Delta\omega_{nat} = 2\gamma_{12} = A_{12}. \tag{3.52}$$

If the lower level of the transition under consideration is not the ground one, its broadening has to be taken into account as well. Let $2\gamma_n \equiv \sum_{m<n} A_{mn}$ be the total probability of a spontaneous transition from level n to all lower levels, then

$$\gamma_{mn} = \gamma_m + \gamma_n. \tag{3.53}$$

In reality, Δf_{nat} is on the order of MHz for the case of visible-range allowed transitions, and $T_2 \sim 10^{-6}$ s.

For sufficiently dense gases, natural broadening is masked by the collision one, and T_2 coincides, within an order of magnitude, with the mean time τ between the collisions of atoms with each other. As a result, $\Delta\omega \approx 2/\tau$, and the line width scales as the pressure p, provided, of course, that the Doppler broadening is smaller than the collision one. For rough estimates, one can assume that at $p = 1$ mmHg, $\Delta f \sim 10...100$ MHz. Note, however, that under certain conditions, the increase of the pressure leads to line narrowing, $\Delta\omega \sim 1/p$, called collision, or dynamical, narrowing. One of the models of this effect is considered in Ref. [Akhmanov (1981)].

Interaction of atoms with the thermostat leads not only to the damping of the states, but also to a certain shift $\delta\omega$ of the transition frequency. In the case where the thermostat is a vacuum, this shift is called the *Lamb* shift. Both effects can be formally taken into account by substituting a complex value for the transition frequency ω_{mn},

$$\tilde{\omega}_{mn} = \omega_{mn} + \delta\omega_{mn} - i\gamma_{mn}. \tag{3.54}$$

It is important that transverse relaxation is not always related to the energy transfer to the thermostat. For instance, elastic collisions in a gas lead to random changes in the phases of complex state amplitudes for separate atoms, $b_m^{(i)}$, and their pair products, $b_m^{(i)} b_n^{(i)*}$. If these phases are initially equal, $\rho_{mn} \neq 0$, then, after a certain time interval $T_2 = 1/\gamma_{mn}$, which is on the same order of magnitude as the mean time interval between collisions, the phase will be uniformly distributed within the interval $0 - 2\pi$, so that $\rho_{mn} \to 0$. A similar effect is caused by the dipole-dipole interaction of neighboring dopant atoms in crystals. Perturbations of this kind, which do not change populations, are called *adiabatic*. Certainly, non-adiabatic perturbations, such as non-elastic collisions, also contribute to the relaxation of non-diagonal elements, as they change both the amplitudes and the phases of the coefficients $b_m^{(i)}$.

Let us now consider the relaxation of diagonal density matrix elements, i.e., populations. Kinetic equations (3.49) contain a set of phenomenological coefficients w_{mn} with the dimensionality 1/s. The w_{21} coefficient defines the rate of transition from state 1 into state 2 due to the effect of the thermostat. (Recall that in quantum mechanics, transition indices are read from right to left.) The role of the thermostat can be played, for instance, by lattice vibrations in crystals, translational degrees of freedom of atoms in gases, and electromagnetic radiation.

In the case of two-level systems, one denotes

$$T_1 = (w_{12} + w_{21})^{-1}. \tag{3.55}$$

T_1 determines the relaxation time of populations, i.e., of the mean energy, and is called the time of *spin-lattice*, or *longitudinal*, relaxation. The time of longitudinal relaxation depends on the thermostat temperature and varies within broad limits, from 10^{-12}s in the case of nonradiative optical transitions in condensed matter, to hours and days in the case of nuclear magnetic resonance. (In this case, interaction with lattice is weak due to the small value of the nuclei magnetic moment, $\mu \sim 10^{-23}$ CGS.) Note that adiabatic perturbations, like dipole-dipole interaction, do not change the populations; therefore, usually $T_1 > T_2$. In experiment, longitudinal relaxation manifests itself in the saturation effect (Sec. 4.3).

Equations (3.48), (3.49) should also involve the case of thermodynamic equilibrium, where $\rho = \rho^{(0)}$ and $\dot{\rho}^{(0)} = 0$; hence, the following relation should hold true:

$$\sum_n (w_{mn}\rho_n - w_{nm}\rho_m) = 0. \tag{3.56}$$

This equality is satisfied, in particular, if one assumes the *principle of detailed equilibrium*,

$$w_{mn}\rho_n = w_{nm}\rho_m. \tag{3.57}$$

Hence, taking into account the Boltzmann distribution, we find the relation between the probabilities of relaxation transitions and thus reduce the number of independent parameters in Eq. (3.49) by a factor of two,

$$w_{mn}/w_{nm} = \exp(\hbar\omega_{nm}/\kappa T), \tag{3.58}$$

with T being the temperature of the thermostat. This condition provides dynamical equilibrium for the populations. Thus, $w_{12} > w_{21}$, in contrast to the case of stimulated transitions in a classical field, where, according to (2.26), $W_{12} = W_{21}$. When the thermostat is at low temperature, it has no excitations (photons, phonons, etc.) with high energy, $\hbar\omega > \kappa T$. Therefore, it can only absorb energy from the system under consideration, and transitions 'up' are practically absent. An extreme example is realized for transitions between nuclei levels in the γ range. Even in condensed matter, such transitions usually occur only due to spontaneous emission, with the probability $w_{12} = A = 1/T_1$. In the case of nuclear isomers, A is extremely small because the transitions are forbidden, and T_1 can be as large as days, similarly to the NMR case.

The parameters w_{mn} can be calculated, in principle, using some model of the thermostat. An example where the role of the thermostat is played by the field has been considered in Sec. 2.5. Then,

$$w_{21} = B\rho, w_{12} = B\rho + A, \tag{3.59}$$

$$1/T_1 = A\coth(\hbar\omega_{21}/2\kappa T) = A/\Delta^{(0)}, \tag{3.60}$$

where A, B are the Einstein coefficients for spontaneous and stimulated transitions, $\rho \equiv \rho^{(0)}(\omega_{21})$ is the spectral density of the equilibrium field, given by the Planck formula, and $\Delta^{(0)}$ is the equilibrium relative population difference (see Eq. (3.29)).

3.3.6 *Interaction picture*

Usually, one has to solve the von Neumann equation for the density matrix by means of the perturbation theory, i.e., a sequence of iterations. The only exception is the case of a two-level system, which will be considered in Sec. 4.3. As in the case of solving the Schrödinger equation in the energy representation (Sec. 2.1), we will assume that the influence of the external alternating field on the electrons in an atom is much weaker than the effect of the nucleus constant field, which determines the unperturbed stationary states of a bound electron. As we will show in what follows, a more precise formulation of the condition for the perturbation theory to be valid has the form $\Omega \ll \tilde{\omega}$, where $\Omega = |\mathbf{d}_{mn} \cdot \mathbf{E}_0|/\hbar$ is the Rabi frequency, i.e., the matrix element of the perturbation energy in frequency units, and $\tilde{\omega}$ is the mismatch between the field frequency and the closest atom frequency, i.e., the energy deficit $|\omega - \omega_{mn}|$ in the virtual state (Sec. 6.2) or the transition bandwidth γ_{mn} given by the relaxation.

Before solving the von Neumann equation, it is convenient to transfer the trivial time dependence of the unperturbed density matrix to the operators. To do this, let us introduce the following notation for the matrix elements of an arbitrary operator in the energy basis:

$$f'_{mn} = f_{mn}\exp(i\omega_{mn}t) = \int d\mathbf{r}\Phi^*_m(\mathbf{r}, t)f\Phi_n(\mathbf{r}, t), \tag{3.61}$$

where the functions $\Phi_n = \varphi_n \exp(-i\mathcal{E}_n t/\hbar)$ satisfy the equation $i\hbar\dot{\Phi}_n = \mathcal{H}_0\Phi_n$.

Transformation of the matrix elements (3.61) corresponds to the following *unitary transformation* of the operators:

$$f'(t) = U_0^+ f U_0, \tag{3.62}$$

$$U_0(t) \equiv \exp(-i\mathcal{H}_0 t/\hbar), \;\; U_0^+ U_0 = I. \tag{3.63}$$

The unitary operator U_0 is called the *unperturbed evolution operator*; it is diagonal in the energy representation and has the eigenvalues $\exp(-i\mathcal{E}_n t/\hbar)$, so that

$$\Phi_n(t) = U_0(t)\varphi_n, \tag{3.64}$$

$$\varphi_n = U_0^+(t)\Phi_n(t). \tag{3.65}$$

In the Dirac notation, the time evolution of a state vector, for the case $\mathcal{V} = 0$, is described as

$$|t\rangle = U_0(t - t_0)|t_0\rangle. \tag{3.66}$$

The inverse transformation has the form

$$|\,\rangle' \equiv |t_0\rangle = U_0^+|t\rangle. \tag{3.67}$$

Let us now substitute into the von Neumann equation (3.42), for ρ_{mn} and $\mathcal{V}_{mn}$, the primed values, according to (3.62), and take into account that $\omega_{mn}+\omega_{nk} = \omega_{mk}$. As a result, we get the equation for the density matrix in the *interaction picture*, also called the *Dirac picture*,

$$i\hbar\dot{\rho}'_{mk} = \sum_n (\mathcal{V}'_{mn}\rho'_{nk} - \rho'_{mn}\mathcal{V}'_{nk}). \tag{3.68}$$

In the invariant notation, it is written as

$$i\hbar\dot{\rho}' = [\mathcal{V}', \rho']. \tag{3.69}$$

Note that the time dependence of an arbitrary operator f is given by the Heisenberg equation,

$$i\hbar\dot{f} = [f, \mathcal{H}]. \tag{3.70}$$

Here, it is assumed that f does not depend on the time directly, i.e., $\partial f/\partial t = 0$.

Transformations of operators of the form (3.63) accompanied by transformations of state vectors of the form (3.67) means passing to the interaction picture, and in the case $\mathcal{V} = 0$, to the *Heisenberg picture*. These transformations are similar to passing to a rotating frame of reference.

In the initial *Schrödinger picture*, state vectors, and also, according to definition (3.4), the density matrix elements are functions of time. Operators can depend on time only due to a varying external force (as, for instance, the energy operator $\mathcal{V}(t) = -\mathbf{d} \cdot \mathbf{E}(t)$ in dipole interaction). On the contrary, in the Heisenberg picture all time dependence is transferred to the operators and their matrix elements, except the density matrix operator, and the state vectors are constant. The interaction picture is intermediate, and it has all values time-dependent.

However, it is important that *the observables do not depend on the choice of the picture*,

$$\langle f \rangle = \mathrm{Tr}(f\rho) = \mathrm{Tr}(f'\rho'). \tag{3.71}$$

This can be proven using definition (3.63), the unitarity, $U_0 U_0^+ = I$, and the invariance of the trace to cyclic permutations, $\mathrm{Tr}(abc) = \mathrm{Tr}(bca)$.

3.3.7 °*Perturbation theory*

It is not difficult to find the formal solution to the von Neumann equation (3.69) using the iteration method. For this, let us represent the density operator as a series expansion (the primes will be temporarily omitted),

$$\rho(t) = \rho^{(0)} + \rho^{(1)}(t) + \rho^{(2)}(t) + \ldots \tag{3.72}$$

and substitute it in (3.69). Here, $\rho^{(0)} = \rho(t_0)$ is the initial condition. By setting equalities between the terms of the same order in the perturbation $\mathcal{V}$, we find the relation

$$i\hbar\dot{\rho}^{(k)} = [\mathcal{V}, \rho^{(k-1)}]. \tag{3.73}$$

Integration yields

$$\rho^{(k)}(t) = (i\hbar)^{-k} \int_{t_0}^{t} dt_k \ldots \int_{t_0}^{t_2} dt_1 [\mathcal{V}(t_k), \ldots [\mathcal{V}(t_1), \rho^{(0)}] \ldots]. \tag{3.74}$$

From this, we find the mean value of an arbitrary operator,

$$\langle f(t) \rangle = \sum_{k=0}^{\infty} (i\hbar)^{-k} \int_{t_0}^{t} dt_1 \ldots \int_{t_0}^{t_{k-1}} dt_k \times \langle [\ldots [f'(t), \mathcal{V}'(t_1)], \ldots, \mathcal{V}'(t_k)] \rangle^{(0)}. \tag{3.75}$$

In the last expression, averaging is over the initial (unperturbed) density matrix $\rho^{(0)}$; the initial time moment t_0 is usually assumed to be $-\infty$. Eq. (3.75) was derived using the property $\mathrm{Tr}(ab) = \mathrm{Tr}(ba)$, which leads to the following equalities under the Tr operation:

$$a[b, c] = [a, b]c, \; a[b, [c, d]] = [[a, b], c]d. \tag{3.76}$$

Relation (3.75) determines the response (reaction) of a quantum system to an external perturbation. For instance, assuming $f = d_\alpha, \mathcal{V}(t) = -\mathbf{d} \cdot \mathbf{E}(t)$ one can find the mean dipole moment of an atom, i.e., the charge displacement due to a given electric field, in the form

$$\langle \mathbf{d}(t) \rangle = \hat{\alpha}\mathbf{E} + \hat{\beta}\mathbf{E}^2 + \hat{\gamma}\mathbf{E}^3 + \ldots, \tag{3.77}$$

where $\hat{\alpha}, \hat{\beta}, \hat{\gamma}$ are some integral operators whose structure is clear from (3.75). Expansion of $\mathbf{d}(t)$ and $\mathbf{E}(t)$ in Fourier integrals or series determines the *polarisability* tensors $\alpha(\omega), \beta(\omega, \omega'), \ldots$ of an atom. Further, by multiplying atom polarisabilities by the density N of the atoms, one can find the macroscopic *susceptibility* tensors of the matter, $\chi^{(1)}(\omega), \chi^{(2)}(\omega, \omega'), \ldots$

As a result of such calculations, some of which will be demonstrated in Secs. 4.2 and 6.2 below, polarization $\mathbf{P} = N\langle \mathbf{d} \rangle$ of the matter can be expressed in terms of the external field and the parameters of the atoms, dipole matrix elements $\mathbf{d}_{mn}$ and the transition frequencies ω_{mn}.

Chapter 4

The Susceptibility of Matter

In classical electrodynamics, the interaction between matter and a field is conventionally studied in two steps, the microscopic one and the macroscopic one. The microscopic part is focused on the behaviour of charged material particles in a given external field. As a result, one finds the averaged, macroscopic, parameters of the matter, such as the susceptibility tensor χ, which determines the polarization $\mathbf{P} = \chi \cdot \boldsymbol{E}$ of homogeneous matter caused by the field, or the dielectric function $\epsilon = 1 + 4\pi\chi$. As a result of frequency Fourier expansion of the field, these quantities become complex functions of the frequency, $\epsilon(\omega) = \epsilon'(\omega) + i\epsilon''(\omega)$.

In the macroscopic approach, the susceptibility of the matter is assumed to be known, and the emission and propagation of the field are studied using Maxwell's macroscopic equations.

The present chapter considers first the definition and the general properties of the susceptibility tensor (Sec. 4.1). Further, in Sec. 4.2 the susceptibility is calculated in the framework of the simplest model of identical, motionless and non-interacting molecules. Both classical and quantum theory is used in this case. Section 4.3 considers saturation, the most important effect in nonlinear optics, in which populations of two levels get balanced due to a strong resonant field. Finally, in Sec. 4.4 the Bloch equations, which are widely used in quantum electronics, are derived.

4.1 Definition and general properties of susceptibility

By definition, linear dielectric susceptibility $\chi(\omega)$ is the proportionality factor between a monochromatic macroscopic field $\mathbf{E}(\omega)$ at frequency ω and the polarization $\mathbf{P}(\omega)$ emerging in a homogeneous medium due to its effect. In an anisotropic medium, polarization can be non-parallel to the field, so that in the general case, susceptibility is a tensor,

$$P_\alpha(\omega) = \sum_\beta \chi_{\alpha\beta}(\omega)E_\beta(\omega), \tag{4.1}$$

with $\alpha, \beta = x, y, z$. Usually, the summation sign is omitted. In more compact notation,

$$\mathbf{P}(\omega) = \chi(\omega)\mathbf{E}(\omega). \tag{4.2}$$

4.1.1 *Symmetry*

Here, $\mathbf{E}(\omega)$ denotes the Fourier component of the field $\mathbf{E}(t)$. As it is common in physics, a function of time $f(t)$ and its Fourier transform $f(\omega)$ are denoted by the same symbol and only their arguments differ,

$$f(t) = \int d\omega e^{-i\omega t} f(\omega), \;\; f(\omega) = \int dt e^{i\omega t} f(t)/2\pi. \tag{4.3}$$

The absence of the integration limits means that they are $\pm\infty$. Note that if $f(t)$ is real, then (4.3) leads to the following symmetry property for $f(\omega)$:

$$f(-\omega) = f^*(\omega). \tag{4.4}$$

Thus, the real part, $f'(\omega)$, is an even function while the imaginary part, $f''(\omega)$, is an odd function, so it is sufficient to know $f(\omega)$ for positive frequencies. From the definition (4.2), it follows that all components of the susceptibility tensor also satisfy (4.4),

$$\chi(-\omega) = \chi^*(\omega). \tag{4.5}$$

The $\chi(\omega)$ tensor and its inverse Fourier transform, $\chi(t)$, called *the Green function or the response function*, have one more general property, typical for arbitrary physical systems: they are symmetric, $\chi = \tilde{\chi}$, or

$$\chi_{\alpha\beta} = \chi_{\beta\alpha}. \tag{4.6}$$

Here, $\tilde{\chi}$ is the *transposed* tensor, $\tilde{\chi}_{\alpha\beta} \equiv \chi_{\beta\alpha}$. This equality is an example of the general *Onsager symmetry principle for kinetic indices*. It is also confirmed by the microscopic theory (see (4.59)). The symmetry of χ is only violated in the case of optical activity, either natural or caused by a constant magnetic field. In the latter case, instead of (4.6) we have $\chi(\mathbf{H}_0) = \tilde{\chi}(-\mathbf{H}_0)$.

Additional relations between different components of $\chi_{\alpha\beta}$ are imposed by the symmetry of the medium. For instance, in crystals with cubic symmetry, $\chi_{\alpha\beta} = \chi\delta_{\alpha\beta}$, as in isotropic media.

4.1.2 *The role of causality*

Dependence of susceptibility on the frequency, $\chi(\omega)$, cannot be arbitrary. As we will see, its real and imaginary parts, $\chi'(\omega)$ and $\chi''(\omega)$, are related via the Hilbert transformation.

Consider polarization induced in a dielectric by a very short field pulse, $E(t) \sim \delta(t)$, so that the pulse duration is much less than the period of the most high-frequency eigenmode of the matter. In the spectrum of such a pulse, all frequencies are distributed uniformly, $E(\omega) = \text{const}$, and the spectrum of the polarization, according to (4.1), repeats the shape of $\chi(\omega)$: $P(\omega) \sim \chi(\omega)$. (For simplicity, the medium is considered as isotropic.) Hence, the polarization pulse $P(t)$ repeats the shape of the Fourier transform of the complex susceptibility,

$$P(t) \sim \int d\omega e^{-i\omega t} \chi(\omega) \equiv 2\pi\chi(t).$$

Apparently, the system cannot respond before the external force is 'turned on'; hence, $\chi(t)$ should turn to zero at $t < 0$,

$$\int d\omega e^{-i\omega t} \chi(\omega) \sim \theta(t), \tag{4.7}$$

where $\theta(t)$ is the Heaviside step function, which is unity at $t > 0$ and zero at $t < 0$.

The causality principle, according to (4.7), restricts considerably the allowed class of $\chi(\omega)$ functions. Indeed, it follows that $\chi(\omega)$, considered as a function of a complex frequency $\omega = \omega' + \omega''$, should be analytical in the upper semiplane. Let us calculate the integral in (4.7) using the residue theory. The integrand contains the factor $e^{\omega'' t}$; therefore, at $t > 0$, the integral should run along a contour in the lower semiplane (see Fig. 4.1), while at $t < 0$, the contour should be in the upper semiplane. However, due to the causality principle, at $t < 0$ the integral should turn into zero. Therefore, the $\chi(\omega)$ function cannot have poles in the upper semiplane (see, for instance, [Vinogradova (1979); Landau (1982, 1964)]).

Further, according to the integral Cauchy formula, the real and imaginary parts of an analytical function are related via the Hilbert transformations,

$$\pi\chi'(\omega) = PV \int d\omega_1 \frac{\chi''(\omega_1)}{\omega_1 - \omega}, \quad \pi\chi''(\omega) = PV \int d\omega_1 \frac{\chi'(\omega_1)}{\omega - \omega_1}, \tag{4.8}$$

where 'PV' denotes the principal value of an integral. These integral equations are called the *Kramers-Kronig relations*. They allow, for instance, the real part of susceptibility to be calculated from the measured imaginary part. The above-given derivation can be extended to the case of an anisotropic medium. Then, equations (4.8) will be valid for all components of the χ tensor.

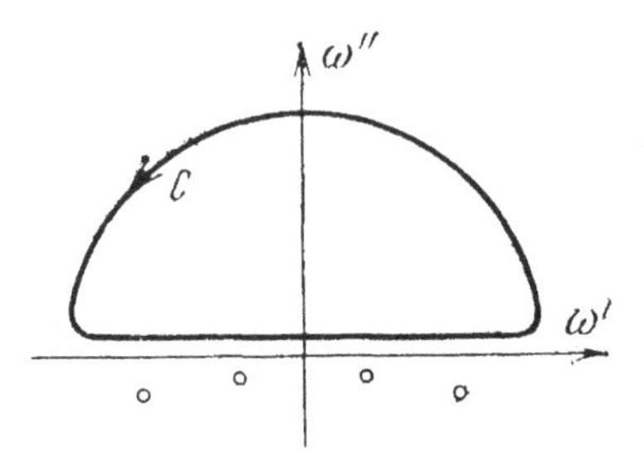

Fig. 4.1 Proof that the susceptibility of a dielectric $\chi(\omega)$ is an analytical function of complex frequency in the upper semiplane: at $t < 0$, the response function $\chi(t)$ turns to zero due to the causality principle. At the same time, it is equal to the integral of $\chi(\omega)e^{-i\omega t}$ along the C contour. Hence, according to the Cauchy theorem, C should not contain poles of $\chi(\omega)$.

4.1.3 *Absorption of a given field*

In the linear optics approximation, susceptibility χ completely determines emission, propagation, and absorption of a macroscopic field in a homogeneous medium, as well as the properties of surface waves, refraction and diffraction at the boundaries. Moreover, according to the fluctuation-dissipation theorem (FDT), χ also determines the equilibrium thermal field in matter (Sec. 7.7).

Let us show that the imaginary part of the susceptibility, $\chi''(\omega)$, determines the power of radiation absorbed or, at $\chi'' < 0$, emitted by the matter. We start from macroscopic Maxwell's equations for a linear non-magnetic medium with $\mathbf{D} = \mathbf{E} + 4\pi\mathbf{P} = (\mathbf{I} + 4\pi\chi)\cdot\mathbf{E}$ and $\mathbf{B} = \mathbf{H}$,

$$c\,\mathrm{rot}\mathbf{H} - \dot{\mathbf{D}} = 4\pi\mathbf{j}, \tag{4.9}$$

$$c\,\mathrm{rot}\mathbf{E} + \dot{\mathbf{H}} = 0, \tag{4.10}$$

$$\mathrm{div}\mathbf{D} = 4\pi\rho, \tag{4.11}$$

$$\mathrm{div}\mathbf{H} = 0, \tag{4.12}$$

where $\mathbf{j}$ and ρ are external (given) densities of current and charge.

At $\mathbf{j} = 0$, the power $\mathcal{P}$ absorbed by a unit volume of the matter, due to the energy conservation law, should be opposite to the divergence of the energy flux density $\mathbf{S}$,

$$\mathcal{P}(t) = -c\,\mathrm{div}(\mathbf{E}\times\mathbf{H})/4\pi = c(\mathbf{E}\cdot\mathrm{rot}\mathbf{H} - \mathbf{H}\cdot\mathrm{rot}\mathbf{E})/4\pi.$$

From (4.9, 4.10), it follows that

$$\mathcal{P}(t) = (\mathbf{E}\cdot\dot{\mathbf{D}} + \mathbf{H}\cdot\dot{\mathbf{H}})/4\pi.$$

In the case of a monochromatic field, this expression contains terms oscillating at a double frequency. After time averaging, it turns into zero, so that the mean power per period is

$$\mathcal{P} \equiv \overline{\mathcal{P}(t)} = \overline{\mathbf{E} \cdot \dot{\mathbf{P}}} = \omega \mathrm{Im}(\mathbf{E}_0^* \cdot \mathbf{P}_0)/2 = i\omega(E_{0\alpha}\chi_{\alpha\beta}^* E_{0\beta}^* - E_{0\alpha}^*\chi_{\alpha\beta}E_{0\beta})/4 \quad (4.13)$$

Let us interchange the α, β indices in the first term and take into account that in a non-gyrotropic material, susceptibility (4.6) is a symmetric tensor, then

$$\mathcal{P} = \omega\chi''_{\alpha\beta}E_{0\alpha}^* E_{0\beta}/2 \equiv \omega \mathbf{E}_0^* \cdot \chi'' \cdot \mathbf{E}_0/2. \quad (4.14)$$

In the case of a gyrotropic material, the imaginary part of χ in (4.14) should be replaced by the anti-Hermitian part, $(\chi - \chi^+)/2i$. For an isotropic medium or a cubic crystal, (4.14) takes the form

$$\mathcal{P} = \omega\chi''|E_0|^2/2. \quad (4.15)$$

4.1.4 °*Susceptibility of the vacuum*

Further, let us find the field generated in a homogeneous medium by external sources, i.e., a given polarization with a harmonic variation in time and space,

$$P = (1/2)P_0 e^{i\mathbf{k}\cdot\mathbf{r} - i\omega t} + \text{c.c.} \quad (4.16)$$

Here, $\mathbf{k}$ and ω are independent variables. In a homogeneous medium, $\mathbf{P}$ induces a plane monochromatic wave with amplitudes $\mathbf{E}_0, \mathbf{H}_0$. Let us substitute (4.16) in (4.9, 4.10) and take into account that $\mathbf{j} = \dot{\mathbf{P}}$. We get a system of algebraic equations for $\mathbf{E}_0, \mathbf{H}_0$ ($\mathbf{n} \equiv c\mathbf{k}/\omega$),

$$\mathbf{n} \times \mathbf{H}_0 + \epsilon \cdot \mathbf{E}_0 = -4\pi \mathbf{P}_0, \quad (4.17)$$

$$\mathbf{n} \times \mathbf{E}_0 - \mathbf{H}_0 = 0. \quad (4.18)$$

Excluding $\mathbf{H}_0$, we get

$$\mathbf{n} \times (\mathbf{n} \times \mathbf{E}_0) + \epsilon \cdot \mathbf{E}_0 = -4\pi \mathbf{P}_0. \quad (4.19)$$

Double vector product in (4.19) projects the $-\mathbf{E}_0$ vector onto the plane orthogonal to the propagation direction $\mathbf{k}$. Let us denote this projection operation by Π. Apparently, the Π tensor has the components

$$\Pi_{\alpha\beta} = \delta_{\alpha\beta} - k_\alpha k_\beta / k^2. \quad (4.20)$$

As a result, Eq. (4.19) takes the form

$$(n^2\Pi - \epsilon) \cdot \mathbf{E}_0 = 4\pi \mathbf{P}_0. \quad (4.21)$$

Thus, the problem is reduced to solving a system of two linear non-homogeneous algebraic equations. The solution can be expressed, in a standard way, in terms of the minors and determinant of the matrix $(n^2\Pi - \epsilon)_{\alpha\beta}$. Instead of solving the system directly, we express $\mathbf{E}_0$ in terms of $\mathbf{P}_0$ using the formalism of the inverse matrix or tensor. By definition, $\mathbf{A} \cdot \mathbf{A}^{-1} = \mathbf{A}^{-1} \cdot \mathbf{A} = \mathbf{I}$; therefore, it follows from (4.18) and (4.21) that

$$\mathbf{E}_0 = \mathbf{G} \cdot \mathbf{P}_0, \ \mathbf{H}_0 = \mathbf{n} \times (\mathbf{G} \cdot \mathbf{P}_0), \ \mathbf{G} \equiv 4\pi(n^2\Pi - \epsilon)^{-1}. \tag{4.22}$$

The $G_{\alpha\beta}(\mathbf{k}, \omega)$ tensor is called the *spectral Green function* for Maxwell's equations. It determines a macroscopic field induced by polarization (4.16), i.e., the response of the electromagnetic 'vacuum in matter' to an external excitation. The tensor function $\mathbf{G}(\mathbf{k}, \omega)$, similarly to χ or ϵ, satisfies the Kramers-Kronig relations. Its Fourier transform, $\mathbf{G}(\mathbf{r}, t)$, determines the fields emerging in a homogeneous medium due to an arbitrary distribution of the polarization $\mathbf{P}(\mathbf{r}, t)$ or a current.

Consider the case of a homogeneous medium. Let the z axis be along $\mathbf{k}$, then from (4.20) and (4.22) we find

$$E_{0x,y} = 4\pi P_{0x,y}/(n^2 - \epsilon), \tag{4.23}$$

$$E_{0z} = -4\pi P_{0z}/\epsilon; \tag{4.24}$$

recall that here $n \equiv ck/\omega$. The last equation for the field E_{0z}, which is longitudinal with respect to the propagation direction, shows that it is independent of $\mathbf{k}$: we usually exclude the effects of *spatial dispersion* where $\epsilon = \epsilon(\mathbf{k}, \omega)$. According to (4.24), longitudinal field created by this polarization is maximal at frequencies where $|\epsilon(\omega)|$ is minimal; these frequencies are given by the condition $\epsilon'(\omega) \approx 0$. Note that the Green function for a longitudinal field can be also obtained from Eq. (4.11) by assuming that $\rho = \mathrm{div}\mathbf{P}$.

Transverse components of the field, $E_{0x,y}$, considered as functions of k, according to (4.23), have a 'wave resonance' at $k = \omega\sqrt{\epsilon'}/c$, i.e., at $n^2 = \epsilon'$. Then, $G_{xx} = G_{yy} = i/\chi''$, and the radiation power is (c.w. (4.15))

$$\mathcal{P} = \omega(|P_{0x}|^2 + |P_{0y}|^2)/2\chi''. \tag{4.25}$$

4.1.5 °*Thermodynamic approach*

Between the ranges of strong absorption in matter, there are transparency 'windows' where one can neglect the energy dissipation, i.e., assume $|\chi''/\chi'| \ll 1$. (We consider the matter to be non-gyrotropic.) In the absence of dissipation, the vibrational energy of the particles caused by the external field is conserved; hence, the work of polarization can be defined as a function of the field amplitude, $A(E_0)$. This work consists of displacing the charges and is performed by the sources of alternating fields. Note that for A to be defined, a finite time is necessary for a stationary amplitude $\mathbf{P}_0$ of polarization oscillations to be formed, which is only possible in the presence of some finite absorption. After introducing the notion of work $A(\mathbf{E}_0)$ we can consider $\mathbf{E}_0$ as one of the thermodynamic parameters defining the state of the matter, in addition to entropy S, density ρ, etc. In the framework of this approach, one can formulate thermodynamical definitions for the polarization $\mathbf{P}_0(S,\rho,\mathbf{E}_0)$ and susceptibility $\chi(S,\rho,\mathbf{E}_0)$ as functions of the state of the matter.

In transparency windows, dispersion is small as well; therefore, polarization follows the field almost instantaneously,

$$\mathbf{P}(t) \approx \chi(\omega)\mathbf{E}(t), \tag{4.26}$$

where ω is the central frequency of a quasi-monochromatic field. In the case of an optical field, $\chi(\omega)$ is certainly defined without accounting for the inertial mechanisms of polarization, for instance, orientation of the molecules by an alternating field (Sec. 6.2). Such mechanisms only contribute to the static and radio-frequency susceptibilities.

Let us first ignore the dispersion completely. Then the state of the matter has a time dependence only via the field $\mathbf{E}(t)$. Then, in (4.13) one can assume $\dot{\mathbf{P}} = \chi\cdot\dot{\mathbf{E}}$, so that the variation rate of the macroscopic field (per unit volume) takes the form[a]

$$\mathcal{P}(t) = (\mathbf{E}\cdot\dot{\mathbf{E}} + \mathbf{H}\cdot\dot{\mathbf{H}})/4\pi + \mathbf{E}\cdot\chi\cdot\dot{\mathbf{E}} = \frac{d}{dt}\left(\frac{E^2+H^2}{8\pi} + \frac{1}{2}\mathbf{E}\cdot\chi\cdot\mathbf{E}\right). \tag{4.27}$$

In the last equation, we have used the symmetry of the χ tensor. The expression in brackets is obviously the energy density of the macroscopic field, the first term being energy of the field in the vacuum, at the same $\mathbf{E}, \mathbf{H}$, and the second one having the meaning of additional work A performed by a field source in the presence of the matter. Additional energy of the matter in a given field has the opposite sign,

$$v = -\boldsymbol{E}\cdot\chi\cdot\boldsymbol{E}/2. \tag{4.28}$$

Strictly speaking, the macroscopic field $\boldsymbol{E}$ inside the matter should be replaced here by the external field $\boldsymbol{E}'$ in the absence of the matter (see Ref. [Landau (1982)], Sec. 11), but for the sake of simplicity we ignore the difference between $\boldsymbol{E}$ and $\boldsymbol{E}'$.

Equations (4.26)–(4.28) assume a linear relation between $\boldsymbol{P}$ and $\boldsymbol{E}$, which is valid only for a sufficiently weak field. An evident generalization of (4.28) is

$$dv = -\boldsymbol{P}(S,\rho,\boldsymbol{E})\cdot d\boldsymbol{E}, \tag{4.29}$$

[a]Taking dispersion into account leads, in the linear approximation, to replacing χ in (4.27) by $d(\omega\chi)/d\omega$ [Landau (1982)].

or

$$v = -\int_0^E \boldsymbol{P}(S,\rho,\boldsymbol{E})\cdot d\boldsymbol{E}. \tag{4.30}$$

Polarization $\boldsymbol{P}$ and, hence, the elementary work of polarization, $-dv$, certainly depend not only on $\boldsymbol{E}$ but on the other parameters defining the state of the matter. Therefore, the integral in (4.30) is along a curve, and in order to find v, this curve should be specified. The work of polarization can be defined at constant entropy S and density ρ, i.e., for a thermally isolated material with a given concentration of molecules $N = \rho/m$. In this case, polarization will not change the internal energy of the matter per unit volume in the absence of the field, $U_0(S,\rho)$ (by definition, $dU_0 = TdS + \mu d\rho$, μ being the chemical potential). Therefore, the internal energy of the matter in the presence of the field is

$$U(S,\rho,\boldsymbol{E}) = U_0(S,\rho) + v(S,\rho,\boldsymbol{E}), \tag{4.31}$$

where E plays the role of an external thermodynamical parameter.

Now, one can define polarization and susceptibility thermodynamically as functions of the state of the matter,

$$P_\alpha(S,\rho,\boldsymbol{E}) \equiv -\partial U/\partial E_\alpha, \tag{4.32}$$

$$\chi_{\alpha\beta}(S,\rho) \equiv -(\partial^2 U/\partial E_\alpha \partial E_\beta)_{E=0}. \tag{4.33}$$

Thus, *by defining the χ tensor in terms of the thermodynamical potential one can provide its symmetry.* In (4.27)–(4.32), one can assume $\boldsymbol{E} = \boldsymbol{E}(\boldsymbol{r},t)$ if the dispersion is neglected; hence, the state of the matter depends on time and coordinate as parameters.

Further, doing the Taylor expansion of the internal energy $U(\boldsymbol{E})$ or the energy of adiabatic polarization $v(\boldsymbol{E})$, near the $\boldsymbol{E} = 0$ point one can define the nonlinear polarization and the nonlinear susceptibility tensors $\chi^{(n)}$ (Sec. 6.1).

It is often convenient to use, instead of U, other thermodynamic potentials such as, for instance, the free energy $F(T,\rho,\boldsymbol{E})$. The field part of F, which has the meaning of the work of polarization, v_F, should be calculated at constant temperature, so that in the general case $v_F \neq v_U$. However, in weak fields, the field parts of all potentials are the same (see Ref. [Landau (1964)], Sec. 15) and equal to $v(\boldsymbol{E})$. As a result, various macroscopic effects in electromagnetic field, such as electrostriction, electrocaloric effect etc., are determined by partial derivatives of χ in density, temperature, and so on (Sec. 6.2).

Let now the field be quasi-monochromatic, then χ should be replaced by $\chi(\omega)$. Thus, transmission of light by transparent matter leads to an increase of thermodynamic potentials by a value of

$$v(t) = -[\boldsymbol{E}_0 \cdot \chi(\omega) \cdot \boldsymbol{E}_0^* + \boldsymbol{E}_0 \cdot \chi(\omega) \cdot \boldsymbol{E}_0 e^{-i2\omega t} + \text{c.c.}]/8. \tag{4.34}$$

Only the constant, or slowly varying, part of the potential is of practical interest,

$$v = -\boldsymbol{E}_0 \cdot \chi(\omega) \cdot \boldsymbol{E}_0^*/4. \tag{4.35}$$

This expression for the effective potential of the matter in a monochromatic field describes, according to the known thermodynamic equations, the effect of light on the state of the matter. Variation of the state (temperature, density, etc.) of the matter, in its turn, influences χ and the transmitted light, i.e., causes a nonlinear optical effect (Sec. 6.2).

Note that, according to (4.35), one can define the polarization amplitude and the susceptibility in terms of the effective potential,

$$P_{0\alpha} = -4\partial v/\partial E_{0\alpha}^*,\ \chi_{\alpha\beta} = -4\partial^2 v/\partial E_{0\alpha}^* \partial E_{0\beta}. \tag{4.36}$$

Let the density of the molecules be N, then in the approximation of non-interacting molecules, polarizability of a single molecule is $\alpha = \chi/N$, and from (4.35) one can find the effective potential of a molecule in an alternating field,

$$\mathcal{V} = -\frac{1}{4}\boldsymbol{E}_0 \cdot \alpha(\omega) \cdot \boldsymbol{E}_0^*. \tag{4.37}$$

This potential defines the mean force of light pressure acting on a molecule in a monochromatic field in terms of the molecule polarizability,

$$\boldsymbol{F} = -\nabla\mathcal{V} = \nabla(\boldsymbol{E}_0 \cdot \boldsymbol{\alpha}(\omega) \cdot \boldsymbol{E}_0^*)/4. \tag{4.38}$$

This expression can be transformed as

$$\begin{aligned} F_\alpha &= \frac{1}{4}\frac{\partial}{\partial x_\alpha}(E_{0\beta}\alpha_{\beta\gamma}E_{0\gamma}^*) = \frac{1}{4}\frac{\partial E_{0\beta}}{\partial x_\alpha}d_{0\beta}^* + \text{c.c.} \\ &= \frac{1}{4}\frac{\partial}{\partial x_\alpha}d_0^* \cdot E_0(\boldsymbol{r}) + \text{c.c.} = \frac{\partial}{\partial x_\alpha}\overline{\boldsymbol{d}(t) \cdot \boldsymbol{E}(\boldsymbol{r}, t)}. \end{aligned} \tag{4.39}$$

The factor $1/2$ is absent here since we assume that the ∇ operator does not act on the dipole moment $\boldsymbol{d} = \boldsymbol{\alpha} \cdot \boldsymbol{E}$ of the molecule. The force (4.39) corresponds to the potential $\mathcal{V}(\boldsymbol{r}) = -\boldsymbol{d} \cdot \boldsymbol{E}(\boldsymbol{r})$. Light pressure will be considered in more detail in Sec. 6.2.

4.2 Dispersion theory

4.2.1 *Dispersion law*

In the transparency windows, $\epsilon'' = 0$ and, according to (4.23), the Green function turns to infinity at $n = \sqrt{\epsilon}$. Usually, it is this 'resonant' value of the ratio ck/ω that is denoted by n and called the refractive index.

The same condition provides the existence of a nontrivial solution ($E_0 \neq 0$) to homogeneous ($P_0 = 0$) Maxwell's equations. Therefore, *a medium without sources can only support propagation of waves with a certain relation between the wavelength, $\lambda = 2\pi/k$, and the frequency*. This relation,

$$k(\omega) = \omega[\epsilon(\omega)]^{1/2}/c, \tag{4.40}$$

or the inverse of it, $\omega(k)$, is called *the dispersion law*, and the waves satisfying it are called *free, or normal*, ones. The condition for normal longitudinal field to exist is $\epsilon(\omega) = 0$.[b] It follows from (4.40) that the phase velocity of transverse normal waves is $\sqrt{\epsilon}$ times as small as the speed of light. The group velocity, as we know, is given by the derivative $d\omega/dk \equiv u$. It follows that longitudinal waves do not propagate, since, according to (4.24), $\omega(k) = \text{const}$ and $u = 0$. (Here, we again neglect the effect of spatial dispersion.)

In an anisotropic medium, the condition for normal waves to exist, or for the Green function to turn to infinity, according to (4.22), has the form

$$\det(n^2\mathbf{\Pi} - \boldsymbol{\epsilon}) = 0. \tag{4.41}$$

This condition is called the Fresnel equation. With the frequency ω and the wavevector direction $\boldsymbol{k}/k$ fixed, Eq. (4.41) only has solutions for two[c] particular directions of the polarization vector $\boldsymbol{e}_\nu$ ($\nu = 1, 2$). In the general case, the polarization vector is not orthogonal to $\boldsymbol{k}$ and may be complex, which corresponds to the elliptical polarization of the normal wave (for more details, see Ref. [Landau (1973)]). The two normal waves have different dispersion laws $\omega_\nu(\boldsymbol{k})$, which leads to birefringence. In an anisotropic medium, the group velocity vector $\boldsymbol{u}$ is equal to $\nabla\omega_\nu(\boldsymbol{k})$ and, in the general case, is not parallel to the phase velocity vector.

4.2.2 *The effect of absorption*

With an account for absorption, the Fresnel equation has solutions only for complex ω and/or $\boldsymbol{k}$. The choice depends on the particular problem. A stationary experiment corresponds to a real frequency and a complex propagation constant. If the wave vector is complex, a free monochromatic wave gets either damped or amplified in the course of propagation. Let us make a replacement in (4.40), $k \to \tilde{k} \equiv k + i\alpha/2$, then the dispersion law of a normal transverse wave in an isotropic medium takes the form $(k + i\alpha/2)^2 = (\epsilon' + i\epsilon'')\omega^2/c^2$. Hence,

$$k^2 - \alpha^2/4 = \omega^2\epsilon'/c^2, \; \alpha k = \omega^2\epsilon''/c^2,$$

[b] We neglect the effects of spatial dispersion, which can be described by a $\epsilon(k)$ dependence (see, for instance, Refs. [Vinogradova (1979); Landau (1982)]).

[c] Effects of spatial dispersion may double the number of normal waves at a given frequency. The corresponding waves are called 'new' ones (Fig. 4.5).

or

$$k = \frac{\omega}{c}\mathrm{Re}\,\sqrt{\epsilon} = \frac{\omega}{c}\left(\frac{|\epsilon| + \epsilon'}{2}\right)^{1/2}, \tag{4.42}$$

$$\alpha/2 = \frac{\omega}{c}\mathrm{Im}\,\sqrt{\epsilon} = \frac{\omega}{c}\left(\frac{|\epsilon| - \epsilon'}{2}\right)^{1/2}. \tag{4.43}$$

The sign by the square root is chosen from physical considerations. These equations define the positions of the two poles of the Green functions $G(\tilde{k}, \omega)$ in the $\tilde{k}$ plane.

In the case of weak absorbtion, $\alpha^2 << k^2$, Eqs. (4.42), (4.43) take the form

$$k = \omega\sqrt{\epsilon'}/c, \tag{4.44}$$

$$\alpha = k\epsilon''/\epsilon'. \tag{4.45}$$

It should be stressed again that the dispersion law $\omega_\nu(\boldsymbol{k})$ and a fixed polarization $\boldsymbol{e}_\nu$ only take place for free waves, i.e., waves generated by distant sources. In the presence of given sources, the spatial and temporal dependencies of 'stimulated' field are determined by the distribution of the currents and can be arbitrary. In particular, thermal fluctuation field inside the matter is created by the chaotic motion of charged particles and the field at a given frequency is a superposition of plane waves of various lengths. Note that waves with maximal amplitudes do not always satisfy (4.44) (see Eq. (4.79)).

Thus, the macroscopic theory enables all basic observable rules of waves emission, propagation, and absorption through the phenomenological function $\chi(\omega)$. The next step is calculating $\chi(\omega)$ in the framework of the microscopic theory. This is a traditional problem of non-equilibrium thermodynamics, and its complete solution is still absent.

4.2.3 *Classical theory of dispersion*

In order to find the order of magnitude and the dispersion dependence for linear dielectric permittivity, let us use the simplest model of the matter as a set of independent, motionless, and identical atoms or molecules. Due to alternating electromagnetic field, the electron cloud of a molecule oscillates (the nuclei are assumed to be motionless), and the molecule gains the dipole moment $\boldsymbol{d}(t) = -e\sum \boldsymbol{r}_i(t)$, which, in the first approximation, scales as the field. Here, $e > 0$ is the electron charge and $\boldsymbol{r}_i$ is the radius vector of the ith electron. As a rule, magnetic dipole moment, quadruple moment and higher-order moments can be ignored since the scale of the spatial field variation, $\lambda > 10^{-5}$ cm, in the optical range exceeds much the typical size of a molecule, $a_0 \sim 10^{-8}$ cm. The product of $\boldsymbol{d}$ and the concentration of molecules N is equal to the dipole moment per unit volume, i.e., polarization, $\boldsymbol{P} = N\boldsymbol{d} = \chi\boldsymbol{E}$.

Thus, the problem of calculating the susceptibility is reduced to calculating the dipole moment of a molecule induced by an external field.

Thermal motion of charges can be taken into account in the framework of the kinetic theory. In the quantum theory, $\boldsymbol{r}_i$ and, hence, $\boldsymbol{d}$, are operators; therefore, one should do both quantum and statistical averaging, i.e., use the density matrix formalism.

Consider first the classical Lorentz model, which represents a molecule as an oscillator. The equation of motion of a linear isotropic oscillator has the form

$$\ddot{\boldsymbol{r}} + 2\gamma\dot{\boldsymbol{r}} + \omega_0^2\boldsymbol{r} = e\boldsymbol{E}_{loc}/m, \tag{4.46}$$

where m, ω_0 and e are the effective mass, frequency, and charge of the oscillator, respectively, γ is the phenomenological damping constant, and $\boldsymbol{E}_{loc}$ is the field at the centre of the molecule, known as the *local field*. After multiplying (4.46) by eN, we find the equation of motion for the polarization,

$$\ddot{\boldsymbol{P}} + 2\gamma\dot{\boldsymbol{P}} + \omega_0^2\boldsymbol{P} = \omega_p^2\boldsymbol{E}_{loc}/4\pi, \tag{4.47}$$

where $\omega_p = (4\pi e^2 N/m)^{1/2}$ is the so-called plasma frequency.

The field $\boldsymbol{E}_{loc}$ at the centre of a motionless molecule differs from the space-averaged macroscopic field $\boldsymbol{E}$. According to Lorentz,

$$\boldsymbol{E}_{loc} = \boldsymbol{E} + \frac{4\pi}{3}\boldsymbol{P} = \frac{\epsilon + 2}{3}\boldsymbol{E}, \tag{4.48}$$

so that (4.47) takes the form

$$\ddot{\boldsymbol{P}} + 2\gamma\dot{\boldsymbol{P}} + \tilde{\omega}_0^2\boldsymbol{P} = \omega_p^2\boldsymbol{E}/4\pi, \tag{4.49}$$

$$\tilde{\omega}_0^2 \equiv \omega_0^2 - \omega_p^2/3. \tag{4.50}$$

Hence, assuming the field to be monochromatic, we find

$$\chi = \frac{\omega_p^2/4\pi}{\tilde{\omega}_0^2 - \omega^2 - 2i\gamma\omega}. \tag{4.51}$$

In what follows, we will assume that the eigenfrequency shifts (4.50) due to the Lorentz correction are included into the definition of ω_0.

Suppose now that there are several types of independent oscillators with eigenfrequencies ω_j and concentrations $f_j N$, $\sum f_j = 1$, then

$$\chi = \frac{\omega_p^2}{4\pi}\sum_j \frac{f_j}{\tilde{\omega}_j^2 - \omega^2 - 2i\gamma\omega}. \tag{4.52}$$

The parameter f_j is called the *oscillator strength*. A similar expression, which in many cases describes well the observed dispersion of susceptibility, will be obtained below using the quantum theory.

Note that for very high frequencies or in the case of free electrons in a plasma or metal, one can assume in (4.52) $\omega \gg \omega_j$, so that

$$\epsilon \approx 1 - \frac{\omega_p^2}{\omega(\omega + 2i\gamma)}.$$

4.2.4 *Quantum theory of dispersion*

Let us now start from the kinetic equations for the density matrix (3.48), (3.49) with the phenomenological relaxation parameters γ_{mn}, w_{mn}. In the dipole approximation, the perturbation energy $\mathcal{V} = -\boldsymbol{d} \cdot \boldsymbol{E}$, and its matrix elements in the case of a monochromatic field have the form

$$\mathcal{V}_{mn}(t) = -\hbar\Omega_{mn}e^{-i\omega t}/2 + \text{h.c.}, \tag{4.53}$$

where

$$\Omega_{mn} \equiv \boldsymbol{d}_{mn} \cdot \boldsymbol{E}_0/\hbar. \tag{4.54}$$

Notation 'h.c.' stands for the Hermitian conjugate matrix,

$$-\boldsymbol{d}^*_{nm} \cdot \boldsymbol{E}^*_0 e^{i\omega t}/2 = -\boldsymbol{d}_{mn} \cdot \boldsymbol{E}^*_0 e^{i\omega t}/2.$$

Monochromatic perturbation will cause, in the linear approximation, the same response; therefore, let us seek the density matrix in the form

$$\rho^{(1)}_{mn}(t) = \rho^{(1)}_{mn}(\omega)e^{-i\omega t} + \text{h.c.} \tag{4.55}$$

In the zeroth order of the perturbation theory, the density matrix is diagonal, $\rho^{(0)}_{mn} = \rho^{(0)}_m \delta_{mn}$, so that, after substituting (4.53) and (4.55) in (3.48), (3.49), we find for $m \neq n$

$$\rho^{(1)}_{mn}(\omega) = \frac{\Omega_{mn}\Delta^{(0)}_{nm}/2}{\omega_{mn} - \omega - i\gamma_{mn}}, \tag{4.56}$$

where $\Delta_{nm} \equiv \rho_n - \rho_m$ is the relative population difference for levels n and m. The diagonal elements $\rho^{(1)}_{nn}$, according to (3.48), (3.49), will scale as the inverse frequency ω of the perturbation; if one is only interested in resonance effects, under the condition $\gamma/\omega \ll 1$ one can assume $\rho^{(1)}_{nn} = 0$. Thus, the amplitude of the response to a harmonic perturbation scales as the population difference and reaches its maximum at resonance, $\omega = \omega_{mn}$.

After substituting (4.56) in (3.6), we find the dipole moment of the molecule and the polarization,

$$\boldsymbol{P} = N\langle \boldsymbol{d}(t)\rangle = \frac{1}{2}\boldsymbol{P}_0 e^{-i\omega t} + \text{c.c.}$$

$$\boldsymbol{P}_0 = \frac{N}{\hbar}\sum_{mn} \frac{\Delta_{nm}^{(0)}\boldsymbol{d}_{nm}(\boldsymbol{d}_{mn}\cdot \boldsymbol{E}_0)}{\omega_{mn}-\omega-i\gamma_{mn}}.$$

Hence, according to definition (4.1),

$$\chi_{\alpha\beta} = \frac{N}{\hbar}\sum_{mn} \frac{\Delta_{nm}^{(0)} d_{nm}^{(\alpha)} d_{mn}^{(\beta)}}{\omega_{mn}-\omega-i\gamma_{mn}}, \tag{4.57}$$

where $d^{(\alpha)} \equiv d_\alpha$ is the projection of the molecule dipole moment onto the axis $\alpha = x, y, z$.

One can easily verify that the obtained expression has the necessary symmetry (4.5) and satisfies the causality principle (Fig. 4.1). Note that (4.57) can be represented in a somewhat different form,

$$\begin{aligned} \chi_{\alpha\beta} &= \sum_n N_n \alpha_{\alpha\beta}^{(n)}(\omega), \\ \alpha_{\alpha\beta}^{(n)}(\omega) &\equiv \frac{1}{\hbar}\sum_m \left(\frac{d_{nm}^{(\alpha)} d_{mn}^{(\beta)}}{\omega_{mn}-\omega-i\gamma_{mn}} + \frac{d_{mn}^{(\alpha)} d_{nm}^{(\beta)}}{\omega_{mn}+\omega+i\gamma_{mn}} \right). \end{aligned} \tag{4.58}$$

Here, $\boldsymbol{\alpha}^{(n)}$ has the meaning of the polarizability tensor of a molecule in state n. In the absence of a static magnetic field, unperturbed wave functions and, hence, the matrix elements $\boldsymbol{d}_{mn} = \boldsymbol{d}_{nm}$, can be considered to be real (see Ref. [Landau (1964)]). Then, according to (4.6), (4.58) is invariant to the permutation of α, β indices,

$$\alpha_{\alpha\beta}^{(n)} = \alpha_{\beta\alpha}^{(n)} = \frac{2}{\hbar}\sum_m \frac{\omega_{mn} d_{mn}^{(\alpha)} d_{mn}^{(\beta)}}{\omega_{mn}^2 - (\omega + i\gamma_{mn})^2}. \tag{4.59}$$

Using (4.57) and (4.14), one can easily show that the contribution of each pair of levels (m,n) into the field energy is positive or negative depending on the sign of $\omega_{mn}\Delta_{nm}$, i.e., amplification of the field requires population inversion, see also (4.60).

In the case of a gas, (4.57) should be averaged over random orientations and velocities of the molecules. Due to orientation averaging, non-diagonal elements of the $d_\alpha d_\beta$ turn into zero, and the diagonal ones become $|d_{mn}^{(\alpha)}|^2 = |\boldsymbol{d}_{mn}|^2/3$. As a result, the susceptibility tensor (4.57) becomes a scalar,

$$\chi = \frac{2N}{3\hbar}\sum_{m>n} \frac{\omega_{mn}\Delta_{nm}^{(0)}|\boldsymbol{d}_{mn}|^2}{\omega_{mn}^2 - (\omega + i\gamma_{mn})^2}, \tag{4.60}$$

In the last equation, we have taken into account that the double sum contains twice each term with $m \neq n$,

$$\sum_{mn} a_{mn} = \sum_{n} a_{nn} + \sum_{m>n} (a_{mn} + a_{nm}), \tag{4.61}$$

and the diagonal terms in (4.60) are zero since $\Delta_{nn} = 0$.

4.2.5 °*Oscillator strength*

In order to compare (4.60) with the classical expression (4.52), let us define dimensionless *oscillator strengths*,

$$f_{mn} \equiv 2m\omega_{nm}|\boldsymbol{d}_{mn}|^2/3\hbar e^2. \tag{4.62}$$

Note that the oscillator strength can be also defined phenomenologically, in terms of χ'', see Ref. [Landau (1982)].

Let us number possible pairs of states (m, n) by a single index $j \equiv \{m, n\}$, assuming $m > n$. If we neglect the γ^2 terms in the denominator of (4.60) and put $f_j \equiv f_{nm}\Delta_{nm}^{(0)}$, Eq. (4.60) takes the form of (4.52). Therefore, quantum calculation confirms the Lorentz model: *in the first approximation in the amplitude of the external field, matter behaves like a set of linear oscillators with damping.* However, f_j may now take negative values, which manifests itself in the effects of quantum amplification ($\chi'' < 0$) and negative dispersion ($\partial\chi'/\partial\omega < 0$ outside of the resonance).

Recall that, according to (4.7), a field δ-pulse causes a pulse of polarization shaped as the Fourier transform of $\chi(\omega)$. According to (4.60), the poles of $\chi(\tilde{\omega})$, understood as the function of a complex frequency, are at points $\tilde{\omega}_j = \pm\omega_j - i\gamma_j$ in the lower semi-plane; therefore, the polarization pulse is a sum of damped harmonic oscillations,

$$\chi(t) = \theta(t)\frac{\omega_p^2}{4\pi}\sum_j \frac{f_j}{\omega_j}\exp(-\gamma_j t)\sin(\omega_j t). \tag{4.63}$$

This expression defines the Green function for the polarization of the matter in terms of the eigenfrequencies and the oscillator strengths of the transitions.

The oscillator strengths satisfy the *sum rules*. For instance, for single-electron transitions,

$$\sum_m f_{nm} = 1. \tag{4.64}$$

This equation can be obtained from the commutation rule $[x, p] = i\hbar$, $p \equiv p_x$. Let $\mathcal{H}_0 = p_\alpha^2/2m + \mathcal{V}(\boldsymbol{r})$, then

$$[x, \mathcal{H}_0] = i\hbar p/m, \; p_{mn} = im\omega_{mn}x_{mn}; \tag{4.65}$$

hence,

$$\omega_{mn}|x_{mn}|^2 = ix_{mn}p_{nm}/m = -ip_{mn}x_{nm}/m,$$

$$\sum_m \omega_{mn}|x_{mn}|^2 = i[p,x]_{nn}/2m = \hbar/2m.$$

From the last equation, we obtain (4.64).

For most strong optical transitions in atoms, $|f_{mn}| \approx 1$. For instance, for the 'resonance' line of atomic hydrogen, $f = 0.416$ (the $1s$–$2p$ transition, $\lambda = 0.12\mu$). Hence, according to (4.62),

$$|x_{mn}| = (\lambda\!\!\!^{-}_c \lambda\!\!\!^{-}_{mn} f_{mn}/2)^{1/2} = 10^{-8}\ \text{cm}, \tag{4.66}$$

which corresponds to $|d_{mn}| = 4.8 \cdot 10^{-18}$CGSE $= 4.8$ D. Here, $\lambda\!\!\!^{-}_c \equiv \hbar/mc \approx 4 \cdot 10^{-11}$ cm is the Compton wavelength. For allowed transitions between rotational levels in the millimeter range, d_{mn} is also on the order of 1D, but in this case, according to (4.62), $f_{mn} \approx 10^{-5}$. Note that the sum in (4.64) should also include an integral over the continuous spectrum of ionized states; for instance, in hydrogen the ionized state for $n = 1s$ has a contribution of $f = 0.43$.

4.2.6 *Isolated resonance*

In the vicinity of a narrow isolated resonance, only a single term in the double sum of (4.60) has to be taken into account,

$$\chi = \chi_\infty + \frac{2\gamma\omega_0\Delta\chi}{\omega_0^2 - \omega^2 - 2i\gamma\omega}. \tag{4.67}$$

Here, $\Delta\chi \equiv \Delta N d^2/3\hbar\gamma = f\omega_p^2/8\pi\gamma\omega_0$ and χ_∞ is the contribution of other resonances, which is real. In the denominator of (4.67), the term γ^2 has been omitted; similarly to the Lorentz correction (4.50), it can be incorporated into the definition of ω_0. The parameter $\Delta\chi$, scaling as the product of the active particle density, ΔN, and the squared dipole moment of the transition, d^2, determines the maximal value of χ'' and the amplitude of χ' variation (Fig. 4.2).

In the optical range, the *Q-factor of a resonance* is usually high, $\omega_0/2\gamma \gg 1$; therefore, in the close vicinity of a resonance one can use a simple approximate formula,

$$\chi = \chi_\infty - \frac{\Delta\chi}{x+i}, \tag{4.68}$$

with

$$x \equiv (\omega - \omega_0)/\gamma,\ \omega \sim \omega_0 \gg \gamma > 0.$$

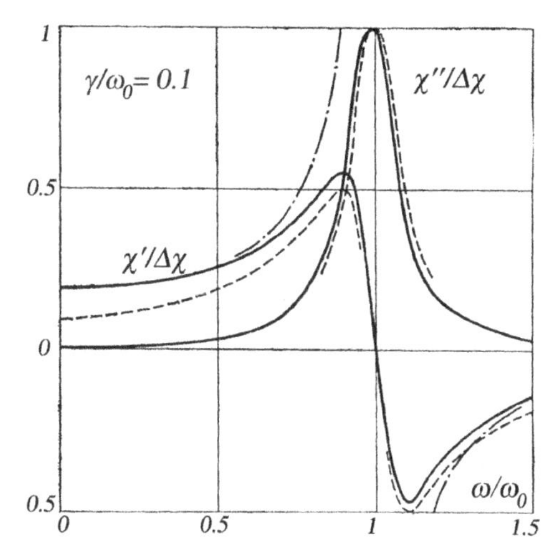

Fig. 4.2 Dispersion dependence of susceptibility $\chi(\omega)$ in the vicinity of an isolated resonance, for the Q-factor $\omega_0/2\gamma = 5$ and $\chi_\infty = 0$: solid lines correspond to (4.67), dashed lines, to approximate formula (4.68), and dash-dotted lines, to approximate formula (4.69).

This equation yields an even dependence for $\chi''(x)$ and an odd dependence for $\chi'(x) - \chi_\infty$ (see Fig. 4.2). It is clear from Fig. 4.2 that even for a resonance with a low Q-factor, this formula provides a good approximation for χ'' and somewhat worse one for χ'.

Note that far from the resonance, $|\chi' - \chi_\infty|$ decays much slower than $|\chi''|$; therefore, in the transparency windows, where $\chi'' \ll 1$, the refractive index n can still noticeably differ from unity. At a sufficient distance from the resonance, absorption can be neglected, and (4.67) takes another asymptotic form (Fig. 4.2),

$$\chi - \chi_\infty \approx \frac{f\omega_p^2/4\pi}{\omega_0^2 - \omega^2} = \frac{\chi_0 - \chi_\infty}{1 - \omega^2/\omega_0^2}, \tag{4.69}$$

where $\chi_0 \equiv \chi(0)$.

In the description of optical experiments, instead of χ or $\epsilon = 1 + 4\pi\chi$, one uses parameters that are more close to experiment, the refractive index and the index of absorption,

$$n \equiv kc/\omega = \mathrm{Re}\,\sqrt{\epsilon}, \;\; \kappa \equiv \alpha c/2\omega = \mathrm{Im}\,\sqrt{\epsilon}. \tag{4.70}$$

The value κ^{-1} has the meaning of the length of the wave penetration into the matter, in $\lambda\!\!\!^{-}/2 \equiv c/2\omega$ units. Figure 4.3 shows the dispersion dependence of these parameters, according to (4.68), (4.69) and (4.42)–(4.45), in the vicinity of an isolated resonance.

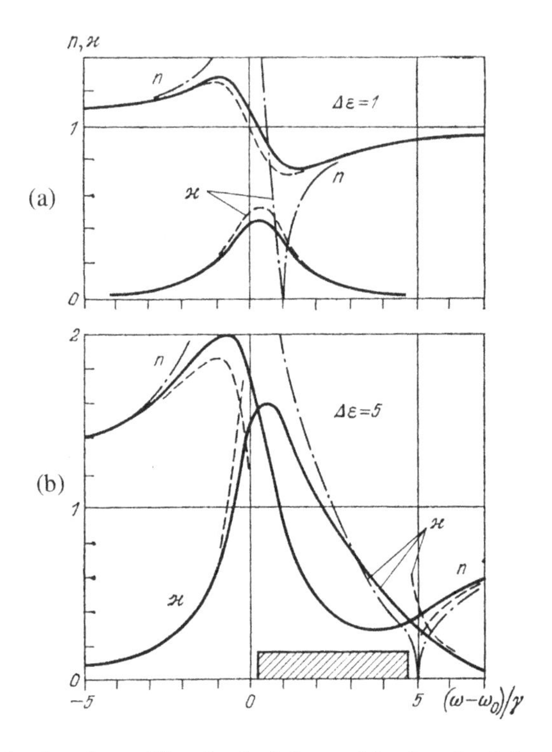

Fig. 4.3 Dispersion dependence of the refractive index n and the absorption index κ in the vicinity of an isolated resonance, for the resonance 'amplitude' $\Delta\epsilon$ being 1 (a) and 5 (b): solid lines correspond to (4.42), (4.43), and (4.68); dashed lines, to (4.44), (4.45) (i.e., weak absorption approximation), and dash-dotted lines, to (4.69) (i.e., approximation of $\epsilon'' = 0$, when $n = \sqrt{\epsilon}$ at $\epsilon > 0$ and $\kappa = \sqrt{-\epsilon}$ at $\epsilon < 0$); shading denotes the energy gap, where $\epsilon' < 0$.

In laser media, as a rule, $|\alpha| \ll 1\ \text{cm}^{-1}$, so that $|\epsilon''/\epsilon'| \approx |\alpha|/k < 10^{-4}$, and approximations (4.44), (4.45) are certainly valid. Substituting into them (4.68), we find

$$n \approx n_\infty\left(1 - \frac{\Delta\epsilon}{2\epsilon_\infty}\frac{x}{1+x^2}\right), \tag{4.71}$$

$$\alpha \approx \frac{\omega\Delta\epsilon/cn_\infty}{1+x^2}, \tag{4.72}$$

where $n_\infty \equiv \epsilon_\infty^{1/2} = (1 + 4\pi\chi_\infty)^{1/2}$ is the refractive index at $x \gg 1$ and

$$\Delta\epsilon \equiv 4\pi\Delta\chi = \frac{f\omega_p^2}{2\gamma\omega_0} = \frac{4\pi\Delta N d^2}{3\hbar\gamma} \tag{4.73}$$

is the resonance 'amplitude' for ϵ. Note that (4.72) coincides with the result of a 'probabilistic' calculation, (2.65), and that population inversion leads to a sign change in χ', χ'', and α. In this case, the dispersion dependence of the refractive index looks opposite to the usual one: n decreases with the growth of the frequency outside of the resonance and increases in the absorption range. This effect is called *negative dispersion*.

In condensed matter, narrow resonances often have large amplitudes, $\Delta\epsilon > 1$. This is especially typical for the ϵ dispersion in the infrared range, near lattice eigenvibrations of ion crystals. The corresponding elementary excitations (*quasi-particles*) are called *optical phonons*. Let, for instance, $d = 1\text{D}$, $\Delta N = 10^{20}\ \text{cm}^{-3}$, and $\Delta\omega \equiv 2\gamma = 1\ \text{cm}^{-1}$; then, according to (4.73), $\Delta\epsilon = 4$.[d] If, in addition, $f = 1$ and $\lambda_0 = 1\ \text{cm}^{-1}$, then the plasma frequency is much larger than $\Delta\omega$ but still much smaller than ω_0: $\omega_p = (\omega_0 \Delta\omega \Delta\epsilon)^{1/2} = 200\ \text{cm}^{-1}$.

From homogeneous Maxwell's equations, it follows that longitudinal oscillations are possible, their dispersion dependence being $\epsilon(\omega, k) = 0$. In the neglection of dissipation and spatial dispersion, such an oscillation has a fixed frequency ω_l and an arbitrary wave vector, i.e., zero group velocity $u = d\omega/dk$. According to (4.69), at $f = \epsilon_\infty = 1$,

$$\omega_l = \sqrt{\omega_0^2 + \omega_p^2} \approx \omega_0 + \omega_p^2/2\omega_0 = \omega_0 + \gamma\Delta\epsilon. \tag{4.74}$$

Hence, at $\Delta\epsilon \gg 1$, the splitting between the longitudinal and transverse frequencies, ω_l and ω_0, is much larger than the damping constant γ. The same condition defines whether the eigenfrequency shifts of molecules due to their Coulomb interaction, (4.50), are high, so ω_0 in (4.74) should be understood as $\omega_0 - \gamma\Delta\epsilon/3$.

In the interval $\omega_0 - \omega_l$, according to (4.69), $\epsilon < 0$, and the wave number $\tilde{k} = \omega\sqrt{\epsilon}/c$ is purely imaginary, so that the field is not a wave any more. Thus, *this interval is a 'forbidden zone' where the module of Fresnel's reflectivity* $R = (\sqrt{\epsilon} - 1)/(\sqrt{\epsilon} + 1)$ *becomes a unity, and the dielectric behaves as a metal. Note that a metal, in its turn, is at* $\omega \gg \omega_p$ *similar to a dielectric.*

4.2.7 °*Polaritons*

In the case of low absorption, macroscopic field in the matter can be treated quantum mechanically (Sec. 7.4). In this case, the notion of *a photon in matter*, or a *polariton*, emerges. (It should not be confused with a polaron, an electron in a dielectric considered together with the polarization it induces.) A polariton (sometimes also called a *light exciton*) is an elementary excitation of a macroscopic field and the molecules interacting with it, having an energy $\hbar\omega$ and propagating with a velocity $u = d\omega/dk$. As ω approaches ω_0, more and more of the polariton energy is contained in the internal energy of the molecules.

[d] Here, the scaling factor $2\pi c$ between ω and frequency in cm^{-1} has been omitted.

When a photon enters a medium from a vacuum, it becomes a polariton, with a probability of $1 - |R|^2$; after passing an average distance of α^{-1}, the polariton gets absorbed. The momentum $\hbar k$ of a photon in matter differs n times from the momentum $\hbar\omega/c$ of a vacuum photon with the same energy.

Polaritons can be also excited through thermal energy, in which case their mean number per mode is given by the Planck function $\mathcal{N}(\omega)$. According to FDT (Sec. 7.7), the ωk spectrum of equilibrium field fluctuations in matter scales as $\mathcal{N}(\omega)G''(\omega, k)$, where G is the Green's function for macroscopic Maxwell's equations (Sec. 4.1).

Above, we discussed the dispersion law, i.e., the relation between the frequency and the wavelength, for free waves created by a distant source. One can suggest other definitions for the functions $n(\omega)$ or $\omega(k)$, for instance, given by the maximum of the Green's function imaginary part $G''(\omega, k)$. The corresponding dispersion dependence manifests itself in experiments on light scattering by polaritons (Sec. 6.5).

Let us substitute $\epsilon(\omega)$ in the single-pole approximation (4.68) into Eqs. (4.23), (4.24). Then, for $\omega > 0$,

$$G_x = \frac{4\pi}{y + \Delta\epsilon/(x+i)}, \tag{4.75}$$

$$G_z = \frac{4\pi}{-\epsilon_\infty + \Delta\epsilon/(x+i)}, \tag{4.76}$$

where

$$x \equiv (\omega - \omega_0)/\gamma, \;\; y \equiv (ck/\omega)^2 - \epsilon_\infty.$$

Hence, the spectra of transverse and longitudinal field fluctuations are described by the functions (Fig. 4.4)

$$G_x'' = \frac{4\pi\Delta\epsilon}{(\Delta\epsilon + xy)^2 + y^2}, \tag{4.77}$$

$$G_z'' = \frac{4\pi\Delta\epsilon}{(\Delta\epsilon - x\epsilon_\infty)^2 + \epsilon_\infty^2}. \tag{4.78}$$

If the y dependence on ω is neglected, then the spectra of fluctuations at fixed k have Lorentzian shapes with the central frequencies given by the equations

$$\tilde{\epsilon}(\omega) = (ck/\omega)^2, \; \tilde{\epsilon}(\omega_l) = 0, \tag{4.79}$$

where $\tilde{\epsilon}(\omega)$ coincides with (4.68) under the condition $\gamma = 0$,

$$\tilde{\epsilon} \equiv \epsilon_\infty - \Delta\epsilon/x \approx \epsilon_\infty + f\omega_p^2/(\omega_0^2 - \omega^2). \tag{4.80}$$

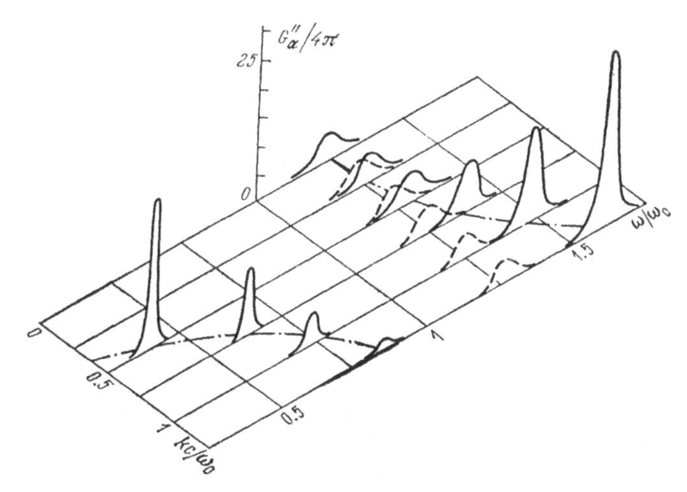

Fig. 4.4 Spectral density of equilibrium field, $\langle E_\alpha^2\rangle_{\omega k}$, normalized to $\hbar N/2\pi^3$, as a function of the frequency and wave vector in the vicinity of a dielectric function resonance for $\omega_0/\gamma = 20$ and $\Delta\epsilon = 5$. Solid lines refer to transverse (with respect to $\boldsymbol{k}$) oscillations , dashed lines, to longitudinal ones. It is clear from the figure that the frequency spectrum of thermal field fluctuations is described by the dispersion dependence (dash-dot) without the anomalous part.

Thus, the *dispersion law for field equilibrium fluctuations (4.79) differs from the dispersion law for free waves (4.42)* $ck/\omega = Re\sqrt{\epsilon(\omega)}$ *by the absence of the damping parameter* γ.

The dispersion law (4.79) corresponds to the condition $\partial G''/\partial x = 0$. At the same time, the condition for the G_x'' maximum at a fixed frequency, $\partial G_x''/\partial y = 0$, leads, according to (4.77), to the known dispersion law (4.44), $\mathrm{Re}\epsilon(\omega) = (ck/\omega)^2$. Hence, it follows that *the dispersion law observed near a resonance depends on the experimental conditions.*

Dispersion properties of a medium can be qualitatively represented as a graph showing the relation between ω and k, instead of the $\epsilon(\omega)$ or $n(\omega)$ dependencies. Figure 4.5 shows such a relation in different approximations and provides the commonly used names of the corresponding quasi-particles.

If the coupling between the transverse field and the oscillations of charges is negligible, which is possible at $k \gg \omega_0/c$, the elementary excitation, i.e., the energy quantum of the matter, in the case of polar oscillations of ions in a crystal lattice, is called an *exciton*, or an *optical phonon*. Excitons, similarly to photons, depending on the wave packet describing them, can be either localized within a certain area in the crystal, or spread over the whole space. In the $\Delta\epsilon = 0$ approximation, dispersion dependencies of photons and excitons overlap without interaction, which means that the incident field does not excite the excitons.

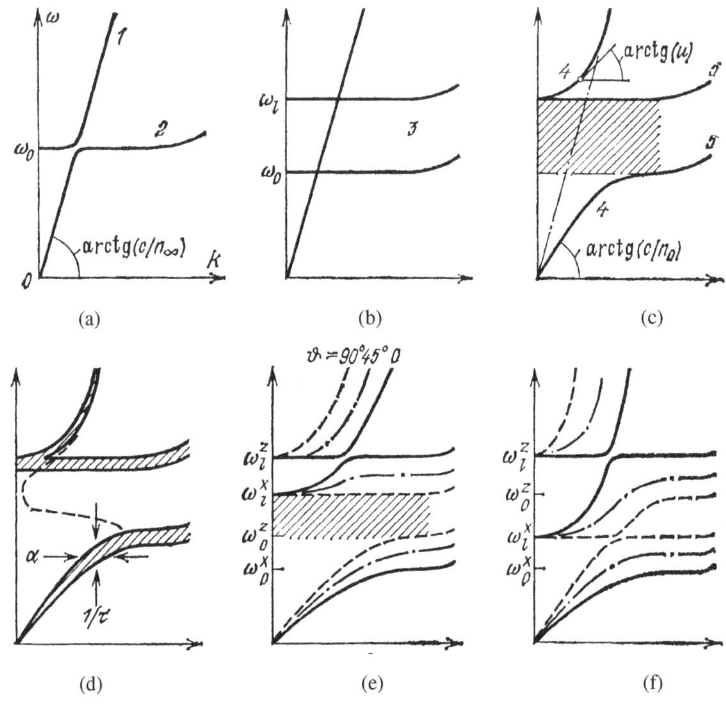

Fig. 4.5 Dispersion in various approximations. The numbers correspond to the following quasiparticles: 1, *photon*; 2, *mechanical exciton*, or *optical phonon*; 3, *Coulomb exciton* (longitudinal and transverse); 4, *polariton (light exciton)*; 5, 'new' waves. The following cases are shown: (a) oscillator strength or the particle density is small, and dispersion dependencies 'anti-intersect' with a small gap; (b) Coulomb interaction between the molecules lifts the degeneracy between the frequencies of longitudinal and transverse excitons; (c) interaction between the molecules and the transverse field leads to polariton effects, namely, to the energy gap (shaded area), to the dispersion of the polariton phase velocity ω/k outside the gap and to its group velocity u turning into zero at both boundaries of the gap; (d) dissipation leads to the smearing of the dispersion dependence, to finite life time τ and finite free path length $1/\alpha = u\tau$ for polaritons; in this case, waves excited externally have anomalous dispersion (dashed line); (e,f) anisotropy of the matter leads to the dependence of exciton and polariton frequencies on the wavevector direction $\boldsymbol{k}/k$. The figure shows frequency and angular dispersion, $k(\omega, \vartheta)$, for the extraordinary wave in a uniaxial crystal in the cases of weak (e) and strong (f) anisotropy; ϑ is the angle between the wave vector and the z axis of the crystal, ω_0^z and ω_l^z are frequencies corresponding to ϵ_{zz} being ∞ and 0, respectively; ω_0^x and ω_l^x are the same frequencies with respect to $\epsilon_{xx} = \epsilon_{yy}$.

At small $\Delta\epsilon$ and negligible dissipation, there is 'anti-crossing', or 'repulsion', of the dispersion dependencies, which look, in the interaction area, like two hyperbolas separated by a small gap (Fig. 4.5(a)).

At $\Delta\epsilon \gg 1$, oscillations of the charges and the field strongly influence each other, and the dispersion dependence changes considerably. There appears the longitudinal branch and the energy gap, near which $u \to 0$ (Figs. 4.4, 4.5(c)).

The electromagnetic wave is followed by an in-phased ($\omega < \omega_0$) or anti-phased ($\omega > \omega_0$) polarization wave, whose contribution into the total energy density is considerable. Note once again that *longitudinal oscillations, in the $u = 0$ approximation, do not propagate, i.e., these are waves with a fixed frequency ω_l and arbitrary wavelengths.*

Certainly, the simplest models considered here provide only a qualitative description of the field dispersion in realistic media. For the description of spatial dispersion effects, one has to take into account the dependence of $\Delta\epsilon$ ad ω_0 on k [Vinogradova (1979); Landau (1982)]. Doppler broadening can be included into the model by integrating (4.67), (4.68), (4.69) w.r.t. ω_0 over the Maxwell distribution in the case of gases and over the Fermi-Dirac distribution in the case of inter-band transitions in condensed matter. Transitions between narrow exciton bands in semiconductors and molecular crystals are described by dispersion dependencies (4.67), (4.68), (4.69) with $\Delta\epsilon$, ω_0, and k depending on ω, k. Calculation of these parameters is an interesting problem in solid state theory.

Note, in conclusion, that sometimes, instead of calculating ϵ, it is simpler to calculate directly the Green function G or the spectral density of equilibrium fluctuations [Zubarev (1971)], which is related to it through the Green-Kubo formula (Sec. 7.7).

4.3 Two-level model and saturation

Susceptibility χ calculated above determines the response of the matter to an alternating field only within the first order of the perturbation theory. This calculation does not take into account saturation of populations and other effects nonlinear in the field. Nonlinear effects are most pronounced under resonance conditions, when the field frequency is close to the eigenfrequencies of the matter.

4.3.1 *Applicability of the model*

This section considers two-level model, which is widely used in quantum electronics and spectroscopy. The model is based on the assumption that the field is quasi-monochromatic and a resonance $\omega \sim \omega_{21} \equiv \omega_0$ takes place only for a single pair of non-degenerate levels of a molecule. Such a situation is typical for magnetic resonance effects, nuclear (NMR) or electronic (EPR) ones. If a molecule has a single uncoupled electron, or the nucleus has a spin $I = 1/2$, then an external magnetic field H_0 splits each level in two Zeeman's sublevels, with the transition frequency $\omega_0 = \gamma H_0$, γ being the *gyromagnetic ratio.* Usually, ω_0 is in the microwave range and differs much from all other frequencies of the molecule.

However, in the absence of a constant magnetic field the Hamiltonian, as a rule, is invariant to certain symmetry operations, such as rotations etc., and therefore all energy levels are degenerate. Nevertheless, even in this case the two-level model provides a qualitatively correct description.

Note that in some cases, the two-level approximation is not applicable at all. For instance, it is not valid for NMR at $I > 1/2$ or for transitions between vibrational levels of molecules with weak anharmonicity. Such a molecule behaves as an oscillator with nearly equidistant levels, so that a resonance takes place for many pairs of levels simultaneously.

4.3.2 *Kinetic equations*

A two-level system is described by density-matrix equations (3.44) with $m, n = 1, 2$ and phenomenological relaxation times w_{mn} and $\gamma_{21} \equiv 1/T_2$ (see (3.48), (3.49)). Let $\mathcal{V}_{nn} = 0$, then

$$\dot{\rho}_{21} = -(i\omega_0 + 1/T_2)\rho_{21} - i(\rho_{11} - \rho_{22})\mathcal{V}_{21}/\hbar, \tag{4.81}$$

$$\dot{\rho}_{11} = w_{12}\rho_{22} - w_{21}\rho_{11} + i(\mathcal{V}_{21}\rho_{12} - \rho_{21}\mathcal{V}_{12})/\hbar. \tag{4.82}$$

In the case of two levels, $\rho_{11} + \rho_{22} = 1$, hence $\dot{\rho}_{11} = -\dot{\rho}_{22}$. Denote

$$\rho_{11} - \rho_{22} \equiv \Delta, \quad w_{12} + w_{21} \equiv 1/T_1. \tag{4.83}$$

With the perturbation 'switched off', $\mathcal{V} = 0$, populations should take their equilibrium values $\rho_{nn}^{(0)}$; therefore, the relaxation rates w_{12} and w_{21} are related by (3.58), which leads to

$$(w_{12} - w_{21})/(w_{12} + w_{21}) = \Delta^{(0)}. \tag{4.84}$$

With an account for (4.83) and (4.84), Eq. (4.82) takes the form

$$\dot{\Delta} = (\Delta^{(0)} - \Delta)/T_1 + 4\mathrm{Im}(\rho_{21}\mathcal{V}_{12})/\hbar. \tag{4.85}$$

Let the field be quasi-monochromatic and have the mean frequency $\omega > 0$ close to ω_0,

$$\boldsymbol{E}(t) = (1/2)\boldsymbol{E}_0(t)\exp(-i\omega t) + \text{c.c.}, \tag{4.86}$$

where $\boldsymbol{E}_0(t)$ is the slowly varying amplitude of the field. Then, in the dipole and resonance approximations, one can assume in (4.81) that

$$\mathcal{V}_{21} \approx -(1/2)\boldsymbol{d}_{21} \cdot \boldsymbol{E}_0(t)\exp(-i\omega t) \equiv -(1/2)\hbar\Omega\exp(-i\omega t + i\varphi). \tag{4.87}$$

Here, we have omitted the non-resonant negative-frequency term[e] proportional to $e^{i\omega t}$ and introduced the Rabi frequency $\Omega \equiv |\boldsymbol{d}_{21} \cdot \boldsymbol{E}_0|/\hbar$ and the interaction phase $\varphi(t)$. For a linearly polarized field and real wave functions, φ is simply the phase of the wave. Let us also define the slowly varying 'envelope' of the non-diagonal density matrix element,

$$\rho_{21}(t) = \rho_0(t)\exp(-i\omega t). \tag{4.88}$$

As a result, kinetic equations for a two-level system take the form

$$\dot{\rho}_0 = [(i(\omega - \omega_0) - 1/T_2]\rho_0 + i\Omega\Delta e^{i\varphi}/2, \tag{4.89}$$

$$\dot{\Delta} = (\Delta^{(0)} - \Delta)/T_1 - 2\Omega \mathrm{Im}(\rho_0 e^{-i\varphi}). \tag{4.90}$$

This system of three equations for three real functions $\rho'(t), \rho''(t)$, and $\Delta(t)$ determines the evolution of a two-level system due to the thermostat and the external field. According to (4.89), the effect of the field on ρ_0 scales as the population difference Δ, which, in its turn, is related to the field through ρ_0, according to (4.90). It is this relation that leads to the nonlinearity of the two-level system response. Under stationary conditions, this nonlinearity manifests itself in the *saturation effect*, i.e., Δ tending to zero provided that $\Omega^2 \gg 1/T_1T_2$. Non-stationary effects caused by the anharmonicity of a two-level system will be considered in Chapter 5.

4.3.3 *Saturation*

Consider now a stationary response of a two-level system to a monochromatic field, with Ω, ρ_0, φ, and Δ being constant. Then it follows from (4.89) (compare with (4.56)) that

$$\rho_0 = \frac{\Omega\Delta/2}{\omega_0 - \omega - i/T_2} e^{i\varphi}, \tag{4.91}$$

Substituting this expression in (4.90) allows one to find the stationary population difference,

$$\Delta = \Delta^{(0)}/(1 + 2T_1 W), \tag{4.92}$$

where

$$W \equiv \frac{\Omega^2 T_2/2}{1 + (\omega_0 - \omega)^2 T_2^2}. \tag{4.93}$$

---

[e] We also neglect higher-order harmonics of the density matrix, which oscillate with frequencies $n\omega$ and are on the order of magnitude of $(\Omega/\omega)^n$.

These equations describe *saturation*, the decrease of population difference due to a strong resonance field. Note that (4.93) coincides with the transition rate calculated in Sec. 2.2 for the case of a Lorentzian shape with the unsaturated width $\Delta\omega_0 = 2/T_2$. The second term in the denominator of (4.92) is called the *saturation factor*,

$$s \equiv 2WT_1 = 2W/(w_{12} + w_{21}). \tag{4.94}$$

It is clear from (4.94) *that saturation results from the competition between the transition rates due to the noise field of the thermostat,* $(w_{12} + w_{21})/2$, *and the external monochromatic field,* W.

According to (4.93), saturation is most pronounced at exact resonance, where the saturation factor takes its maximal value,

$$s_0 = \Omega^2 T_1 T_2 = 2\sigma_0 F T_1 \equiv F/F_s. \tag{4.95}$$

Here, we have introduced the transition cross section $\sigma \equiv W/F$, photon flux density F, and the photon flux density corresponding to a two-fold decrease in the population difference,

$$F_s \equiv 1/(2\sigma_0 T_1) = \hbar c/(8\pi\omega_0 |\boldsymbol{d}_{12}^{(e)}|^2 T_1 T_2). \tag{4.96}$$

Thus, in a resonant field the population difference decreases as

$$\Delta = \frac{\Delta^{(0)}}{1 + F/F_s}. \tag{4.97}$$

4.3.4 °*Lineshape in the presence of saturation*

By substituting (4.92) and (4.93) into (4.91), one finds the value of ρ_0 with an account for population saturation. The dipole moment of a two-level system (we consider a non-polar molecule, $d_{nn} \equiv 0$) is

$$\langle \boldsymbol{d}(t) \rangle = \boldsymbol{d}_{12}\rho_{21} + \text{c.c.}, \tag{4.98}$$

so that the susceptibility of a medium consisting of N two-level molecules takes the form (see (4.57))

$$\begin{aligned} \chi_{\alpha\beta}(E_0) &= N d_{12}^{(\alpha)} d_{21}^{(\beta)} \Delta/\hbar(\omega_0 - \omega - i/T_2) \\ &= \hbar^{-1}\Delta^{(0)} N d_{12}^{(\alpha)} d_{21}^{(\beta)} \frac{\omega_0 - \omega + i/T_2}{(\omega_0 - \omega)^2 + (1 + s_0)/T_2^2}. \end{aligned} \tag{4.99}$$

Recall that here, frequencies ω and ω_0 are positive and the values of χ at $\omega < 0$ are determined by (4.99) with ω, i replaced by $-\omega, -i$. Note that at $s_0 \neq 0$ it follows from (4.99) that the susceptibility $\chi(\omega, E_0)$, considered as a function of the complex variable $\tilde{\omega} = \omega' + i\omega''$, has poles both in the upper and lower semi-planes,

$$\tilde{\omega} = \pm\omega_0 \pm i(1 + s_0)^{1/2}/T_2.$$

As a result, the function $\chi(\omega, E_0)$ does not satisfy the Kramers-Kronig relations (4.8) at $E_0 \neq 0$.

Let us find, using (4.99) and (4.14), the power of absorbed, or, at $\Delta^{(0)} < 0$, emitted radiation:

$$\mathcal{P} = \hbar\omega\Delta^{(0)} N W_0/[1 + s_0 + (\omega_0 - \omega)^2 T_2^2] \equiv \mathcal{P}_{max} s_0/[1 + s_0 + (\omega_0 - \omega)^2 T_2^2], \quad (4.100)$$

where $W_0 \equiv |\boldsymbol{d}_{21} \cdot \boldsymbol{E}_0/\hbar|^2 T_2/2$ is the stimulated transition rate at exact resonance. It follows from this equation that at saturation, the spectral line maintains its Lorentzian shape of the form $1/(1 + x^2)$, but its width increases $\sqrt{1 + s_0}$ times (Fig. 4.6),

$$\Delta\omega = 2\sqrt{1 + s_0}/T_2. \quad (4.101)$$

This effect is called *radiation-induced broadening* or *field-induced broadening*.

Let $s_0 \gg 1 + (\omega_0 - \omega)^2 T_2^2$ (strong saturation), then it follows from (4.100) that

$$\mathcal{P} = \hbar\omega\Delta^{(0)} N/2T_1 \equiv \mathcal{P}_{max}. \quad (4.102)$$

Thus, *at strong saturation the power absorbed by the matter is no more dependent on the intensity and frequency of the field and is only determined by the*

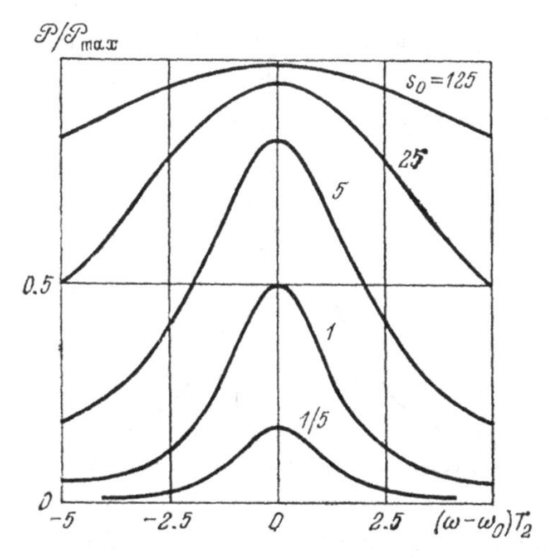

Fig. 4.6 The absorbed power $\mathcal{P}$ as a function of the frequency ω at different saturation factors s_0, according to (4.100).

rate of energy transfer from the molecules to the thermostat. In the case of strong saturation, the susceptibility real part, according to (4.99), scales linearly with the frequency,

$$\chi' \sim (\omega_0 - \omega)/s_0. \tag{4.103}$$

Let us make a numerical estimate. For a wave with the intensity $I = 1$ W/cm^2 and wavelength $\lambda = 1\mu$, $\hbar\omega = 2 \cdot 10^{-12}$ erg, $F = 5 \cdot 10^{18}$ photon/(s cm^2), and $E_0 = \sqrt{8\pi I/c} = 0.1$G= 30 V/cm. If $d_{12} = 1$D, then $\Omega = 10^8$ s^{-1}, so that $s_0 = 1$ at $T_1 = T_2 = 10^{-8}$ s. The transition rate is $W = 5 \cdot 10^7$ s^{-1}, and if $\Delta^{(0)}N = 10^{19}$ cm^{-3}, as it is the case for a doped crystal or a gas at atmospheric pressure, then it follows from (4.100) that $\mathcal{P} = 50$ MW/cm^3. This estimate shows that in the optical range, saturation is accompanied by strong heating of the matter. On the other hand, for electronic paramagnetic resonance in the $\lambda = 1$ cm range and at $T_1 = 10^{-3}$ s it follows from (4.102) that $\mathcal{P} = 0.1$ W/cm^3, so that stationary saturation is possible. This is used in paramagnetic amplifiers.

Saturation is very important in quantum electronics. It is used for creating population inversion by means of auxiliary radiation (pump) in lasers based on doped solids and in paramagnetic amplifiers. It is also applied for producing short strong light pulses via Q-switching and mode locking. Saturation stabilizes the amplitude of quantum oscillators and limits the dynamical range of quantum amplifiers.

In the case of inhomogeneous broadening, for instance, due to the Doppler effect, saturation affects not all the line but only its part, with the width on the order of collision or natural bandwidth. This effect, leading to the *Bennett hole burning* in the velocity distribution of the molecules (Fig. 4.7), is used for frequency stabilization of lasers and in saturation spectroscopy (Sec. 6.4).

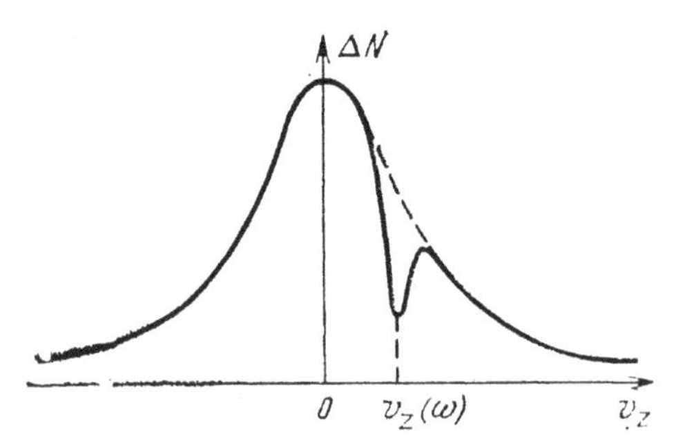

Fig. 4.7 Bennett hole burning. For a line inhomogeneously broadened due to the Doppler effect, saturation only affects those molecules whose velocities have given projections $v_z(\omega) = (\omega_0 - \omega)/k$ on the direction $\boldsymbol{k}$ of the wave propagation. $\Delta N(v_z)$ is the velocity distribution of active particles.

4.4 °Bloch equations

4.4.1 *Kinetic equations for the mean values*

In the previous section, we first solved kinetic equations for the density matrix and then, with the help of the obtained solution $\rho(t)$, we found the necessary mean (observed) value according to the formula $\langle f(t)\rangle = \mathrm{Tr}\{f\rho(t)\}$. It seems natural to try to exclude ρ and to find kinetic equations for the observables directly. Such equations can be obtained from the equations for ρ, but here we will find them in a different way.

In the case of a closed system, equations for the observables can be found by averaging the Heisenberg equations, which determine the time dependence of operators in the Heisenberg picture,

$$i\hbar df/dt = [f(t), \mathcal{H}(t)]. \tag{4.104}$$

(We assume that f has no direct time dependence.) The averaging is over the initial density matrix, which is usually considered as equilibrium,

$$i\hbar d\langle f\rangle/dt = \mathrm{Tr}\{[f(t), \mathcal{H}(t)]\rho(t_0)\}. \tag{4.105}$$

In the case of multi-particle systems, the derivative $d\langle f\rangle/dt$ depends, as a rule, not only on the mean value $\langle f(t)\rangle$ but also on the second moments or correlation functions, $\langle f(t)g(t')\rangle$. One can write the Heisenberg equations for the second moments, but after averaging, the result will contain the third moments, and so on. In order to 'unlink' this infinite chain of equations for the moments, one has to neglect the correlation of some values at a certain point, $\langle fg\rangle \approx \langle f\rangle\langle g\rangle$.

As a result, after excluding 'excessive' variables, one can obtain relatively simple kinetic equations for the observables of a single particle. In these equations, interaction with other particles and with the thermostat is taken into account with the help of a few phenomenological parameters, like relaxation times T_1, T_2 for a two-level system or kinetic transfer coefficients. It is noteworthy that, according to FDT (Sec. 7.7), kinetic coefficients for the first moments provide information about the second moments, i.e., fluctuations.

Approaches based on the density-matrix equations are certainly equivalent to the ones based on the equations for the mean values, and they should yield similar results. Note that in classical statistical physics, similarly, there are two basic methods of describing the kinetics: the first one is based on the distribution function (Liouville's equation, Boltzmann's distribution, Fokker-Planck equation), and the other one, on the moments (diffusion and transfer equations).

Macroscopic Maxwell's equations are, in fact, kinetic equations for the first field moments $\langle E\rangle, \langle H\rangle$, with the phenomenological function $\epsilon(\omega)$. The relaxation time τ of a monochromatic field is apparently equal to the ratio of the energy density, $\mathcal{E} = \epsilon'|E_0|^2/8\pi$, and the loss power, $\dot{\mathcal{E}} = \omega\epsilon''|E_0|^2/8\pi$, i.e., $\tau = \epsilon'/\omega\epsilon''$.

The same result is obtained if we set equality between τ and the absorption length $1/\alpha$ divided by the velocity of the wave, c/n.

Below, we will consider equations for the observables of a two-level system in two typical cases, namely, for electric dipole and magnetic dipole interactions, the energy of interaction with the field being $-\boldsymbol{d}\cdot\boldsymbol{E}$ and $-\boldsymbol{\mu}\cdot\boldsymbol{H}$, respectively. In the latter case, the observable is the magnetic moment $\langle\mu\rangle$ of the particle or magnetization $\boldsymbol{M} = N\langle\boldsymbol{\mu}\rangle$; kinetic equations for these observables are called the *Bloch equations*. In the case of electric dipole interaction, kinetic equations for $\langle\boldsymbol{d}\rangle$ and the population difference Δ, i.e., energy in $\hbar\omega_0$ units, are similar to the Bloch equations and are called the *optical Bloch equations*.

4.4.2 *Pauli matrices and expansion of operators*

Description of two-level systems is most convenient in terms of two-dimensional *Pauli matrices*, which are defined as

$$\sigma_x \equiv \begin{pmatrix} 0 & 1 \\ 1 & 0 \end{pmatrix}, \quad \sigma_y \equiv \begin{pmatrix} 0 & -i \\ i & 0 \end{pmatrix}, \quad \sigma_z \equiv \begin{pmatrix} 1 & 0 \\ 0 & -1 \end{pmatrix}. \tag{4.106}$$

These matrices $\boldsymbol{\sigma}_{\alpha mn} \equiv \langle m|\sigma_\alpha|n\rangle$ *represent* certain operators $\boldsymbol{\sigma}_\alpha$, whose eigenvalues are $\lambda = \pm 1$. (Recall that the eigenvalues of a matrix f_{mn} are defined as the roots of the characteristic equation, $\det\{f_{mn} - \lambda\delta_{mn}\} = 0$.) The matrix representation (4.106) is the *eigenrepresentation* for the $\boldsymbol{\sigma}_z$ operator. From (4.106) and the rules of matrix multiplication, we find the multiplication table for the Pauli operators,

$$\begin{aligned} \sigma_x\sigma_y &= -\sigma_y\sigma_x = i\sigma_z, \\ \sigma_\alpha^2 = I, \quad \sigma_y\sigma_z &= -\sigma_z\sigma_y = i\sigma_x, \\ \sigma_z\sigma_x &= -\sigma_x\sigma_z = i\sigma_y. \end{aligned} \tag{4.107}$$

Thus, the Pauli matrices *anti-commute* with each other ($\sigma_\alpha\sigma_\beta + \sigma_\beta\sigma_\alpha = 2\delta_{\alpha\beta}$), and their commutation relations coincide with the ones for the Cartesian components of the angular momentum s.

It is also convenient to introduce elementary matrices called *diadic tensors*, or *outer products* of vectors,

$$\sigma^{(1)} \equiv |1\rangle\langle 1| = \begin{pmatrix} 1 & 0 \\ 0 & 0 \end{pmatrix}, \quad \sigma^{(+)} \equiv |1\rangle\langle 2| = \begin{pmatrix} 0 & 1 \\ 0 & 0 \end{pmatrix}$$

$$\sigma^{(-)} \equiv |2\rangle\langle 1| = \begin{pmatrix} 0 & 0 \\ 1 & 0 \end{pmatrix}, \quad \sigma^{(2)} \equiv |2\rangle\langle 2| = \begin{pmatrix} 0 & 0 \\ 0 & 1 \end{pmatrix}.$$

In the general case, two arbitrary vectors $|a\rangle$ and $|b\rangle$ can always compose a diadic tensor, with the matrix elements given by the products of the corresponding components of the vectors,

$$\sigma^{(ab)} \equiv |a\rangle\langle b|, \ \sigma^{(ab)}_{mn} = \langle m|a\rangle\langle b|n\rangle.$$

A symmetric diadic operator $\sigma^{(n)} \equiv |n\rangle\langle n|$ is called a projector, since its action on a vector consists of projecting it onto the $|n\rangle$ direction,

$$\sigma^{(n)}|a\rangle = |n\rangle\langle n|a\rangle = \text{const} \cdot |n\rangle.$$

Usually, $|n\rangle$ is a unit vector, $\langle n|n\rangle = 1$. Note that the mean value of a diadic operator $\sigma^{(mn)}$ coincides with the corresponding element of the transposed density matrix,

$$\langle \sigma^{(mn)} \rangle = \text{Tr}\{\rho|m\rangle\langle n|\} = \sum_{kl} \rho_{kl}\langle l|m\rangle\langle n|k\rangle = \rho_{nm}.$$

The Pauli operators are related to the diadic operators as

$$\begin{aligned}
&2\sigma^{(1)} = I + \sigma_z, \ 2\sigma^{(2)} = I - \sigma_z, \ 2\sigma^{(\pm)} = \sigma_x \pm i\sigma_y, \\
&\sigma_x = \sigma^{(+)} + \sigma^{(-)}, \ \sigma_y = i(\sigma^{(-)} - \sigma^{(+)}), \ \sigma_z = \sigma^{(1)} - \sigma^{(2)}, \\
&I = \sigma^{(1)} + \sigma^{(2)}.
\end{aligned}$$

One can easily find the multiplication table and the commutation relations for the diadic operators,

$$\begin{aligned}
\sigma^{(1)^2} = \sigma^{(1)}, \ \sigma^{(\pm)^2} = 0, \ \sigma^{(+)}\sigma^{(-)} = \sigma^{(1)}, \\
\sigma^{(-)}\sigma^{(+)} = \sigma^{(2)}, \ [\sigma^{(+)}, \sigma^{(-)}] = \sigma_z, \ [\sigma^{(\pm)}, \sigma_z] = \mp 2\sigma^{(\pm)}.
\end{aligned}$$

Note that $\sigma^{(\pm)}$ are non-Hermitian operators: $(\sigma^{(\pm)})^+ = \sigma^{(\mp)}$. They can be called *creation and annihilation operators* for an energy quantum. Indeed, let $|2\rangle$ be the ground-state wave function of the system. The $\sigma^{(+)}$ operator turns it into the excited-state wave function, $\sigma^{(+)}|2\rangle = |1\rangle$. Similarly, $\sigma^{(-)}|1\rangle = |2\rangle$.

One can easily verify that any Hermitian operator acting in the Hilbert space of a two-level system can be represented as a sum (sometimes we will keep the 'hats' over the operators),

$$\hat{f} = a\hat{I} + b\hat{\sigma}_x + c\hat{\sigma}_y + d\hat{\sigma}_z, \tag{4.108}$$

where a, b, c, d are real numbers. Indeed, combining (4.106) and (4.108), we find the relations that define the coefficients in the expansion (4.108) in terms of the matrix elements f_{mn},

$$\hat{f} = \begin{pmatrix} a + d & b - ic \\ b + ic & a - d \end{pmatrix}. \tag{4.109}$$

Recall that quantum mechanics, similarly to vector calculus, operates with three types of variables: usual complex *numbers* (c-numbers, scalars), complex *vectors* (wave functions of discrete or continuous variables), given by n numbers, and *operators* (matrices, tensors), given by n^2 numbers. (Here, n is the dimensionality of the vector space, equal to the number of states of the system.) For a given operator, one can find, according to certain known rules, the corresponding scalars (eigenvalues and trace), which are invariant with respect to a change of representation, i.e., to a rotation of the basic vectors. The Pauli vector, $\hat{\boldsymbol{\sigma}} \equiv \{\hat{\sigma}_x, \hat{\sigma}_y, \hat{\sigma}_z\}$, and the orbital momentum $\hat{\mathbf{s}}$, which is proportional to it, are vectors in a real three-dimensional space and, at the same time, operators in the abstract *space of states*, having two complex basic vectors $|1\rangle, |2\rangle$.

Note that the coefficients in the Pauli-matrix expansion (4.108) of an arbitrary operator $\hat{f}$ have a clear physical meaning. They determine two values that the observable f can take at single measurements. By writing the equation for the eigenvalues of matrix (4.109), one can see that the *spectrum* $f_{nn} \equiv f_n$ consists of two numbers,

$$f_{1,2} = a \pm (b^2 + c^2 + d^2)^{1/2}. \tag{4.110}$$

Let the basic vectors for the representation (4.106) be the energy states of the system. Then the operator $\hat{\mathcal{H}}_0$ is diagonal and, for $f \equiv \hat{\mathcal{H}}_0$, it follows from (4.109) that $a = b = c = 0$, $d = -\hbar\omega_0/2$. Therefore, the Hamiltonian of the system scales as the $\hat{\sigma}_z$ operator,

$$\hat{\mathcal{H}}_0 = -\hbar\omega_0\hat{\sigma}_z/2, \tag{4.111}$$

where $\hbar\omega_0 \equiv \mathcal{H}_{022} - \mathcal{H}_{011}$. The relative population difference is in this case equal to the mean value of $\hat{\sigma}_z$,

$$\langle\sigma_z\rangle = \rho_{11} - \rho_{22} \equiv \Delta. \tag{4.112}$$

Let the eigenfunctions $|m\rangle$ of the $\hat{\mathcal{H}}_0$ operator and, hence, the matrix elements of the electric dipole moment $\boldsymbol{d}_{mn}$ be real, $\boldsymbol{d}_{12} = \boldsymbol{d}_{21} = \boldsymbol{d}_0$. Let us also assume that the diagonal elements are absent (the molecule is non-polar), $d_{nn} = 0$, then the operator $\hat{\boldsymbol{d}} = -e\hat{\boldsymbol{r}}$ scales as $\hat{\sigma}_x$,

$$\hat{\boldsymbol{d}} = \boldsymbol{d}_0\hat{\sigma}_x, \tag{4.113}$$

and the perturbation operator takes the form

$$\hat{\mathcal{V}} = -(\boldsymbol{d}_0 \cdot \boldsymbol{E})\hat{\sigma}_x. \tag{4.114}$$

4.4.3 *The Bloch vector and the Bloch sphere*

Let us define the *Bloch vector* $\boldsymbol{R}$ (also called *pseudo-spin*) as the mean value of the Pauli operator $\boldsymbol{\sigma}$. Using (4.106), $\boldsymbol{R}$ can be also written in terms of the density matrix,

$$\boldsymbol{R} \equiv \langle \boldsymbol{\sigma} \rangle = \{2\rho'_{21}, 2\rho''_{21}, \Delta\}. \tag{4.115}$$

Thus, the $\boldsymbol{R}$ vector, similarly to the density matrix ρ_{mn}, fully determines the state of the system. In other words, an arbitrary state of a two-level system is given by three real numbers, which can be explicitly shown by a point or a radius vector in some three-dimensional space. In the case of a spin-1/2 particle, this vector is parallel to the mean orbital momentum. In the case of an electric dipole two-level system, $\boldsymbol{R}$ does not correspond to any observable vector, but its x and z components, according to (4.111), (4.113), have a clear physical meaning.

Let us find the length of the $\boldsymbol{R}$ vector. According to (4.115) and (3.18),

$$R^2 = (\rho_{11} - \rho_{22})^2 + 4|\rho_{21}|^2 \leq 1. \tag{4.116}$$

In the case of a pure state, by definition (3.4), $|\rho_{21}|^2 = \rho_{11}\rho_{22}$ and $\boldsymbol{R}$ is a unit vector. Thus, an arbitrary pure state of a system can be shown by a point on a sphere called the *Bloch sphere*. If the state is mixed, then (see (3.18)), $|\rho_{21}|^2 < \rho_{11}\rho_{22}$, and $R < 1$.

During time evolution, the depicting point (the $\boldsymbol{R}(t)$ vector) moves along some trajectory on the unit sphere. This trajectory, for an arbitrary perturbation $\mathcal{V}(t)$ (4.114), can be found using the Heisenberg equation (4.104) for σ_α and the commutation rules (4.107). At $\mathcal{V} = 0$, (3.46) immediately yields

$$\begin{aligned} R_x &= R_{x0} \cos \omega_0 t + R_{y0} \sin \omega_0 t, \\ R_y &= -R_{x0} \sin \omega_0 t + R_{y0} \cos \omega_0 t, \\ R_z &= R_{z0}. \end{aligned} \tag{4.117}$$

Thus, in the case of a closed system, the end of the $\boldsymbol{R}$ vector circles around the z axis, similarly to the precession of a spinning top around the gravity force direction (Fig. 4.8). According to (4.111) and (4.113), the energy is constant in this case, while the dipole moment oscillates with the transition frequency. In particular, for a pure coherent state, with $c_n = \exp(i\varphi_n)/\sqrt{2}$, it follows from (3.4) that

$$\begin{aligned} R_x &= \cos(\omega_0 t + \varphi_1 - \varphi_2), \\ R_y &= -\sin(\omega_0 t + \varphi_1 - \varphi_2), \\ R_z &= 0, \end{aligned} \tag{4.118}$$

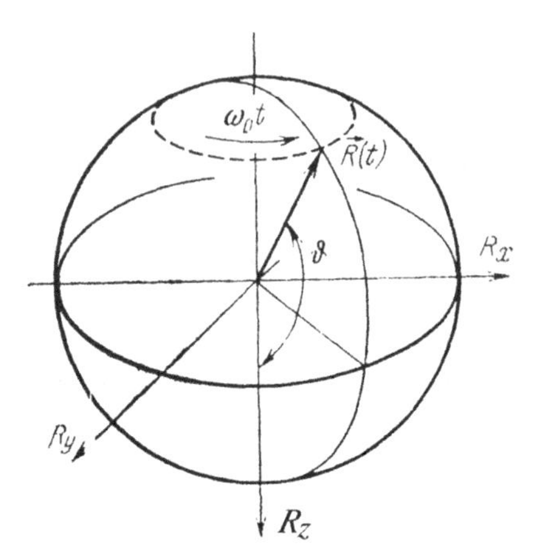

Fig. 4.8 The state of a two-level system represented geometrically using the Bloch vector $\boldsymbol{R}$ whose components determine the dipole moment $\langle d \rangle = d_0 R_x$ and the population difference $\Delta = R_z$. The z axis is directed downwards, so that points denoting the excited states of the system are above the point denoting the ground state. Under free evolution of the system, $\boldsymbol{R}$ undergoes precession with the frequency ω_0, the angle of precession θ being determined by the initial conditions.

i.e., the depicting point moves along the 'equator' of a unity sphere, and the *precession angle* is $\vartheta \equiv \arctan(R_\perp / R_z) = \pi/2$. For an energy state, the point rests at one of the poles ($c_n = 0$ or 1, $\vartheta = 0$ or π). A weak resonant perturbation causes a slow variation of the precession angle with the Rabi frequency, the so-called *nutation* (Sec. 5.1).

4.4.4 *Higher moments and distributions*

Recall that the density matrix (or, according to (4.115)), the vector $\boldsymbol{R} = \langle \boldsymbol{\sigma} \rangle$) provides full statistical information about the system, i.e., allows one to find the higher moments $\langle f^k \rangle$ and the probability distributions $P(f)$ of an arbitrary observable f. Moments can be easily expressed in terms of ρ or $\langle \sigma_\alpha \rangle$ using (4.108) and the multiplication formulas (4.107). In particular, it follows from (4.107) that

$$\sigma_\alpha^{2k+1} = \sigma_\alpha, \ \sigma_\alpha^{2k} = I.$$

Hence, the main measure of fluctuations, the variance, is

$$\Delta\sigma_\alpha^2 \equiv \langle \sigma_\alpha^2 \rangle - \langle \sigma_\alpha \rangle^2 = 1 - R_\alpha^2. \tag{4.119}$$

For an energy state, where $R_z = \pm 1$ and $R_{x,y} = 0$, the energy variance is equal to zero, while the variance of the dipole moment is a unity (in d_0 units). In the case of a coherent state, according to (4.118), the energy fluctuates with the unity variance (in $\hbar\omega_0/2$ units), while the variance of the transverse components, $\sigma_{x,y}$, depends on the time instance of the measurement, as it oscillates with $2\omega_0$ frequency between 0 and 1.

Consider the uncertainty relations containing the variances of the Pauli vector components and limiting the accuracy of their simultaneous measurement. According to (4.119),

$$\Delta\sigma_\alpha^2 \Delta\sigma_\beta^2 = 1 + R_\alpha^2 R_\beta^2 - R_\alpha^2 - R_\beta^2,$$
$$\Delta\sigma_\alpha^2 + \Delta\sigma_\beta^2 = 2 - R_\alpha^2 - R_\beta^2.$$

For an arbitrary state, the length of the $\boldsymbol{R}$ vector does not exceed unity (see (4.116)); therefore, the following inequalities hold true:

$$\Delta\sigma_\alpha^2 \Delta\sigma_\beta^2 \geq R_\gamma^2 + R_\alpha^2 R_\beta^2, \tag{4.120}$$

$$\Delta\sigma_\alpha^2 + \Delta\sigma_\beta^2 \geq 1 + R_\gamma^2, \tag{4.121}$$

where $\gamma \neq \alpha, \beta$. In the case $R_\alpha = 0$, inequality (4.120) takes the form of a standard uncertainty relation,

$$\Delta f \Delta g \geq |\langle [f, g] \rangle|/2. \tag{4.122}$$

Let us now find the probability distributions. Let $P_\alpha(\pm 1)$ be the probability of σ_α taking values ± 1, then

$$\langle \sigma_\alpha \rangle = P_\alpha(1) - P_\alpha(-1) = 2P_\alpha(1) - 1.$$

Hence,

$$P_\alpha(\pm 1) = (1 \pm R_\alpha)/2.$$

For instance, for a coherent state (4.118),

$$P_x(1) = \cos^2[(\omega_0 t + \varphi_2 - \varphi_1)/2].$$

Similarly, for an arbitrary observable of a two-level system,

$$P(f_1) = 1 - P(f_2) = (\langle f \rangle - f_2)/(f_1 - f_2). \tag{4.123}$$

4.4.5 *Bloch equations*

Let us find the equations of motion for the Pauli vector. From (4.104), (4.107), (4.111), and (4.114), it follows that

$$\begin{aligned}
\dot{\sigma}_x &= \omega_0 \sigma_y, \\
\dot{\sigma}_y &= -\omega_0 \sigma_x + \Omega(t)\sigma_z, \\
\dot{\sigma}_z &= -\Omega(t)\sigma_y,
\end{aligned} \tag{4.124}$$

where $\Omega(t) \equiv 2\boldsymbol{d}_0 \cdot \boldsymbol{E}(t)/\hbar$ is the 'instantaneous' Rabi frequency. Let us introduce the vector $\boldsymbol{A}(t) \equiv \{\Omega(t), 0, \omega_0\}$, then (4.124) can be represented as a vector product,

$$\dot{\boldsymbol{\sigma}} = \boldsymbol{\sigma} \times \boldsymbol{A}. \tag{4.125}$$

From (4.125), we immediately find a similar equation for the Bloch vector,

$$\dot{\boldsymbol{R}} = \boldsymbol{R} \times \boldsymbol{A}. \tag{4.126}$$

According to (4.126), *the* $\boldsymbol{R}$ *vector, which represents all properties of a two-level system, is precessing around the instantaneous direction of the effective field vector* $\boldsymbol{A}(t)$.

Further, let us take into account, in the simplest approximation, the interaction between the particle and its environment. Assume that there is an exponential relaxation with two positive parameters T_1, T_2, which characterize the rate of approaching thermodynamic equilibrium after the perturbation is off. As a result, the Heisenberg equations (4.125) turn into the so-called *optical Bloch equations*,

$$\dot{\boldsymbol{R}} = \boldsymbol{R} \times \boldsymbol{A} - \boldsymbol{R}_\perp/T_2 - \hat{z}(R_z - \Delta^{(0)})/T_1, \tag{4.127}$$

where $\hat{z}$ is a unit vector along the z axis.

One can easily verify that these equations, with an account for (4.115), coincide with the density-matrix equations (4.81), (4.85), and hence all results of Sec. 4.3 are still valid. However, now the system behavior has an obvious geometric interpretation.

It should be stressed that the Bloch equations (4.127) are kinetic equations, describing only the first moments of the observables $\boldsymbol{R} = \langle \boldsymbol{\sigma} \rangle$; they provide no information about fluctuations and higher moments. The latter can be only found by choosing some particular stochastic model of relaxation.

In the case of a monochromatic field, a settled stimulated motion of $\boldsymbol{R}$ is precession with the field frequency ω around the z axis (Fig. 4.8). The precession angle ϑ, according to (4.91), is given by ($\Omega \equiv |\boldsymbol{d}_0 \cdot \boldsymbol{E}_0|/\hbar$)

$$\tan\vartheta \equiv \frac{R_\perp}{R_z} = \frac{2|\rho_{21}|}{\Delta} = \frac{\mathrm{sign}(\Delta)\Omega T_2}{[1 + (\omega_0 - \omega)^2 T_2^2]^{1/2}}, \tag{4.128}$$

and the length of the $\boldsymbol{R}$ vector, according to (4.92), is

$$R = \Delta/\cos\vartheta = \Delta^{(0)}/[\cos\vartheta(1 + \xi\tan^2\vartheta)], \tag{4.129}$$

where $\xi \equiv T_1/T_2$. Let $T_1 = T_2 \equiv \tau$ and $\omega = \omega_0$, then the $\boldsymbol{R}$ vector, due to the resonance field, shrinks $(1 + \Omega^2\tau^2)^{1/2}$ times and precesses at an angle $\vartheta = \arctan(\Omega\tau\mathrm{sign}(\Delta))$.

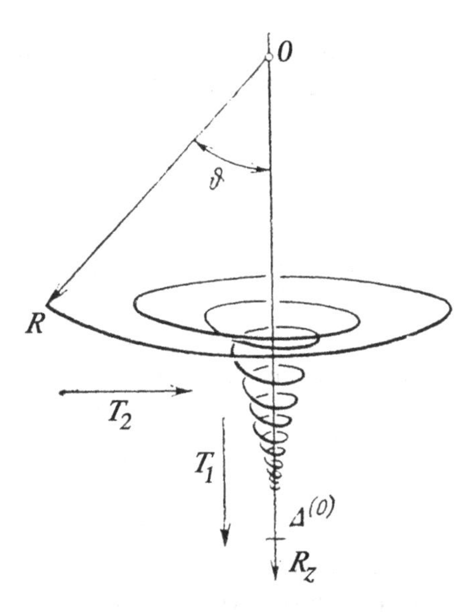

Fig. 4.9 Relaxation of the Bloch vector $\boldsymbol{R}$. After the perturbation is switched off, the transverse component goes down to zero at a rate of T_2^{-1}, while the longitudinal component takes its equilibrium value $\Delta^{(0)}$ at a rate of T_1^{-1}.

If the perturbation is suddenly switched off, the $\boldsymbol{R}$ vector, according to (4.127), will simultaneously precess with the Bohr frequency ω_0 and undergo relaxation to the equilibrium value $\{0, 0, \Delta^{(0)}\}$, i.e., move along a spiral (Fig. 4.9). If $T_1 \gg T_2$, then the first to disappear is the transverse component, $\boldsymbol{R}_\perp$, i.e., the non-diagonal element of the density matrix, ρ_{21}, and after that R_z takes its equilibrium value.

4.4.6 *Equation for polarization*

In the framework of the two-level approximation, the obtained equations (4.127) fully determine the optical properties of the medium, both stationary (Sec. 4.3) and non-stationary (Chapter 5) ones. To make it more obvious, let us pass from the variables R_x, R_y to polarization $\boldsymbol{P} = N\boldsymbol{d}_0 R_x$ [Pantell (1969)], which enters Maxwell's equations. From (4.127), it follows that

$$\dot{P} + P/T_2 = \omega_0 N d_0 R_y. \tag{4.130}$$

Taking the second derivative, we get

$$\ddot{P} + \dot{P}/T_2 = \omega_0 N d_0(-R_y/T_2 - \omega_0 R_x + \Omega(t) R_z). \tag{4.131}$$

In practice, always $\omega_0 T_2 \gg 1$, so that at $\Omega \ll \omega \sim \omega_0$, according to (4.130), one can assume in the right-hand side of (4.131) that

$$\omega_0 N d_0 R_y \approx \dot{P}. \tag{4.132}$$

As a result, we find that polarization satisfies the second-order linear differential equation (we assume that $\boldsymbol{d}_0 || \boldsymbol{E}$)

$$\ddot{P} + \frac{2}{T_2}\dot{P} + \omega_0^2 P = \frac{2\omega_0 d_0^2}{\hbar} E \Delta N, \tag{4.133}$$

where $\Delta N \equiv N R_z$ is the population difference per unit volume. Substituting (4.132) into the equation for R_z, we find the equation for ΔN,

$$\Delta\dot{N} + \frac{1}{T_1}(\Delta N - \Delta^{(0)} N) = -\frac{2}{\hbar\omega_0} E\dot{P}. \tag{4.134}$$

This equation has a simple meaning: according to (4.13), $E\dot{P}$ is the power absorbed in the matter.

Thus, a two-level system behaves as a harmonic oscillator with the damping $1/T_2$ whose coupling with the external force E depends on the force itself, with the inertia time T_1. In the case of a weak field, with the saturation factor $\Omega^2 T_1 T_2 \ll 1$, the system is equivalent to a linear oscillator.

4.4.7 *Magnetic resonance*

As it was already mentioned, two-level approximation is most applicable to a spin-1/2 particle. In a magnetic field, an electron acquires additional energy $\mathcal{H} = -\boldsymbol{\mu} \cdot \boldsymbol{H}$, with $\boldsymbol{\mu}$ being the magnetic moment of the electron, anti-parallel to its mechanical moment (*spin*) $\boldsymbol{s}$, $\boldsymbol{\mu} = -(g\mu_0/\hbar)\boldsymbol{s}$. Here, $g = 2.002$ is the so-called g *factor* of a free electron, $\mu_0 = e\hbar/2mc = 0.927 \cdot 10^{-20}$erg/G is the atomic unit of magnetic dipole moment (*Bohr's magneton*). The spin is usually given in $\hbar$ units: $\boldsymbol{s}' \equiv \boldsymbol{s}/\hbar$. The operators of spin projections, s_α, scale as the Pauli matrices, i.e., $\boldsymbol{s}' = \boldsymbol{\sigma}/2$. Thus, if we ignore the difference between g and 2, then $\boldsymbol{\mu} = -\mu_0 \boldsymbol{\sigma}$, and

$$\mathcal{H} = \mu_0 \boldsymbol{H} \cdot \boldsymbol{\sigma}. \tag{4.135}$$

Now, using commutation relations for σ_α (4.107), we can easily find equations of motion for any operator. For instance,

$$i\hbar\dot{\sigma}_x = \mu_0([\sigma_x, \sigma_y]H_y + [\sigma_x, \sigma_z]H_z) = 2i\mu_0(H_y\sigma_z - H_z\sigma_y). \tag{4.136}$$

Derivatives for other components can be written in a similar way. As a result, equation of motion for the Pauli vector takes a simple form (compare with (4.125)),

$$\dot{\boldsymbol{\sigma}} = \gamma \boldsymbol{\sigma} \times \boldsymbol{H}, \tag{4.137}$$

where $\gamma \equiv -2\mu_0/\hbar = -2\pi \cdot 2.8$ MHz/G is the *gyromagnetic ratio*. Since the $\boldsymbol{\sigma}$ vector scales as the magnetic and mechanical moment of an electron, equations for $\boldsymbol{\mu}$ and $\boldsymbol{s}$ have a similar form, for instance, $\dot{\boldsymbol{s}} = \gamma \boldsymbol{s} \times \boldsymbol{H}$.

The last equation has the same form as the classical equation for rotational motion, according to which the rate of angular momentum variation is equal to the torque of the forces acting on a dipole in a magnetic field, $\boldsymbol{\mu} \times \boldsymbol{H} = -2\mu_0 \boldsymbol{s} \times \boldsymbol{H}/\hbar$. Thus, an electron in a magnetic field behaves similarly to a spin top under the action of two forces. In the case of a constant magnetic field, (4.137) describes *precession*, the motion of the momentum vector along a cone around $\boldsymbol{H}_0$ (Fig. 4.8). The precession frequency, $\omega_0 = |\gamma| H_0$, coincides with Bohr's transition frequency, $(\mathcal{H}_{22} - \mathcal{H}_{11})/\hbar$.

However, in contrast to a classical spin top, the observable angular momentum of an electron can have only a single absolute value, $s \equiv (\sum s_\alpha^2)^{1/2}$, equal to $\hbar\sqrt{3}/2$, since $\boldsymbol{\sigma}_\alpha^2 = 1$. Also, a single measurement of its projection onto an axis can only yield one of the two values, $\pm\hbar/2$. Note that $s^2 \neq \langle \boldsymbol{s} \rangle^2 = \sum \langle s_\alpha \rangle^2$; for instance, in a mixed state with equal populations ($\Delta = \rho_{21} = 0$), all three projections are equal to zero, so that $\langle \boldsymbol{s} \rangle^2 = 0$.

Let the magnetic field have, in addition to the constant component $\boldsymbol{H}_0$, a time-varying component $\boldsymbol{H}_\perp(t)$, orthogonal to $\boldsymbol{H}_0$. Let the x axis be parallel to $\boldsymbol{H}_\perp$ and the z axis be anti-parallel to $\boldsymbol{H}_0$. Then the electron energy (4.135) can be written in terms of the Pauli operators,

$$\mathcal{H} = \mathcal{H}_0 + \mathcal{V}(t) = -\hbar(\omega_0 \sigma_z + \Omega(t)\sigma_x)/2, \tag{4.138}$$

where, this time, $\Omega(t) = \gamma H_\perp$. We have chosen H_{0z} to be negative for the subscript 1 to denote the lower level. Once again, the Heisenberg equations take the form (4.124).

Let us average these equations over the initial density matrix, pass to magnetization $\boldsymbol{M} = -\mu_0 N \langle \boldsymbol{\sigma} \rangle$, and add relaxation. We get

$$\begin{aligned} \dot{M}_x &= -M_x/T_2 + \omega_0 M_y, \\ \dot{M}_y &= -M_y/T_2 - \omega_0 M_x + \gamma M_z H_\perp, \\ \dot{M}_z &= (M_z^{(0)} - M_z)/T_1 - \gamma M_y H_\perp, \end{aligned} \tag{4.139}$$

where $M_z^{(0)} \equiv -\mu_0 N \Delta^{(0)}$ is the static equilibrium magnetization.

This system of equations, determining the magnetization kinetics of a paramagnetic material (electronic or nuclear) in a constant or variable field, is called the *Bloch equations*. The Bloch equations describe magnetic resonance, i.e., resonant absorption of radio waves. This effect forms the basis of the most important directions in radio spectroscopy, *electronic paramagnetic resonance* (EPR), *nuclear magnetic resonance* (NMR), and *ferromagnetic resonance*. One can easily verify that the Bloch equations are equivalent to the density-matrix equations in the case of a two-level system; therefore, all above-made conclusions are also valid for the case of magnetic resonance, with $-\boldsymbol{d} \cdot \boldsymbol{E}(t)$ replaced by $\mu_0 \boldsymbol{\sigma} \cdot \boldsymbol{H}_\perp(t)$.

As we have already mentioned, often $T_1 > T_2$, since the relaxation of the population energy $\langle \sigma_z \rangle$ is only caused by non-adiabatic interactions of the particle with the environment, for instance, non-elastic collisions in a gas or spin-lattice (spin-phonon) interaction in a crystal. At the same time, variation of the transverse components, $\langle \sigma_\perp \rangle$ (or ρ_{21}), does not require energy transfer, and hence the 'life time' of $\langle \sigma_\perp \rangle$ reduces as a result of both adiabatic (spin-spin) and non-adiabatic perturbations. One can say that perturbations 'disturb' the precession phase $\varphi_1 - \varphi_2$ in (4.118), which is the argument of ρ_{21}, and hence the time and ensemble mean values tend to zero, $\langle \sigma_\perp \rangle \to 0$.

Chapter 5

Non-Stationary Optics

In previous chapters, we mostly considered settled, stationary cases of the interaction between a field and matter. With a harmonic field on, transient processes decay due to the relaxation, and the oscillation of charges goes on with a stationary amplitude, given by the susceptibility χ. Susceptibility with an account for saturation, $\chi(E_0)$ (see (4.99)), determines the response of the matter, namely, of the polarization amplitude P_0, to a quasi-monochromatic perturbation, provided that either the amplitude E_0 is constant or the time of its variation is much greater than the relaxation time, $\tau_E \gg T_{1,2}$.

If the field is weak and there is no saturation, the population relaxation time T_1 does not play any role, and the stationarity condition has the form $\tau_E \gg T_2$. It is important that in the absence of saturation, the equations of motion of matter are linear, and hence the susceptibility $\chi(\omega)$ also determines non-stationary, transient processes. For instance, the response of matter to a short ($\tau_E \ll T_2$) weak pulse scales as the Fourier transform of $\chi(\omega)$, i.e., Green's function $\chi(t)$, and has the form of a set of oscillations decaying with the times T_{2mn} (see (4.63)). The response to a weak pulse of an arbitrary shape, $E(t)$, is given by the convolution of $\chi(t)$ and $E(t)$.

A question arises: what will be the response of matter to short and strong pulses? In this chapter, it is shown that under such conditions, several unusual optical effects can be observed. Among these effects, called *coherent*[a] ones, there are, for instance, *self-induced transparency, optical echo, and superradiance.* Some of these essentially non-stationary effects have been observed in radio spectroscopy relatively long ago; however, in the optical range they have only been observed after the invention of the lasers. Apart of their general theoretical importance, optical non-stationary effects are of practical interest from the

[a]In this book, instead of the ambiguous term 'coherent' we will use the term 'non-stationary', which is more accurate for the description of the effects considered.

viewpoint of spectroscopic applications and for the optimization of lasers. They also form the basis of *coherent* (non-stationary) spectroscopy [Steinfeld (1978); Manykin (1984)]. More detailed description of non-stationary effects can be found in Refs [Allen (1975); Macomber (1976)]. In Sec. 5.1, using the Bloch equations, we find the variation of the state of a two-level system under a short resonant light pulse. Further, we consider emission from a single atom (Sec. 5.2) and an ensemble of N atoms (Sec. 5.3) under different initial conditions.

5.1 Stimulated non-stationary effects

Consider the reaction of a two-level system to a quasi-harmonic perturbation of finite duration. For simplicity, assume that the pulse envelope is rectangular, with the duration $\tau_E \ll T_2$, so that one can neglect relaxation and use the relations for the density matrix ρ_{mn} or the equivalent Bloch equations for $\mathbf{R} = \{2\rho'_{21}, 2\rho''_{21}, \rho_{11} - \rho_{22}\}$ at $T_{1,2} = \infty$. Then, the Bloch equations (4.126) have the same form as the Heisenberg equations for the σ operator (4.125). In other words, at time intervals much shorter than the typical times of the interaction between the system and its environment, it can be considered as isolated and described, instead of kinetic equations, in terms of the Schrödinger equation for the wave function, the von Neumann equation for $\rho(t)$, or the Heisenberg equations for the operators.

We have already solved this problem using the Schrödinger equation in Chapter 2, where we have found the amplitudes of the perturbed states $c_{mn}(t)$ and the transition probability $P_{mn} = |c_{mn}|^2$ for a multi-level system in the first order of the perturbation theory. However, in the case of a two-level system, there is no need to represent the solution as a perturbative series.

5.1.1 *Atom as a gyroscope*

Let us turn to the geometric representation of the instant state of the system using the Bloch vector $\mathbf{R}$. According to (4.126), it behaves similarly to the orbital momentum of a body with a single fixed point,

$$\dot{\mathbf{R}} = \mathbf{R} \times \mathbf{A}. \tag{5.1}$$

Here, $\mathbf{A} \equiv \{2d_0E(t)/\hbar, 0, \omega_0\}$ is the effective field vector, $\mathbf{d}_0 \equiv \mathbf{d}_{21} = \mathbf{d}_{12}$ is the dipole moment of the transition, which is assumed to be real and parallel to the field $\mathbf{E}(t)$, and ω_0 is the transition frequency. Equation (5.1) preserves the length of the $\mathbf{R}$ vector, which is unity in the case of a pure state. The end of the $\mathbf{R}$ vector moves then along the surface of a unit sphere.

Consider the trajectory of this motion in a quasi-monochromatic field,

$$\mathbf{E}(t) = (1/2)\mathbf{E}_0(t)e^{-i\omega t + i\varphi_0} + \text{c.c.}$$

Let us pass to a reference frame rotating with the field frequency ω around the z axis,

$$\begin{aligned} \rho_{21} &= \rho_0 e^{-i\omega t}, \\ R_x &= R_{0x}\cos\omega t + R_{0y}\sin\omega t, \\ R_y &= -R_{0x}\sin\omega t + R_{0y}\cos\omega t. \end{aligned} \tag{5.2}$$

Let the conditions $|\omega - \omega_0| \ll \omega_0$, $d_0 E_0 \ll \hbar\omega_0$ hold, then the rotating-wave approximation is applicable and the functions $\rho_0(t)$, $\mathbf{R}_0(t)$ are slow compared to $e^{-i\omega t}$. These functions have a simple physical meaning: they determine the amplitude and the phase of the mean dipole moment, $\langle \mathbf{d} \rangle = 2\mathbf{d}_0\rho'_{21} = \mathbf{d}_0 R_x$. The power absorbed from the external field is proportional to $R_{0y}(t)$, see (5.12).

It is not difficult to verify that in the rotating-wave approximation, equation of motion (5.1) takes the form

$$\dot{\mathbf{R}}_0 = \mathbf{R}_0 \times \tilde{\mathbf{\Omega}}, \tag{5.3}$$

where

$$\begin{aligned} \mathbf{R}_0 &\equiv \{2\rho'_0, 2\rho''_0, \Delta\}, \\ \tilde{\mathbf{\Omega}} &\equiv \{\Omega\cos\varphi_0, \Omega\sin\varphi_0, \omega_0 - \omega\}, \\ \Delta &\equiv \rho_{11} - \rho_{22}, \Omega \equiv d_0 E_0/\hbar, \end{aligned} \tag{5.4}$$

φ_0 is the initial phase of the field at the center of the atom; in the case of a plane wave, $\varphi_0 = kz + \varphi_1$. Recall that for simplicity, we assume the matrix element of the transition to be real. In the general case, $\varphi_0 = \arg(\mathbf{d}_{21} \cdot \mathbf{E}_0)$.

Thus, in a harmonic field the $\mathbf{R}_0$ vector, which represents the state of the system, rotates with the rate $\tilde{\Omega} = [\Omega^2 + (\omega_0 - \omega)^2]^{1/2}$ around the 'effective field' direction $\tilde{\mathbf{\Omega}}$ (Fig. 5.1). This slow rotation is called *nutation*; it adds to the fast stimulated precession with the rate equal to the field frequency ω. In other words, the angle of precession $\vartheta(t)$ (see (4.128)) slowly varies within the pulse duration, so that the depicting point moves along a spiral on a sphere with the fixed radius $R = (\Delta^2 + 4|\rho_{21}|^2)^{1/2}$.

In the resonant case, where $|\omega_0 - \omega| \ll \Omega$, the nutation axis $\tilde{\mathbf{\Omega}}$ is in the equatorial plane. If, in addition, the initial state is an energy one, i.e., $\mathbf{R}(0)$ is at one of the poles and $\vartheta(0) = 0$ or π, then, due to the effect of the field, $\mathbf{R}$ moves along a meridian with the longitude $\varphi_0 \pm \pi/2$.

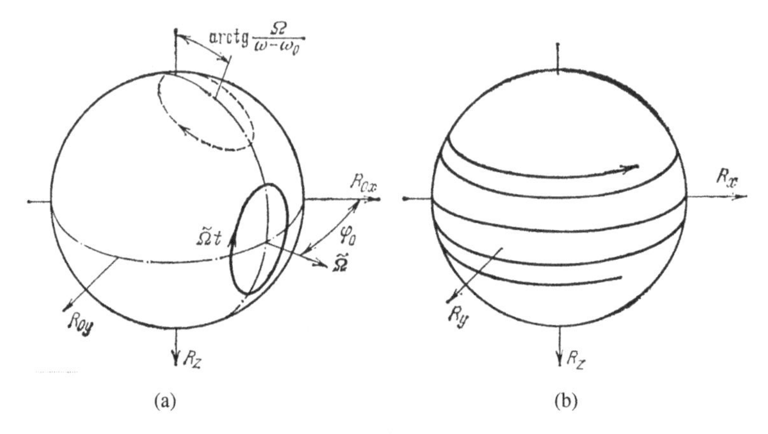

Fig. 5.1 Nutation of the Bloch vector **R**. Due to a monochromatic field, **R**, in addition to precession, rotates around the direction of the effective field $\tilde{\mathbf{\Omega}}$ with the nutation rate $\tilde{\Omega}$: (a) the frame of reference rotates with the field frequency ω; (b) a fixed frame of reference; solid line corresponds to the exact resonance, dashed line, to $\omega \neq \omega_0$.

As to the non-resonant case, $|\omega_0 - \omega| \gg \Omega$, the nutation axis is almost parallel to the z axis, so that the **R** vector moves approximately along latitudes, with the angular rate $\tilde{\Omega} \approx \omega_0 - \omega$ in the rotating reference frame, i.e., with the rate ω_0 in a fixed frame.

In practice, of course, the envelope $E_0(t)$ of the field increases and decreases gradually, so that the rates Ω and $\tilde{\Omega}$ are slow functions of time. At exact resonance, $\omega = \omega_0$, variation of the precession angle is determined by the 'area' of the pulse [Allen (1975)],

$$\vartheta(t) - \vartheta(0) = \int_0^t \Omega dt = d_0 \int_0^t E_0 dt/\hbar. \tag{5.5}$$

The phase φ_0 of the field is assumed to be constant, i.e., only amplitude modulation is considered.

Note that equation (5.3) provides a recipe for preparing any arbitrary pure state of the system (with given 'latitude' and 'longitude') with the help of a coherent field: one should first cool the system, i.e., move the depicting point to the South pole, and then apply a resonant pulse with a given area and phase.

5.1.2 *Analytical solution*

In the case $E_0 = \text{const}$, it is not difficult to describe nutation algebraically. Let, for instance, the initial state be stationary, i.e., the depicting point is at $t = 0$ at one of the poles, and $\varphi_0 = 0$. Then (5.3) is satisfied by the following functions of the time [Allen (1975)]:

$$
\begin{aligned}
R_{0x} &= \Delta(0)\frac{2\Omega(\omega_0-\omega)}{\tilde{\Omega}^2}\sin^2(\tilde{\Omega}t/2),\\
R_{0y} &= \Delta(0)\frac{\Omega}{\tilde{\Omega}}\sin(\tilde{\Omega}t),\\
R_z &= \Delta(0)\left[1-\frac{2\Omega^2}{\tilde{\Omega}^2}\sin^2(\tilde{\Omega}t/2)\right].
\end{aligned}
\tag{5.6}
$$

The last equation can be understood in terms of the transition probability. Let the system initially be at the ground level, $\Delta(0) = +1$, then the probability P of a transition up is $\rho_{22} = (1 - R_z)/2$, and (5.6) leads to the *Rabi formula*,

$$P = \left[\frac{\Omega}{\tilde{\Omega}}\sin(\tilde{\Omega}t/2)\right]^2. \tag{5.7}$$

Note that at $\Omega \ll |\omega_0 - \omega|$, (5.7) coincides with the transition probability (2.34), which was found using the perturbation theory,

$$P = \left[\frac{\Omega}{(\omega_0-\omega)}\sin\left(\frac{\omega_0-\omega}{2}t\right)\right]^2. \tag{5.8}$$

Hence, if one finds the transition rate $W \equiv \frac{dP}{dt}$ from (5.7) and averages it over the exponential distribution of interaction times (see the end of Sec. 2.2), the resulting expression will be equivalent to formula (4.100) describing the stationary absorption with an account for saturation (at $T_1 = T_2$).

Thus, *in the presence of a resonant perturbation, a quantum system passes periodically from the ground level to the excited one and back* (Fig. 5.2). The transition time, according to (5.7), is

$$t_\pi = \pi/\Omega = \pi\hbar/d_0E_0. \tag{5.9}$$

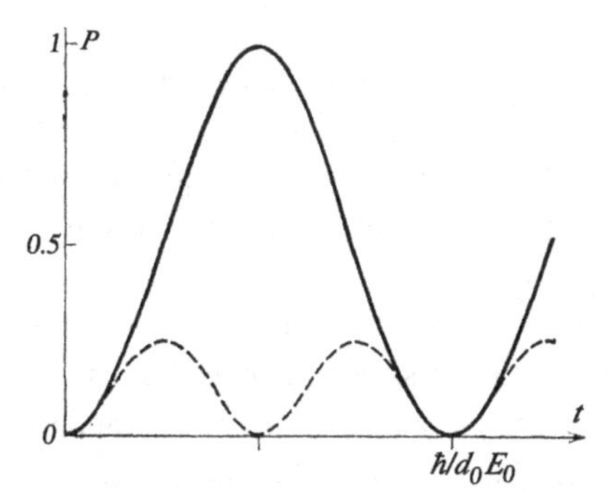

Fig. 5.2 Transition probability as a function of time, according to the Rabi formula, in the case of exact resonance (solid line) and with $\omega - \omega_0 = \sqrt{3}\Omega$ (dashed line).

This picture should be compared with the idea of 'quantum jumps', commonly used at the early stage of quantum mechanics.

Let a transition with $d_0 = 1$D be caused by a strong plane wave with the peak intensity 1 GW/cm^2; then, $E_0 = 1$ MW/cm, and (5.9) leads to $t_\pi = 1$ps. This time is in many cases much smaller than the relaxation time, hence our approximation $T_{1,2}$ is valid.

But for usual optical experiments, $t_\pi \gg T_{1,2}$, and the **R** vector, depicting the state of a two-level particle on the sphere, does not have enough time to move far from the South pole before the next 'collision' of the particle with the neighboring ones, which returns **R** into the equilibrium position with $\mathbf{R}_\perp = 0$. As a result, on the average, dynamical equilibrium takes place, with a certain stationary angle of precession (4.128).

Note that the amplitude of the field required for stationary saturation, $E_0 \geq \hbar/d_0\sqrt{T_1T_2}$ (see (4.95)), does not exceed, on the order of magnitude, the pulse amplitude required for observing coherent effects,

$$E_0 \geq \hbar/d_0T_2. \tag{5.10}$$

In other words, *a field that can cause a transition within a time much shorter than the relaxation time, will cause, on a long-time scale, strong saturation.*

Below, we will consider several experimental methods of observing non-stationary effects.

5.1.3 °*Nutation*

Let an equilibrium two-level system interact with an electromagnetic wave with the amplitude E_0 and the resonant frequency $\omega = \omega_0$. At short time intervals, $t \ll T_2$, the response of the system is described by equation (5.3). According to this equation, the Bloch vector **R**, which denotes the instant state of the system, performs, like a spin top, nutational motion, i.e., periodic variation of the precession angle ϑ with the Rabi frequency $\Omega = d_0E_0/\hbar$. At exact resonance, **R** moves along a meridian from the South pole (the equilibrium state) through the equator (a coherent state) to the North pole (population inversion) and back, with the angular rate Ω.

Evidently, when the **R** vector moves upwards, towards the North, the ensemble-averaged energy of the system increases due to the energy of the wave, and at the time instance when $\vartheta = \Omega t = \pi$, the wave, on the average, loses exactly one energy quantum $\hbar\omega$. Further, when **R** goes downwards, towards the South, the system gives the stored energy back. As a result, the wave becomes amplitude modulated with the nutation frequency Ω (Fig. 5.3). Provided that there are suf-

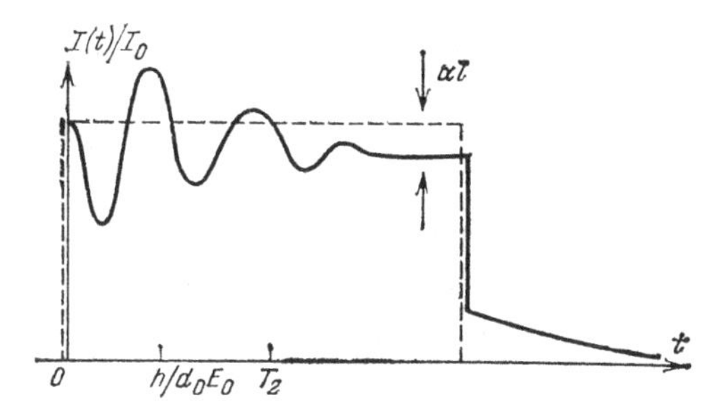

Fig. 5.3 Effects of transient optical nutation and free polarization decay. The front slope of a resonant pulse (dashed line) creates in the matter a transient process, nutation, which leads to the intensity modulation $I(t)$ of the transmitted light (solid line) with the Rabi frequency. Here, α is the absorption coefficient with an account for saturation, l is the layer thickness, T_2 is the transverse relaxation time. The back slope of the observed pulse gets a 'tail' caused by the free polarization precession.

ficiently many atoms, the wave transmitted through the matter can have a 100% modulation. This periodic variation of instantaneous optical density of the matter is called *optical nutation.*

Let us find the power absorbed by the matter. According to (4.98), polarization of a medium with the density N of two-level atoms is

$$P = 2d_0N\rho'_{21} = d_0N(R_{0x}\cos\Omega t + R_{0y}\sin\omega t) \equiv P_c\cos\omega t + P_s\sin\omega t, \qquad (5.11)$$

where the *in-phase*, P_c, and the *quadrature*, P_s, polarization components have been introduced.[b] (We assume that $\varphi_0 = 0$ and $E = E_0\cos\omega t$.) From (4.13), it follows that the loss power averaged over the period $2\pi/\omega$ per unit volume of the matter is determined by the quadrature component of the polarization,

$$\mathcal{P}(t) = \frac{1}{2}\omega E_0 P_s(t) = \omega d_0 E_0 N R_{0y}(t)/2. \qquad (5.12)$$

Hence, with the help of (5.6) we find

$$\mathcal{P}(t) = \hbar\omega\Delta(0)N\Omega\sin(\Omega t)/2. \qquad (5.13)$$

This result agrees with the above-given qualitative consideration assuming that the field and the atoms periodically exchange energy quanta. Note that, because at $t \ll T_{1,2}$ relaxation is too slow to manifest itself, real absorption, i.e., irreversible energy dissipation into the thermostat, is absent here.

Further, at $t > T_{1,2}$, the amplitude of atoms nutation reduces due to relaxation, and a stationary angle of precession is established (4.128). Absorption of the field

[b]Editors' note: the terms come from radio spectroscopy; in nonlinear and quantum optics they would be called amplitude and phase quadratures.

occurs then in a usual way and is described, according to Eq. (4.15), by the imaginary part of the susceptibility. Modulation of the transmitted wave disappears, and its amplitude is given by the Bouguer law with an account for saturation (see Fig. 5.3 and Sec. 6.4).

If then the field is suddenly switched off, the matter will continue to emit light for some time (Fig. 5.3) due to the free precession of the **R** vector; the matter is then in a coherent *superradiance state* (Sec. 5.3). This effect is called *free polarization decay*.

In the presence of inhomogeneous, for instance, Doppler's, broadening, these formulas refer to only a small group of atoms with a certain projection of the velocity. In order to find the total polarization, they should be integrated over Maxwell's distribution.

Optical nutation can be used for finding the transition dipole moment by measuring the modulation frequency Ω and for the study of relaxation processes by observing the modulation decay rate.

5.1.4 *Self-induced transparency*

Further, let us consider interaction of two-level atoms with short resonant pulses with duration $\tau_E \ll T_{1,2}$. According to (5.13), such pulses increase the energy per unit volume of the matter by a value

$$\mathcal{E} = \int_0^{\tau_E} \mathcal{P} dt = \hbar\omega\Delta(0)N \sin^2(\Omega\tau_E/2). \tag{5.14}$$

Let, for instance, $\tau_E = 2\pi/\Omega$, so that $E_0\tau_E = 2\pi\hbar/d_0$ forms the so-called '2π pulse', then $\mathcal{E} = 0$, and the pulse should pass through the matter without absorption! This theoretical prediction seems paradoxical: *a material that is completely opaque in the usual sense has full transmission for sufficiently short and strong pulses.*

This effect, named *self-induced transparency* (SIT), or *self-bleaching*, is indeed observed in experiment. (It should be distinguished from the saturation effect, which also leads to deviations from the Bouguer law.) The effect of the matter on the radiation is manifested here only in the reduction of the propagation velocity and in the pulse distortion. However, it is important that the 'area' of the pulse, $\vartheta = \Omega\tau_E$ (see (5.5)), remains equal to 2π.

SIT can be qualitatively explained as follows. The first half of the pulse incident on the matter ($\vartheta = \pi$) is absorbed by the atoms, which pass into the excited state with the population inversion. However, the second half of the pulse 'eliminates' this inversion, so that the absorbed energy is coherently returned to the field. Note that irradiating the matter with short π pulses is an example of a non-stationary way to create population inversion in two-level atoms.

For the quantitative description of SIT, it is necessary to consider the inhomogeneous wave equation in combination with the equation of matter, i.e., Maxwell's and Bloch's equations (see, for instance, Ref. [Allen (1975)]). Such a consideration shows that the propagation velocity u of a 2π pulse is determined by its 'length', $c\tau_E \equiv l$, and the usual absorption coefficient α_0,

$$c/u = 1 + \alpha_0 l/2. \tag{5.15}$$

Thus, the longer the pulse, the more delayed it gets. For instance, at $\tau_E = 1\text{ns}$ ($l = 30$ cm) and $\alpha_0 = 10^2$ cm^{-1} the pulse is slowed down by three orders of magnitude.

An interesting result of the wave theory in nonlinear materials is the prediction of *solitons*, stable pulses whose shape and amplitude do not change in the course of propagation [Rabinovich (1989)]. In the case of two-level anharmonicity, solitons have a hyperbolic secant shape,

$$E_0(\tau) = (2\hbar/d_0\tau_E)\text{sech}(\tau/\tau_E), \quad \tau \equiv t - z/u. \tag{5.16}$$

One can easily verify that (5.16) is a 2π pulse, i.e., $\int dtE_0 = 2\pi\hbar/d_0$.

It follows from (5.14) that 4π and, generally, $2\pi n$ pulses do not get absorbed either. According to the wave theory, in the course of propagation such pulses split in separate stable 2π solitons.

Note that SIT also takes place for inhomogeneously broadened transitions (for instance, in the presence of the Doppler effect in gases or ingomogeneous static fields in solids), where $2/\Delta\omega \equiv T_2^* \ll T_2$. It is important that the condition $\tau_E \ll T_2^*$ is not necessary. SIT is also observed under inter-band transitions in semiconductors.

5.2 Emission of an atom

Consider an equilibrium two-level system after the incidence of a resonant $\pi/2$ pulse with the duration $\tau_E = \pi/2\Omega$. According to (5.3), in the end of the pulse duration the system is in a coherent state with the coordinates $\vartheta = \varphi = \pi/2$ on the Bloch sphere (the initial field phase is assumed to be $\varphi_0 = 0$). Recall that in a coherent state, the energy of a quantum ensemble has no definite value. When measured, it randomly takes values 0 or $\hbar\omega_0$, the mean value being half of a quantum, $\mathcal{E} \equiv \langle \mathcal{H}_0 \rangle = \hbar\omega_0/2$, if the energy is measured from the ground level.

According to (5.6), after the $\pi/2$ pulse is over, the density matrix of the atom has the following elements:

$$\rho_{21}(t) = i\Delta(0)e^{-i\omega_0 t}/2,\ \ \Delta(t) \equiv \rho_{11} - \rho_{22} = 0. \tag{5.17}$$

Hence, the mean dipole moment of the atom oscillates (see Fig. 2.1) with the transition frequency $\omega_0 = \omega_{21}$, its amplitude being equal to the transition matrix element $d_0 \equiv d_{21}$ (we assume that $d_{21} = d_{12}$ and $\Delta(0) = 1$),

$$\langle \mathbf{d}(t) \rangle = 2\mathbf{d}_0 \rho'_{21}(t) = \mathbf{d}_0 \sin \omega_0 t \tag{5.18}$$

5.2.1 *Emission of a dipole*

Let us now take into account the spontaneous emission of the atom, which we so far neglected. As it was already mentioned in Sec. 2.5, spontaneous transfer of energy from the atom to the vacuum can be considered as relaxation with a typical time $T_1 = T_2/2 = 1/A$, where A is the spontaneous transition rate and $1/A$ is the radiative life time of the excited state. The vacuum plays then the role of a thermostat with a zero temperature and infinite heat capacity. Spontaneous emission is the simplest quantum model of an irreversible process.

Consider the spontaneous emission of an oscillating dipole (5.18) from the classical viewpoint. It follows from Maxwell's equations that the time-averaged power emitted by moving charges is equal, in the dipole approximation, to [Landau (1973)]

$$\mathcal{P} = 2\overline{\ddot{d}_{cl}^2}/3c^3. \tag{5.19}$$

If we identify d_{cl} with $\langle d \rangle$, we will underestimate the power twice compared to the prediction of the quantum mechanics (see below). Assume then that $d_{cl} = \sqrt{2}\langle d \rangle$, then

$$\mathcal{P}_{coh} = 4\overline{\langle \ddot{d} \rangle^2}/3c^3 = 2\omega_0^4 d_0^2/3c^3. \tag{5.20}$$

In addition to estimating the total power of spontaneous emission, the semi-classical model predicts its polarization and directivity diagram. For instance, in the case of an electric dipole transition, the field in the far-field zone is

$$\mathbf{E}(\mathbf{r}, t) = -\langle \ddot{\mathbf{d}}_\perp(t') \rangle / c^2 r, \tag{5.21}$$

where $\mathbf{d}_\perp$ is the projection of the dipole moment on a plane orthogonal to $\mathbf{r}$ and $t' \equiv t - r/c$. Hence, the directivity diagram has a usual 'dipole' shape $\mathcal{P} \sim \sin^2 \vartheta$, where ϑ is the angle between $\mathbf{r}$ and $\langle \mathbf{d} \rangle$.

Note that an atom that is initially in a coherent state emits field with a definite phase. This means that another atom, being in a ground state at a distance r from the first one, with a certain probability, scaling as r^{-2}, can get into a coherent state, with the precession phase shifted by $\omega_0 r/c$.

5.2.2 *Probability of a spontaneous transition*

Emission (5.20) will lead to a constant decrease in the stored energy and the oscillation amplitude of the dipole moment, so that the system will eventually come into the ground state. The Bloch vector **R** will then move along a spiral converging to the South pole (in a fixed reference frame), and the amplitude of the dipole moment $\langle d \rangle$ oscillations in (5.18) and (5.20) will slowly (as $A \ll \omega_0$) decay, see Fig. 4.9. Similar behavior will take place after exciting the system with any pulse, except for πn or $2\pi n$ pulses, which yield $\langle d \rangle = 0$. Thus, *an atom in a non-stationary state has an oscillating dipole moment and, according to the classical electrodynamics, it spontaneously emits a quasi-monochromatic wave with the frequency equal to the transition frequency* ω_0. The angle of precession gradually tends to zero in this case, due to the radiative energy loss.

The semi-classical expression (5.4) for the power $\mathcal{P}_{coh}$ emitted in the case of a coherent state allows one to estimate the probability of a spontaneous transition per unit time, A. To do this, let us postulate that the power of the atom emission decays exponentially,

$$\mathcal{P}(t) = \mathcal{P}(0)e^{-At}, \tag{5.22}$$

then the total energy is

$$\mathcal{E} = \int_0^\infty \mathcal{P}(t)dt = \mathcal{P}(0)/A. \tag{5.23}$$

Assuming that it is equal to the initial energy of the atom, $\hbar\omega_0/2$, we find, with the help of (5.20), that

$$A = 4\omega_0^3 d_0^2/3\hbar c^3, \tag{5.24}$$

which coincides with the result of a more consistent calculation (Sec. 7.7).

In addition to using an arbitrary numerical coefficient,[c] this reasoning has other, more serious, flaws. Namely, it does not result in an exponential decay and predicts the stability of an excited energy state ($\Delta = -1, \rho_{12} = 0$) with $\langle d(t) \rangle = 0$.

One can try to improve the situation by substituting into (5.19) $\langle d^2 \rangle$ instead of $\langle d \rangle^2$. However, the d^2 operator is proportional to the unity operator,

$$d^2 = \begin{pmatrix} 0 & d_0 \\ d_0 & 0 \end{pmatrix}\begin{pmatrix} 0 & d_0 \\ d_0 & 0 \end{pmatrix} = \begin{pmatrix} d_0^2 & 0 \\ 0 & d_0^2 \end{pmatrix} = d_0^2 I, \tag{5.25}$$

[c] Recall that we voluntarily assumed $d_{cl}/\langle d \rangle = \sqrt{2}$.

therefore $\langle d^2 \rangle = d_0^2$ regardless of the state of the atom; this also follows from (4.107). For instance, in this model the atom should emit even in the ground state, which contradicts the energy conservation law. Instead of a stable excited state we have obtained an unstable state. The same problem arises in the well-spread interpretation of spontaneous transitions as caused by zero-point vacuum fluctuations.

A consistent theory of the spontaneous emission, as well as statistical optics in general, should be based on the quantum description of the field (Chapter 7).

5.2.3 °*Normally ordered emission*

However, spontaneous emission can still be described correctly in the framework of semiclassical theory, provided that the squared dipole moment in (5.19) is replaced by the mean value of the *normally ordered* squared dipole moment (Sec. 7.7),

$$\mathcal{P} = (2/3c^3)\langle : (\ddot{d}) :^2 \rangle = (4/3c^3)\langle \ddot{d}^{(-)}\ddot{d}^{(+)} \rangle. \tag{5.26}$$

The colons denote normal ordering, i.e., placing *positive-frequency operators* $f^{(+)}$ on the right of *negative-frequency operators* $g^{(-)}$. By definition, an $f^{(+)}$ operator in the Heisenberg picture contains only positive-frequency harmonics,

$$f^{(+)}(t) = \sum_n f_n \exp(-i\omega_n t), \ \omega_n > 0.$$

Similarly, the Fourier transform of a $g^{(-)}(t)$ is only nonzero for negative frequencies.

Any operator can be represented as a sum of positive- and negative-frequency parts (disregarding the constant component), $d = d^{(+)} + d^{(-)}$; for Hermitian operators, $d^{(+)} = (d^{(-)})^+$. Hence,

$$: d^2 := d^{(+)2} + 2d^{(-)}d^{(+)} + d^{(-)2} = 2d^{(-)}d^{(+)}, \tag{5.27}$$

while according to (5.25),

$$d^2 \equiv d^{(+)2} + d^{(-)}d^{(+)} + d^{(+)}d^{(-)} + d^{(-)2} = d_0^2 I \tag{5.28}$$

(in what follows, we will show that $d^{(\pm)2} = 0$).

During a spontaneous transition, the $d(t)$ dependence in the Heisenberg representation can be approximately considered as unperturbed, since $A \ll \omega_0$,

$$d(t) = d_0 \begin{pmatrix} 0 & \exp(-i\omega_0 t) \\ \exp(i\omega_0 t) & 0 \end{pmatrix}, \tag{5.29}$$

where $d_0 \equiv d_{21} = d_{12}, \omega_0 = \omega_{21} > 0$. Hence, $\ddot{d} = -\omega_0^2 d$ and, by definition,

$$d^{(+)}(t) = d_0 \exp(-i\omega_0 t) \begin{pmatrix} 0 & 1 \\ 0 & 0 \end{pmatrix} \equiv d_0 \sigma^{(+)}(t),$$
$$d^{(-)}(t) = d_0 \exp(i\omega_0 t) \begin{pmatrix} 0 & 0 \\ 1 & 0 \end{pmatrix} \equiv d_0 \sigma^{(-)}(t), \tag{5.30}$$

where $\sigma^{(\pm)} \equiv \sigma_x \pm i\sigma_y$ (Sec. 4.4). By multiplying these matrices, we find

$$: d^2 := 2d_0^2 \begin{pmatrix} 0 & 0 \\ 0 & 1 \end{pmatrix} = d_0^2(I - \sigma_z); \tag{5.31}$$

recall that $d = d_0 \sigma_x$. To justify this procedure, let us point out that it is namely the normally ordered square of the field, $E^{(-)}E^{(+)} \sim d^{(-)}d^{(+)}$ that determines the 'useful' energy, which can be absorbed by the other atom (Sec. 7.7). Also, note that this procedure does not require time averaging over high-frequency oscillations of the power (see (5.19)).

From (5.26), with an account for (5.31), it follows that $\mathcal{P} = 2\omega_0^4 d_0^2 \langle I - \sigma_z \rangle / 3c^3$, or

$$\mathcal{P} = \hbar\omega_0 A \rho_{22}. \tag{5.32}$$

Let us stress that this result is valid for any state of the atom, including a purely energy or coherent one. In the latter case, $\rho_{22} = 1/2$, and (5.32) coincides with (5.20). Thus, *power of the spontaneous emission of a two-level atom at a given time moment scales as the upper level population.*[d]

Due to the energy conservation law, evidently, the following equality should hold true:

$$\hbar\omega_0 \dot{\Delta} = 2\mathcal{P}. \tag{5.33}$$

The factor 2 takes into account that each transition changes the population difference by a value of 2. Hence, replacing ρ_{22} in (5.32) by $(1 - \Delta)/2$, we find the kinetic equation,

$$\Delta = (1 - \Delta)T/T_{1nat}, \quad T_{1nat} \equiv 1/A. \tag{5.34}$$

Thus, we have confirmed the exponential law (5.22) of the excited state decay,

$$\Delta(t) = 1 + [\Delta(0) - 1]e^{-At}. \tag{5.35}$$

A more rigorous description of the interaction between a two-level system and the vacuum can be obtained using the Heisenberg equations (4.125) for the Pauli

[d]This conclusion becomes evident in Dirac's notation, $\sigma^{(+)} = |1\rangle\langle 2|, \sigma^{(-)} = |2\rangle\langle 1|, \sigma^{(-)}\sigma^{(+)} = |2\rangle\langle 2|, \langle \sigma^{(-)}\sigma^{(+)} \rangle = \rho_{22}$.

vector, by substituting, for the external field, the emission field in the zeroth-order approximation, i.e., the *reaction field* (for more detail, see Ref. [Allen (1975)]). Such equations describe the effect of *radiation-induced damping*, well known from classical electrodynamics [Landau (1973)]. For σ_z, they lead to an equation of the form (5.34), while for $\sigma^{(\pm)}$, to harmonic-oscillator equations with the eigenfrequency $\omega_0 + \delta\omega \mp iA/2$, where $\delta\omega$ is the correction to the transition frequency, called the *Lamb shift*, $\delta\omega \sim A$. For instance, for the resonant line L_α of a hydrogen atom, $\delta\omega/2\pi \sim 10^9$ Hz. The imaginary correction to the transition frequency has the meaning of the inverse time $1/T_2$ of transverse relaxation due to the radiation reaction. Thus, for an isolated atom, $T_{2nat} = 2T_{1nat} = 2/A$.

5.2.4 *Relation between spontaneous and thermal emission*

According to the semiclassical theory, spontaneous transitions in atoms lead to the emission of exponential wave 'trains'. Then, the field depends on the time as

$$E(t) = \theta(t)E_0 e^{-At/2}\cos(\omega_0 t + \varphi_0), \tag{5.36}$$

where $\theta(t)$ is the Heaviside step function. The Fourier transform of (5.36), giving the spectrum of the radiation, has a Lorentzian (*dispersion*) shape,

$$E(\omega) = \frac{iE_0}{4\pi}\left(\frac{\exp(-i\varphi_0)}{\omega - \omega_0 + iA/2} + \frac{\exp(i\varphi_0)}{\omega + \omega_0 + iA/2}\right). \tag{5.37}$$

A similar result follows from the quantum theory. The spectral line width $\Delta\omega_{nat} = A$ due to the spontaneous emission is called the *natural* one.

Emission spectra observed in practice are usually created by a large number of atoms excited at random time moments into states with random phases. (The opposite case, leading to superradiance, will be considered later.) For instance, in a heated gas at low pressure the atoms are excited through collisions. During a collision, an atom is in the pulsed field created by the neighboring atom, and this field changes its state according to the general formulas of the perturbation theory (Sec. 2.1). After the collision, atoms, according to the semiclassical theory, emit exponential wave trains of the form (5.36) with random initial phases. The average length of a train is determined by the spontaneous life time $1/A$ or the time τ between collisions.

Usually, radiation cooling of a gas is compensated for by the external heating, so that the superposition of all trains creates the stationary field of thermal radiation. In the case of a small optical density of the gas, $\alpha l \ll 1$, the intensity will scale as the upper level population. If $\alpha l \geq 1$, one has to take into account not only spontaneous transitions but also stimulated ones, leading to the absorption and further re-emission of photons (the *radiation trapping* effect). In the limit $\alpha l \gg 1$, the gas emits equilibrium Planck's radiation. In Sec. 7.1 we will consider a simple qualitative model of this process, also covering the case of amplified spontaneous emission at $\alpha < 0$.

5.2.5 *On the emission of fractions of a photon*

Let us mention here an interesting paradox, namely, the apparent contradiction between our theory and the traditional picture of photons. For a coherent state with the precession angle ϑ, according to (5.14), the energy stored by the atom is $\mathcal{E} = \hbar\omega_0 \sin^2(\vartheta/2)$. As a result of spontaneous emission, this energy after a certain time will be transferred to the field. At the same time, in the 'photon' language this means that the atom emits a fraction of the photon energy, equal to $\mathcal{E}/\hbar\omega_0 = \sin^2(\vartheta/2)$. For instance, after an atom is excited by a $\pi/2$ pulse into a coherent state on the equator, it emits an exponential train with the energy equal to half a photon, in contradiction with the initial postulates by Planck and Bohr.

However, this conclusion by no means contradicts to quantum electrodynamics, which states that *a field contains an integer number of photons N only in pure energy states*. This class of states with a definite photon number is a very special one, and even exotic from the viewpoint of its experimental preparation, with the exception for the vacuum state with $N = 0$. As to coherent, as well as mixed states of the field, $\langle N\rangle = \langle\mathcal{H}\rangle/\hbar\omega_0$ can be any non-negative number.

The paradox, as usual, originates from a terminology confusion: $\mathcal{E} = \hbar\omega_0/2$ is the energy at $t = 0$ averaged over the atomic ensemble, $\langle\mathcal{H}(0)\rangle = \mathcal{E}$, it has no relation to the result of a *single* energy measurement, $\mathcal{E}_i$, which yields either $\mathcal{E}_1 = 0$ or $\mathcal{E}_2 = \hbar\omega_0$ with equal probabilities. At $t \gg 1/A$, the energy of the field is $\hbar\omega_0/2$ only on the average; a single measurement of the photon number, i.e., of the field energy, yields, according to the basic postulate of the quantum mechanics, only an average number of photons $0, 1, 2, \ldots$.

Thus, *the energy exchange between the atoms and the field is still possible via fractions of quanta, but only if 'energy' is understood as the ensemble-averaged one,* $\mathcal{E} = \langle\mathcal{H}\rangle$. As to the energy that is transferred in a single interaction event, it is not definite, as the initial state of the atom is almost always not an energy one.

5.2.6 °*Quantum beats*

In the case of a multi-level system, the natural bandwidth can be explained by the finite lifetimes of all excited levels,

$$\Delta t_m = 1/A_m \equiv 1/\sum_{n<m} A_{nm}, \tag{5.38}$$

due to spontaneous transitions to all lower levels. The probabilities A_{nm} are determined by (5.24) with $\omega_0^3 d_0^2 \rightarrow \omega_{mn}^3 |d_{nm}|^2$. Due to the uncertainty relation $\Delta\mathcal{E}\Delta t = \hbar$, the lifetime Δt_m of the system on a certain level corresponds to the broadening of this level, which is equal, in circular frequency units, to the inverse lifetime,

$$\Delta\omega_m \equiv \Delta\mathcal{E}_m/\hbar = 1/\Delta t_m = \sum_{n<m} A_{nm}. \tag{5.39}$$

Finally, the linewidth of emission or absorption corresponding to a certain pair of levels is obviously given by the sum of the widths of the two levels,

$$\Delta\omega_{mk} = \Delta\omega_m + \Delta\omega_k. \tag{5.40}$$

Hence, it follows that a weak line with $d_{23} \approx 0$ can have a large natural width due to large d_{12} and d_{13}.

Spontaneous emission of a multi–level system excited by coherent radiation has an interesting feature: its power oscillates in time. Let us consider the time dependence of the dipole emission power.

According to the 'upgraded' semiclassical formula (5.26), the power of radiation emitted by an atom at a given time moment scales as the mean value of the normally ordered square of the dipole moment second derivative,

$$\mathcal{P}(t) = \frac{4}{3c^3} \sum_{klm} \rho_{kl} \ddot{d}_{lm}^{(-)}(t) \ddot{d}_{mk}^{(+)}(t). \tag{5.41}$$

(For simplifying the notation, we assume that the dipole moments of all transitions are parallel.) Here, ρ_{kl} is the density matrix of the atom at the initial time moment $t_0 \equiv 0$. It is determined by the excitation, which should be pulsed for observing quantum beats, i.e., the duration of the excitation pulse should be much smaller than the beat period, $2\pi/\omega_{32}$. In the case of a gas, this is achieved using a pulsed laser or discharge while in the case of an atomic beam the latter should be passed through a thin foil. As a result, a considerable number of atoms are in the same pure non-energy state $\hat{\rho}(0)$, so that the beat signal has a definite phase and a large power.

After the excitation pulse is over, the dynamics of a given atom is determined by its unperturbed Hamiltonian $\mathcal{H}_0$. (Relaxation can be neglected provided that the beat frequency is sufficiently large, $\omega_{32}T_2 \ll 1$.) The matrix elements in the energy representation will then depend on time harmonically,

$$d_{mn}(t) = d_{mn} \exp(i\omega_{mn}t), \; \ddot{d}_{mn}(t) = -\omega_{mn}^2 d_{mn}(t). \tag{5.42}$$

Let us number the levels according to increasing energy. Then the matrices of positive- and negative-frequency operators have 'triangular' shapes,

$$d_{mn}^{(+)}(t) = d_{mn}(t)\theta_{nm}, \; d_{mn}^{(-)}(t) = d_{mn}(t)\theta_{mn}, \tag{5.43}$$

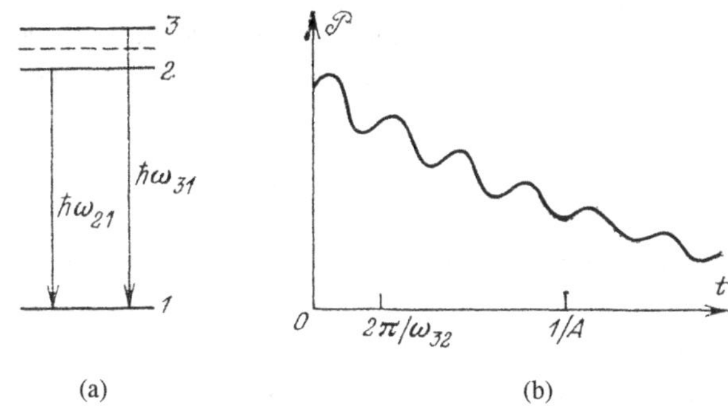

Fig. 5.4 Quantum beats: (a) dipole moment of a three-level system in a coherent state with indefinite energy (dashed line) oscillates with the frequencies ω_{21} and ω_{31}; (b) as a result, the power $\mathcal{P}$ of the spontaneous emission is modulated with the beat frequency $\omega_{32} = \omega_{31} - \omega_{21}$.

where $\theta_{mn} \equiv 1$ at $m > n$ and 0 at $m < n$ (see (5.30) for the case of a two-level system). As a result, the only nonzero terms in the sum (5.41) are the ones with $m < k, m < l$,

$$\mathcal{P}(t) = \frac{4}{3c^3} \sum_{m<k,m<l} \omega_{lm}^2 \omega_{mk}^2 \rho_{kl} d_{lm} d_{mk} \exp(i\omega_{lk} t). \tag{5.44}$$

In particular, the power of radiation emitted by a three-level system contains four terms,

$$\begin{aligned}\mathcal{P}(t) &= \hbar(\omega_{31}A_{13} + \omega_{32}A_{23})\rho_{33} + \hbar\omega_{21}A_{12}\rho_{22} \\ &\quad + 2\hbar(\omega_{31}\omega_{21}A_{13}A_{12})^{1/2}|\rho_{32}|\cos(\omega_{32}t + \varphi),\end{aligned} \tag{5.45}$$

where

$$A_{nm} \equiv 4\omega_{mn}^3 |d_{mn}|^2 / 3\hbar c^3, \; \varphi \equiv \arg(\rho_{23} d_{31} d_{12}).$$

The first three terms in Eq. (5.45) scale as the populations of the excited levels and correspond to usual spontaneous emission with the power constant in the framework of our approximation. To take into account radiative energy losses, one should add factors $\exp(-A_{mn}t)$ to these terms. The last term in (5.45) describes, at $A_{mn} \ll \omega_{32} \ll \omega_{21} \sim \omega_{31}$, *quantum beats*, the periodic modulation of the total power of spontaneous emission from two close levels 2 and 3, with the difference frequency $\omega_{32} = \omega_{31} - \omega_{21}$ (Fig. 5.4).[e]

Let us represent (5.45) in the form

$$\begin{aligned}\mathcal{P}(t) &= \mathcal{P}_0[1 + m\cos(\omega_{32}t + \varphi)], \\ \mathcal{P}_0 &\equiv \hbar\omega_{21}A_{12}\rho_{22} + \hbar\omega_{31}A_{13}\rho_{33}, \\ m &\equiv 2\hbar(\omega_{31}\omega_{21}A_{13}A_{12})^{1/2}|\rho_{32}|/\mathcal{P}_0,\end{aligned} \tag{5.46}$$

[e]Note that if superscripts '±' are omitted in (5.41), then (5.45) will contain beats with the frequency ω_{21}, not observed in experiment.

where we have neglected the weak radiation at the difference frequency ω_{32}, which is usually in the microwave range. If the atom is in a pure state, then $\rho_{mn} = c_m c_n^*$, and

$$m = 2/(\epsilon + 1/\epsilon),\ \epsilon \equiv |\omega_{31}^2 c_3 d_{31}/\omega_{21}^2 c_2 d_{21}|. \tag{5.47}$$

Thus, the modulation coefficient m reaches a unity at $A_{31} \approx A_{21}$, provided that the atom is excited into a coherent state, $|t_0\rangle = (|2\rangle + |3\rangle)/\sqrt{2}$. From the classical viewpoint, the atom emits two sine waves, which interfere, so that at some time moments the radiation is completely canceled while at other ones it is doubled. However, one should keep in mind that it is impossible to observe quantum beats by measuring the field energy in the case of a single atom: multiple measurements are required, with different t. In practice, one uses many atoms in the same state.

Usually, levels 2, 3 correspond to the fine or superfine structure, or Zeeman's sublevels, so that the beat frequency is in the microwave range and the modulation can be discovered by radio electronic methods. (For instance, PMTs enable observing modulation up to frequencies on the order of 100 MHz.) It is important that quantum beats are a single-atom phenomenon; therefore, the Doppler effect has nearly no influence on the beat frequency, which allows one to measure small splitting of levels [Aleksandrov (1972)].

Note that there is another method, developed in the 1960-s, which also uses the spectral analysis of the photocurrent rather than the direct analysis of light, but in a stationary regime. This method, called spectroscopy of *optical mixing* or spectroscopy of *intensity fluctuations*, allows one to observe extremely small frequency splittings on the order of 1 Hz [Cummins (1974)].

5.2.7 °*Resonance fluorescence*

Consider now the case of a cold gas, where $\kappa T \ll \hbar\omega_0$, and the atoms are excited by external directed radiation. The total secondary field of a macroscopic sample can be divided in two parts, the one that is coherent with the incident field, and the scattered one.

The coherent part is determined by the space-averaged atom density, it interferes with the incident primary field, and a joint resulting field propagates through the medium in the same direction. Homogeneous matter, i.e., the constant component in the space Fourier distribution of the matter mass, only slows the wave down and, if the energy dissipation is taken into account, reduces its amplitude. These effects are described by the macroscopic susceptibility of matter (Chapter 4).

The secondary radiation that is scattered sideways is caused by the atomic inhomogeneity of the matter. The fields of different atoms are not coherent with each other; therefore, the total power of the scattered radiation is additive, i.e., equal to the number of atoms times the emission power $\mathcal{P}$ of a single atom (provided that multiple scattering is negligible — the *Born approximation*). Due to the optical anharmonicity of the matter (Sec. 6.2), the spectra of the secondary radiation contain, in addition to the elastic (unshifted) component, new components, providing valuable information about the structure of the matter (fluorescence, Raman scattering, etc.).

Here, using the gyroscopic model, we will consider *resonance fluorescence* (resonance scattering), which is emission of radiation with a frequency close to the transition frequency ω_0, provided that the excitation frequency ω is close to ω_0 as well. This phenomenon, discovered by R. Wood in Na vapor as early as at the beginning of the 20th century, attracts much attention nowadays in connection with its two interesting features.

First, resonance fluorescence of single atoms brings the field into states with unusual statistics, which cannot be described in terms of classical statistical optics. These are states with *photon anti-bunching* and *squeezed states* (for more details, see Chapter 7).

Second, in the case of a large intensity of the incident light (the *pump*), the spectrum of the scattered light contains, near the elastic (*Rayleigh*) component, two satellites with the frequencies ω and $\omega \pm \tilde{\Omega}$ (Fig. 5.5). Here, $\tilde{\Omega} \equiv [\Omega^2 + (\omega_0 - \omega)^2]^{1/2}$ is the generalized Rabi frequency and $\Omega \equiv d_0 E_0/\hbar$. Besides, amplification of weak probe light at these frequencies due to the pump energy is observed.

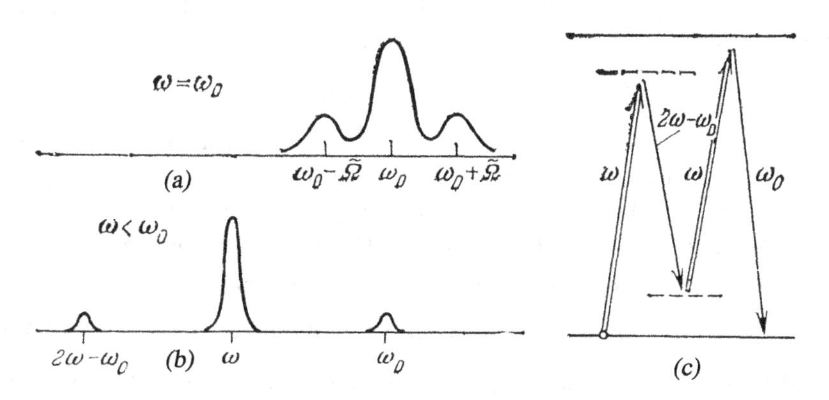

Fig. 5.5 Resonance fluorescence. A monochromatic external field modulates the wave function of a two-level system with the frequencies ω and $\omega \pm \tilde{\Omega}$. Therefore, the spectrum of the scattered field consists of three components. The following cases are shown: (a) small mismatch and strong field; (b) large mismatch; (c) in the case of a large mismatch, satellites appear due to the four-photon process in which two pump photons (thick lines) become two photons with frequencies ω_0 and $2\omega - \omega_0$.

The spectrum of the scattered light is observed under stationary conditions. In this case, a quantitative theory requires an account for the relaxation and quantization of the field (see, for instance, Ref. [Grishanin (1981)]); however, the emerging of the satellites can be easily explained qualitatively in the framework of the semiclassical theory and neglecting the relaxation. According to (5.3), a monochromatic field, in addition to the precession of the $\mathbf{R}(t)$ vector with the field frequency ω, also causes its nutation with the frequency $\tilde{\Omega}$ (Fig. 5.1). As a result, the mean dipole moment of the atom $\langle d(t)\rangle \sim R_x(t)$, and the emitted field should contain frequencies ω and $\omega \pm \tilde{\Omega}$. Therefore, the nonlinearity of a two-level system leads to the modulation of the field scattered by it, with the generalized Rabi frequency $\tilde{\Omega}$. (This should be compared with the Raman scattering, where the modulation frequency coincides with one of the frequencies of the system, so that the satellites have frequencies $\omega \pm \omega_0$, and with the optical nutation, see Sec. 5.1.)

Consider two limiting cases. Let $|\omega_0 - \omega| \ll \Omega$, then $\tilde{\Omega} \approx \Omega$, and the spectrum of resonance fluorescence is a triplet symmetric with respect to the transition frequency, $\omega_0, \omega_0 \pm d_0E_0/\hbar$ (Fig. 5.5(a)). In the opposite case, $\tilde{\Omega} \approx |\omega_0 - \omega|$, and the spectrum consists of the central elastic component ω, a line at the transition frequency ω_0, and the 'mirror' component at $2\omega - \omega_0$ (Fig. 5.5(b)). From the photon viewpoint, radiation at frequencies ω_0 and $2\omega - \omega_0$ is a result of a single four-photon elementary process, i.e., absorption of two photons of the incident light and emission of two secondary photons, according to the scheme $2\hbar\omega \to \hbar\omega_0 + \hbar(2\omega - \omega_0)$ (Fig. 5.5(c)).[f]

Note that in a non-resonance case, emission in the vicinity of a transition is not exactly at frequency ω_0 but shifted. Let $\Omega \ll \omega_0 - \omega > 0$, then the frequency of the anti-Stokes satellite (the right-hand one, on the frequency scale) is

$$\omega_0' = \omega + [\Omega^2 + (\omega_0 - \omega)^2]^{1/2} \approx \omega_0 + (d_0E_0/\hbar)^2/2(\omega_0 - \omega). \tag{5.48}$$

This shift of the observed transition frequency, dependent on the intensity of the exciting radiation, is called the *high-frequency shift*, or the *optical Stark effect*. It follows from (5.48) that on the order of magnitude *the shift is equal to the squared excitation energy in frequency units, Ω, divided by the mismatch.*

Let us estimate the total power of emission. According to (5.7) and (5.32),

$$\mathcal{P}(t) \approx \hbar\omega_0 A\rho_{22}(t) = \hbar\omega_0 A(\Omega/\tilde{\Omega})^2 \sin^2(\tilde{\Omega}t/2). \tag{5.49}$$

Thus, if the relaxation is negligible, $\mathcal{P}$ oscillates with the frequency $\tilde{\Omega}$, i.e., the process is non-stationary and the notion of the spectral density is non-applicable. It is noteworthy that in the non-resonant case, $\mathcal{P}$ scales as $d_0^4/(\omega_0 - \omega)^2$.

---

[f]Editors' note: in modern terms, this process is called *spontaneous four-wave mixing*; for more details, see Chapter 7.

Above, we did not take into account relaxation. It is intuitively clear that it should lead to the broadening of discrete spectral components up to a width of about $2/T_2$ (Fig. 5.5). Hence, the condition for observing the satellites in the case $\omega = \omega_0$ is $\Omega > 2/T_2$, i.e., strong saturation should take place, which is only possible with the help of lasers. (The effect was first observed in 1974 [Delone (1975)].) Note that according to the Bloch equations with an account for relaxation (Sec. 4.4), nutation of the $\boldsymbol{R}$ vector disappears under stationary conditions, so that the above-given explanation of the effect seems to be not valid. However, the kinetic Bloch equations do not describe the fluctuations of $\boldsymbol{R}$ caused by the thermostat and containing the frequency $\tilde{\Omega}$.

If we put aside the rotating-wave approximation, the two-level model will describe other nonlinear effects, for instance, emission at the frequency 3ω. In the case of a polarized atom (or molecule) with nonzero diagonal matrix elements of the dipole moment, $d_{nn} \neq 0$, or in the case of the magnetic resonance, the model will describe two-photon absorption and the Raman effect, i.e., emission at frequencies $\omega \pm \omega_0$. Thus, *according to the simplest two-level model of an atom, the matter can scatter the incident light with changing its spectrum, which is a manifestation of the nonlinear properties of the matter, i.e., the optical anharmonicity.* Another simple nonlinear effect, saturation, has been considered in Sec. 4.3.

5.3 Collective emission

As we have already mentioned, emission from macroscopic matter depends strongly on the conditions of its excitation. Under usual, chaotic (incoherent) excitation, the states of different atoms are statistically independent, the matter is described by a diagonal density matrix, and as a result, the emission power $\mathcal{P}$ depends on the total number of atoms N linearly. (We assume that the linear size of the sample is much smaller than the free path of a photon, $\alpha l \ll 1$.) In the case of a coherent excitation, all N atoms are in the same state, and the sample can be described by a joint wave function. Then, as one can easily verify using the semiclassical theory, $\mathcal{P}$ can depend on N quadratically, which should drastically change the situation. In what follows, we will consider two unusual optical effects originating from the quantum coherence formed in a macroscopic sample.

5.3.1 *Superradiance*

Let at the initial time moment there be N identical atoms in independent coherent states,

$$\psi_j = \{|1\rangle_j + |2\rangle_j \exp[-i(\omega_0 t + \varphi_j)]\}/\sqrt{2}. \tag{5.50}$$

This is a particular case of a possible state of a system containing N particles distinguishable through their coordinates $\boldsymbol{r}_j$, with the wave function of the system being factorable,

$$\psi = \psi_1\psi_2 \ldots \psi_N = \prod_{j=1}^{N} \psi_j. \tag{5.51}$$

The mean dipole moment of the system is

$$\langle d \rangle = \langle \psi | \sum_j d_j | \psi \rangle = \sum_j \langle \psi_j | d_j | \psi_j \rangle = d_0 \sum_j \cos(\omega_0 t + \varphi_j), \tag{5.52}$$

where the transition dipole moments $d_0 \equiv d_{12}$ are considered as real and oriented along a certain axis. Here, the phase φ_j of the j-th atom coherent state has a clear meaning: this is the phase of the Bloch vector precession. If the state is prepared by means of a $\pi/2$ pulse, this phase is determined by the phase of the field at the point $\boldsymbol{r}_j$ of the atom. If $N = 2$, then, according to (5.52), the oscillation amplitude of the total dipole moment is doubled at $\varphi_1 = \varphi_2$ and turns into zero at $\varphi_1 - \varphi_2 = \pi$. In the case of N atoms with the same φ_j,

$$\langle d \rangle = N d_0 \cos(\omega_0 t + \varphi).$$

Suppose then that the linear size of the system is much smaller than the wavelength, $l \ll \lambda_0 = 2\pi c/\omega_0$, so that formula (5.19) for the dipole emission power is valid. Replacing there d_{cl} by $\langle d \rangle$, in the case of equal phases we get $\mathcal{P}_{coh} \sim N^2$, i.e., *the power emitted by dipoles oscillating with equal phases scales as the square of their number.* This effect of coherent (collective, cooperative) spontaneous emission of a multi-atomic system is called *superradiance.* Although its classical interpretation is trivial, the quantum theory has been first considered only in 1954 by Dicke [Fain (1972); Allen (1975)].

A close phenomenon had been observed earlier in the microwave range, in nuclear magnetic resonance (NMR) experiments. In such experiments, a paramagnetic sample is placed between two crossed inductance coils (the *magnetic induction* method). A resonance current in one of the coils excites the stimulated precession of the macroscopic magnetic moment $\boldsymbol{M}$ of the sample around the direction of a constant magnetic field (see (4.139)). The rotating moment $\boldsymbol{M}$ induces the induction electromotive force in the other coil, the receiver one. As a result, the power of the observed signal scales as the squared number of nuclei in the sample.

Coherent emission of phased dipoles is observed in many nonlinear optical effects such as generation of optical harmonics. However, here, similarly to the effect of nuclear induction, emission is at the frequency of the external force, which in optics can be very different from the eigenfrequencies of the transitions. It is important that the condition $l \ll \lambda$ is not necessary: at an appropriate phase of stimulated or free precession along some direction, so that $\varphi_j = kz_j$, where k is the wave vector and z_j is the j-th atom coordinate (*spatial phase matching* condition), the emission along this direction will be coherent. Thus, superradiance of extended systems is not isotropic; it has a maximum along the phase matching direction z.[g]

The semiclassical approach, based on the formula (5.19) for the power of coherent radiation with d_{cl} replaced by $\langle d \rangle$, yields an exact result only at $N \gg 1$. Recall that at $N = 1$, we had to introduce an additional factor of 2 into formula (5.20). As we have mentioned above, in the general case we have to use, instead of d_{cl}^2, the normally-ordered square $2\langle d^{(-)}d^{(+)}\rangle$, which leads to the replacement of N^2 by $N(N+1)$,

$$\begin{aligned}\langle d^{(-)}d^{(+)}\rangle &= \sum_{jk}\langle d_j^{(-)}d_k^{(+)}\rangle \\ &= \sum_{j}\langle\psi_j|d_j^{(-)}d_j^{(+)}|\psi_j\rangle + \sum_{j\neq k}\langle\psi_j|d_j^{(-)}|\psi_j\rangle\langle\psi_k|d_k^{(+)}|\psi_k\rangle \\ &= d_0^2\left[N + \sum_{j<k}\cos(\varphi_j - \varphi_k)\right]/2 = d_0^2 N(N+1)/4. \end{aligned} \tag{5.53}$$

Substituting (5.53) into (5.26), we obtain the superradiance power of N atoms in coherent states,

$$\mathcal{P}_{coh} = \hbar\omega_0 A N(N+1)/4. \tag{5.54}$$

The last equality in (5.53) implies that all phases φ_j are equal, i.e., the states are mutually coherent.

Otherwise, if the precession phases are independent random variables, distributed uniformly within the $0..2\pi$ interval, then all cosines will turn into zero after averaging, and (5.53) will only contain N diagonal terms,

$$\mathcal{P}_{incoh} = \hbar\omega_0 A N/2. \tag{5.55}$$

Here, the subscript 'incoh' indicates that separate atoms, each of them being in a coherent state on the equator of the Bloch sphere, emit incoherently.

Apparently, in addition to the limiting cases (5.53) and (5.56) considered here, there are many other states of an N-atom system. In particular, as one can see from the above-given example for two atoms, *emission-free states* are possible, in which collective effects suppress dipole spontaneous emission.

[g] In Sec. 5.1, we have already come across such a phenomenon, free polarization decay.

Due to the energy losses accompanying the emission, spontaneous superradiance has fast decay. The pulse duration of superradiance can be estimated as the ratio of the energy $\hbar\omega_0 N/2$ stored by the system to the initial power of coherent emission (5.54),

$$\tau_{coh} \approx 2/A(N+1). \quad (5.56)$$

Thus, *at $N \gg 1$ and $l \ll \lambda$ the radiative lifetime of the coherent state of an N-atom system is reduced $N/2$ times due to the collective effect.* However, as it has been already mentioned, superradiance in extended systems is directive, i.e., only part of the field modes 'work' efficiently and take the energy, and this leads to an increase of τ_{coh}. The increase is approximately $4\pi/\Delta\Omega$ times, with $\Delta\Omega$ being the effective solid angle into which the emission is directed. For a stretched sample with the Fresnel number $\alpha^2/\lambda l \sim 1$, this slowing down of the superradiance is on the order of $(a/\lambda)^2 \sim l/\lambda \gg 1$, where a^2 is the sample cross section and l is the length.

Consider now the case where the initial state of an N-atom system is an energy one, $\psi(0) = \prod |2\rangle_j$. The Bloch vector of each atom is directed up, towards the North pole. Initially, the atoms emit independently,

$$\mathcal{P} \sim 2\langle\psi|d^{(-)}d^{(+)}|\psi\rangle = Nd_0^2, \quad (5.57)$$

and gradually pass from the energy state into coherent ones, with some precession angles ϑ_j and phases φ_j.

However, one can expect that the atoms influence each other through the transverse field, and therefore the precession phases should mutually synchronize, provided that the atom density is sufficiently high [Andreev (1980)]. As a result, after a certain delay time t_0 the system will pass by itself into a superradiance coherent state with phased dipoles. At this moment, slow spontaneous emission will turn into a short strong superradiance pulse with a definite phase of the field (Fig. 5.6).

5.3.2 *Analogy with phase transitions*

A consistent theory, as well as the experiment (see Ref. [Andreev (1980)]), performed for the first time in 1973, confirm this qualitative picture. Often, it is this surprising effect that is called superradiance. Spontaneous emerging of an ordered coherent state from an initially chaotic state of a macrosystem is generally of huge interest in physics. Let us mention the analogy between the spontaneous formation of a superradiance state in a system of excited atoms and phase transitions in equilibrium matter. Another example of such a phase transition in a non-equilibrium system is emerging of auto-oscillations with a certain phase and

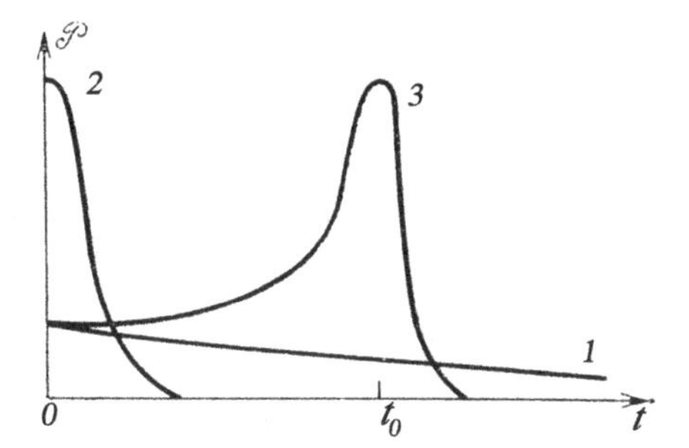

Fig. 5.6 Time dependence of the emission power for atoms excited at time $t = 0$: 1, usual (incoherent) spontaneous emission, observed at low atom density; 2, superradiance of atoms that are initially in a coherent state; 3, superradiance in the case of an energy initial state occurs with a typical delay t_0.

a macroscopic amplitude in a laser operating above the threshold [Haken (1977); Klimontovich (1980); Arecchi (1974)]. In this case, again, the interaction of atoms through the field of the cavity makes the chaotic noise field become a regular one and leads to the formation of regular macroscopic alternating polarization. Note that in equilibrium systems, phase transitions are also caused by the interaction of atoms or molecules through the field, but a static one. A rough description of the ferroelectric transition is provided by the Lorentz model: according to (4.50), the frequency of the polarization resonance turns into zero provided that the atom density is sufficiently high. In this case, susceptibility at zero frequency turns into infinity and a macroscopic static polarization is created in the system.

Superradiance in a system with a population inversion should be distinguished from *superfluorescence* (stationary or pulsed), amplified spontaneous emission (Sec. 7.1). During superfluorescence, emission-free relaxation processes (such as collisions in gases or interaction with phonons in solids) prevent a macroscopic volume from emitting coherently, and therefore the emission depends on the number of particles linearly, and not quadratically.

5.3.3 *Photon echo*

In the case of inhomogeneous broadening, different atoms have slightly different eigenfrequencies due to the Doppler effect, inhomogeneous static fields, etc. Then the observed linewidth $\Delta\omega \equiv 2/T_2^*$ of a transition is much greater than the collision-induced or radiation-induced width $2/T_2$.

Let a short ($\tau_E \ll T_2^*$) resonant $\pi/2$ pulse drive the atoms of the sample from the ground state into the same coherent state on the equator of the Bloch sphere

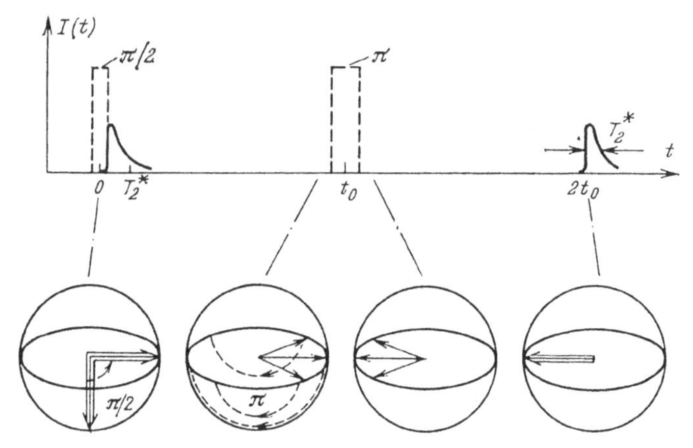

Fig. 5.7 Photon echo. After being excited by two short light pulses (dashed lines, top), the matter emits a flash of light (solid line) delayed by a time t_0 equal to the interval between the pulses. The effect is explained by the fact that at time $2t_0$, different molecules have the same phase of charge oscillations (bottom).

(Fig. 5.7). It creates a polarization

$$P(t) = \sum_{j=1}^{N} \langle d_j(t) \rangle = d_0 \sum_j R_x^{(j)}(t) = d_0 \sum_j \sin(\omega_{21}^{(j)} t). \tag{5.58}$$

Here, N is the particle number density, $\boldsymbol{d}_0 \equiv \boldsymbol{d}_{12}^{(j)}$ the transition dipole moment, $\boldsymbol{R}^{(j)}$ the Bloch vector of the jth atom, with the components $2\rho_{21}^{(j)\prime}, 2\rho_{21}^{(j)\prime\prime}, \rho_{11}^{(j)} - \rho_{22}^{(j)}$.

At the initial stage, when $t \ll T_2^*$, the difference between the eigenfrequencies $\omega_{21}^{(j)}$ can be neglected, so that

$$P(t) = Nd_0 \sin(\omega_0 t), \tag{5.59}$$

with $\omega_0 \equiv \overline{\omega_{21}^{(j)}}$ being the mean transition frequency. The macroscopic polarization (5.59) is accompanied by strong superradiance. However, in a short time T_2^*, the dipoles in (5.58) get out of phase, $P(t)$ becomes close to zero, and only slow spontaneous emission remains, with a typical decay time $T_1 = 1/A$.

The dephasing process can be illustrated with the help of the vector model (Sec. 5.1). In the frame of reference rotating with the frequency ω_0, the $\boldsymbol{R}_0^{(j)}$ vectors will undergo precession clockwise or anti-clockwise, with the angular rates $\Delta\omega_j \equiv \omega_{21}^{(j)} - \omega_0$, which are within the interval $-\Delta\omega \cdots + \Delta\omega$. Prior to the $\pi/2$ pulse, all $\boldsymbol{R}_0^{(j)}$ are directed down along the z axis, while immediately after the pulse, they are directed along the y axis, $R_{0y}^{(j)}(0) = 1$. (We assume the pulse to be very short.)

Let T_2^* be much smaller than the superradiance time τ_{coh} and $t \ll T_1$, then the radiative energy losses can be neglected and each Bloch vector is precessing with its own velocity in the equatorial plane. The Bloch vectors will be unfolding like a fan (Fig. 5.7), and at times $t > T_2^*$ they will uniformly fill the equatorial plane.

The precession of the jth atom can be described analytically in the rotating frame of reference by the formula

$$\rho_0^{(j)} = (i/2)\exp(-i\Delta\omega_j t), \tag{5.60}$$

where $\rho_0^{(j)} \equiv \rho_{21}^{(j)} \exp(i\omega_0 t)$ is the 'slow' amplitude of the density-matrix non-diagonal element.

At first sight, this decay of the macroscopic order is irreversible. However, there exists a smart way to restore the coherent state of the system provided that $t < T_2$, where T_2 is the typical time of irreversible precession dephasing caused by collisions. The idea is that a second resonant pulse, with the 'area' π, is incident on the system at a time t_0.

According to the gyroscopic Bloch equation (5.3), the π pulse 'turns' an $\boldsymbol{R}_0^{(j)}$ vector by 180° around the x axis (5.7). Since $\boldsymbol{R}_0^{(j)}$ is in the equatorial plane, the rotation is equivalent to the mirror reflection with respect to the xz plane or to the change of the $R_{0y}^{(j)}$ sign, i.e., complex conjugation of $\rho_0^{(j)}(t_0)$.

One can easily see that after this operation, the 'fan' or $R_0^{(j)}$ vectors starts folding, and at the time $t = 2t_0$ all vectors will be parallel once again (5.7), this time along $-y : R_{0y}^{(j)}(2t_0) = -1$, $\rho_0^{(j)}(2t_0) = -i/2$. Again, we consider the pulse to be very short.

At this time moment, a superradiance pulse is emitted again, its duration about T_2^* (Fig. 5.7). It is this pulse of emission from the sample, appearing in a time t_0 after the second external pulse, that is called the *photon, or spin, echo*. As the interval t_0 increases, the amplitude of the echo signal reduces as $\exp(-2t_0/T_2)$.

The appearance of echo at $t = 2t_0$ can be illustrated by an analogy with runners at a stadium, who start running at $t = 0$ with different speeds. At $t = t_0$ they simultaneously turn round and run back with the same speeds. Clearly, at $t = 2t_0$ they will cross the start line simultaneously.

The coherence recovery under a π pulse can be also described analytically. Before the pulse,

$$\rho_0^{(j)}(t_0 - 0) = (i/2)\exp(-i\Delta\omega_j t_0), \tag{5.61}$$

while immediately after the pulse,[h]

$$\rho_0^{(j)}(t_0 + 0) = \rho_0^{(j)}(t_0 - 0)^* = -(i/2)\exp(i\Delta\omega_j t_0). \tag{5.62}$$

[h]Note that a similar operation of amplitude complex conjugation, or time reversal, forms the basis of the *phase conjugation* effect (Sec. 6.5).

Further, at $t > t_0$, free precession takes place again, with the initial amplitude given by (5.62),

$$\rho_0^{(j)}(t) = \rho_0^{(j)}(t_0 + 0)\exp[-i\Delta\omega_j(t - t_0)] = -(i/2)\exp[-i\Delta\omega_j(t - 2t_0)]. \quad (5.63)$$

Hence, at $t = 2t_0$ all $\rho_0^{(j)}$ become equal.

This effect, one of the most beautiful phenomena in quantum electronics, has been discovered by Hahn in 1950 in NMR experiments, with inhomogeneous broadening caused by the magnetic field inhomogeneity, and named as the spin echo. In optics, it was first observed in ruby by Kurnit et al. in 1964 using a ruby laser. Photon and spin echo are used for the measurement of relaxation parameters and fine structure of transitions (*echo spectroscopy*, see Ref. [Manykin (1984)]).

Chapter 6

Nonlinear Optics

Some nonlinear optical phenomena have been already considered above (saturation, resonance fluorescence). This chapter will present a more systematic description of these effects.

Although most nonlinear optical phenomena are well described by the semiclassical radiation theory and do not require the quantization of the field, many effects have a clear interpretation and classification in terms of photons. For instance, *frequency doubling* can be represented as resulting from elementary three-photon processes where, due to the interaction between the field and the matter, two photons of the incident light (pump) are destroyed and a photon with the double energy $2\hbar\omega$ is created (Fig. 6.1(a)). It is clear from the figure that, since the process does not change the energy of the matter (such processes are called *parametric* ones), the energy of the created photon is exactly twice as large as the energy of the pump photon.

In the case of a biharmonic pump with the frequencies ω_1, ω_2, the matter emits photons with the combination frequencies, $\omega_0 = \omega_1 \pm \omega_2$. These are *sum- and difference-frequency generation* effects, see Figs. 6.1(b) and 6.1(c). A similar description can be suggested for four-photon and higher-order processes. The processes that take place in a *parametric oscillator* and in the corresponding spontaneous effect, *spontaneous parametric down-conversion*, are inverse to the sum frequency generation. Namely, a pump photon is split in two photons with fractional frequencies, $\omega_0 \to \omega_1 + \omega_2$ (Fig. 6.1(d)).

The efficiency of parametric processes in a macroscopic material is dramatically increased under the condition of photon momentum conservation, $\boldsymbol{k}_1 + \boldsymbol{k}_2 = \boldsymbol{k}_0$. This equality is called the *spatial phase-matching condition*.

In *non-parametric processes*, the matter changes its energy and passes to other energy levels. For instance, in two-photon absorption, two pump photons are annihilated and an atom goes to an excited level (6.1(e)). Such effects lead to nonlinear absorption scaling as the square of the light intensity. In a *Raman two-photon transition*, a photon is annihilated and another one, with a different energy, is created (6.1(g) and 6.1(h)).

All these effects, both parametric and non-parametric, are widely used in nonlinear spectroscopy, as well as for the variation of laser radiation frequency. An important role is played by nonlinear optical effects in laser thermonuclear synthesis, laser separation of isotopes, laser chemistry, and many other fields of quantum electronics. Also, note that *the possibility of quantum amplification is principally related to the nonlinearity of the material*, since in a linear system the levels are equidistant and the stimulated emission is always compensated for by the absorption.

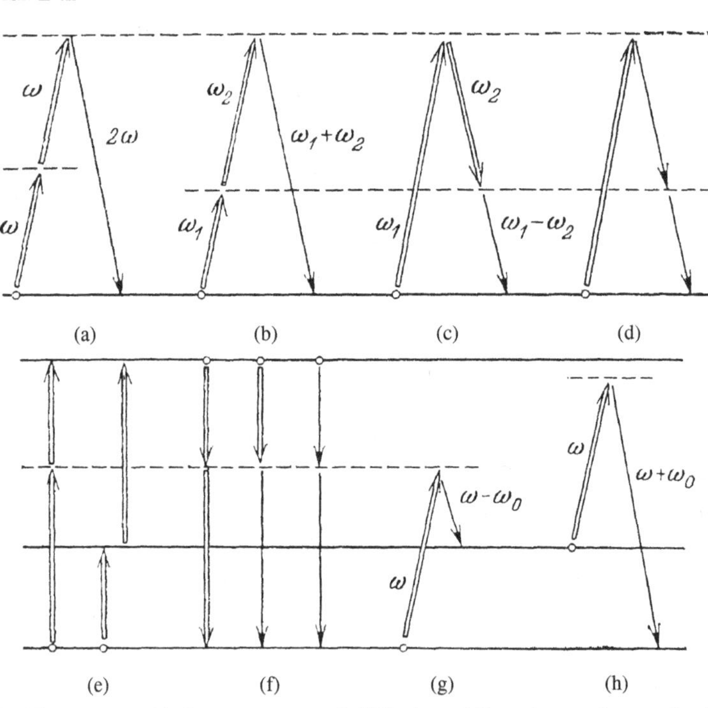

Fig. 6.1 Elementary multi-photon processes. Solid horizontal lines show real energy levels of the matter; dashed horizontal lines denote virtual levels. Arrows pointing upwards are absorbed photons; arrows pointing downwards, emitted ones. Bold arrows correspond to the photons of the primary radiation (the pump), thin ones, to the secondary or spontaneous radiation. Top row: three-photon parametric (coherent) processes; bottom row: two-photon non-parametric processes; (a) second harmonic generation; (b, c) sum- and difference-frequency generation; (d) parametric down-conversion; (e) non-resonance and resonance (cascaded) two-photon absorption; (f) two-photon emission (stimulated, spontaneous-stimulated, and spontaneous); (g, h) Stokes and anti-Stokes Raman processes.

In the quantitative description of nonlinear optical effects, similar to the case of linear optics, the theory is divided in two parts, the microscopic and the macroscopic ones. The microscopic theory is aimed at calculating the polarization $\boldsymbol{P}(\boldsymbol{E})$ induced by a given field $\boldsymbol{E}$. (In the general case, the magnetic field $\boldsymbol{H}$ should be taken into account as well.) The equations of motion of charged particles are nonlinear; therefore, the function, or, rather, the functional $\boldsymbol{P}(\boldsymbol{E})$ can be written as a series expansion containing the quadratic term $\chi^{(2)}\boldsymbol{E}^2$, the cubic one, $\chi^{(3)}\boldsymbol{E}^3$, etc. At the macroscopic stage of calculations, one substitutes the function $\boldsymbol{P}(\boldsymbol{E})$ into the Maxwell equations, and a self-consistent solution $\boldsymbol{E}, \boldsymbol{H}$ is searched at given field sources and boundary conditions. According to this scheme, Sec. 6.1 discusses the general properties of $\chi^{(n)}$, Sec. 6.2 considers various models of the optical anharmonicity of the matter, which allow one to estimate $\chi^{(n)}$. Sections 6.3–6.5 describe the basic problems of macroscopic nonlinear optics, as well as the ways to solve them, and some observable effects.

It should be stressed that nonstationary problems of nonlinear optics are solved by writing joint equations for the field and the matter, without using the susceptibilities. For instance, a quantitative analysis of self-induced transparency (Sec. 5.1) requires a joint solution to Maxwell's and Bloch's equations.

Nonlinear optics can be studied in more detail by reading Refs. [Vinogradova (1979); Akhmanov (1981); Fain (1972); Apanasevich (1977); Landau (1982); Akhmanov (1964); Bloembergen (1965); Kielich (1980); Fabelinsky (1965); Klyshko (1980); Letokhov (1975); Akhmanov2 (1981); Delone (1978); Dmitriev (1982); Butylkin (1977); Schubert (1973); Zernike (1973); Walther (1976); Letokhov (1983)].[a] Currently, there is also a new rapidly developing field, *nonlinear optics of the surface* [Chang (1981)]. One of the most interesting phenomena in this field is *giant Raman scattering of light* by molecules adsorbed on a rough metal surface. Cross section of this scattering is by orders of magnitude greater than the cross section of usual bulk scattering per one molecule.

6.1 Nonlinear susceptibilities: definitions and general properties

Before analyzing various models of optical anharmonicity, it is reasonable to find out the general properties, such as the symmetry of the nonlinear response of the matter, which do not depend on the choice of the model. For this, let us generalize the notion of the phenomenological susceptibility (Sec. 4.1) to the nonlinear case.

[a]Editors' note: See also [Agrawal (2007); Boyd (2008)].

6.1.1 *Nonlinear susceptibilities*

Let a field with a discrete spectrum,

$$\boldsymbol{E}(t) = (1/2)\sum_n \boldsymbol{E}_n \exp(-i\omega_n t) + \text{c.c.}, \; n = 1, 2, \ldots, \tag{6.1}$$

be incident on the matter. Equations of motion for the charged particles of the matter are nonlinear. As a result, charge displacements induced by the field (6.1) and, hence, the polarization $\boldsymbol{P}(t)$ will contain not only the Fourier components with the frequencies of the external force ω_n but also the ones with the combinations of these frequencies, $\omega_n \pm \omega_m$, including multiple frequencies $2\omega_n$ and the zero frequency, $\omega_n - \omega_n = 0$.

Let us first reduce the consideration to the case of the nonlinearity quadratic in the external field. Then, the phenomenological relation between the spectral components of the polarization and the electric field has the form

$$\boldsymbol{P}_0^{(2)} = \boldsymbol{\chi}^{(2)}(\omega_1, \omega_2)\boldsymbol{E}_1\boldsymbol{E}_2, \tag{6.2}$$

where $\boldsymbol{P}_0$ is the complex amplitude of polarization oscillations with the frequency $\omega_0 = \omega_1 + \omega_2$, and we are so far assuming $\omega_1 \neq \omega_2$. The *quadratic susceptibility* (or *quadratic polarizability*) $\boldsymbol{\chi}^{(2)}$ defined this way sets a relation between three vectors and is therefore a third-rank tensor. Notation (6.2) is not based on a certain frame of reference (often, the relation is written with a colon, $\boldsymbol{P}_0 = \boldsymbol{\chi}^{(2)} : \boldsymbol{E}_1\boldsymbol{E}_2$). If some Cartesian frame of reference is chosen, (6.2) takes the form

$$P_{0\alpha}^{(2)} = \sum_{\beta\gamma} \chi_{\alpha\beta\gamma}^{(2)}(\omega_1, \omega_2)E_{1\beta}E_{2\gamma}. \tag{6.3}$$

In what follows, as usual, we will omit the summation over 'dumb' indices β, γ.

Each of the 27 components $\chi_{\alpha\beta\gamma}^{(2)}(\omega, \omega')$ of the $\boldsymbol{\chi}^{(2)}$ tensor is a function of two independent arguments ω, ω' taking values from $-\infty$ to $+\infty$. Since the Fourier components of the field and the polarization are complex, $\chi_{\alpha\beta\gamma}^{(2)}$ is also complex, and in total, there are 54 real functions of these variables. However, as we will show below, there are many relations between these functions, and the number of independent variables is reduced.

By analogy with (6.3), the nonlinear susceptibility of an arbitrary order is defined as

$$\begin{aligned} P_{0\alpha}^{(m)} &= \chi_{\alpha\alpha_1\ldots\alpha_m}^{(m)}(\omega_1, \ldots, \omega_m)E_{1\alpha_1}\ldots E_{m\alpha_m}, \\ \omega_0 &= \omega_1 + \omega_2 + \cdots + \omega_m. \end{aligned} \tag{6.4}$$

For instance, the cubic susceptibility $\chi_{\alpha\beta\gamma\delta}^{(3)}(\omega_1, \omega_2, \omega_3)$ is a fourth-rank tensor, and each of its 81 complex components depends on three continuous variables. Note that the Cartesian indices $\alpha, \beta, \ldots$ can be also considered as arguments of $\chi^{(m)}$, each of them taking three discrete values.

The relation between the polarization and the field can be written symbolically as a power series,

$$P(E) = \sum_{m=0}^{\infty} \chi^{(m)} E^m. \tag{6.5}$$

In the spectral representation, this relation is an algebraic one, while in the time representation, $\chi^{(m)}$ should be understood as integral operators. Their kernels, $\chi^{(m)}(t_1, \ldots, t_m)$, called *multi-time Green's functions* or *response functions* of the matter, are defined in terms of the spectral susceptibility $\chi^{(m)}(\omega_1, \ldots, \omega_m)$ using the m-fold Fourier transformation (for $m = 1$, see (4.7)). By analogy with the linear case (Sec. 4.1), the causality principle leads to integral relations between the real and imaginary parts of $\chi^{(m)}$, similar to the Kramers-Kronig relations.

The effect of the magnetic field can be taken into account by doing a double power-series expansion in (6.5), $\sum \chi^{(mn)} E^m H^n$. Some nonlinear optical effects reveal spatial dispersion, which can be described as the dependence of $\chi^{(m)}$ not only on $\omega_1, \ldots, \omega_m$ but also on the wavevectors $k_1, \ldots, k_m$.

6.1.2 °*Various definitions*

Often, one uses the definition of spectral amplitudes $\tilde{E}_n$ that differs from (6.4) by the absence of the $1/2$ factor,

$$E(t) = \sum_n \tilde{E}_n \exp(-i\omega_n t) + \text{c.c.}, \quad \tilde{E}_n \equiv E_n/2. \tag{6.6}$$

Similarly, at $\omega_0 \neq 0$, $\tilde{P}_0 = P_0/2$, so (6.2) leads to the following relation for the quadratic nonlinearity:

$$\tilde{P}_0^{(2)} = (1/2)\chi^{(2)} E_1 E_2 = 2\chi^{(2)} \tilde{E}_1 \tilde{E}_2 \equiv \tilde{\chi}^{(2)} \tilde{E}_1 \tilde{E}_2. \tag{6.7}$$

Thus, two different definitions for the field and polarization spectral amplitudes correspond to two differently defined m-th-order susceptibilities, related as

$$\tilde{\chi}^{(m)} = 2^{m-1} \chi^{(m)}, \ \omega_0 \neq 0. \tag{6.8}$$

An exception from this rule are even-order susceptibilities at $\omega_0 = 0$, describing the *optical rectification* effect. For this case, $\tilde{P}_0 = P_0$, and

$$\tilde{\chi}^{(m)} = 2^m \chi^{(m)}, \ \omega_0 = 0.$$

Assume now that the inertia of the material response can be neglected. This is the so-called *inertia-free*, or *Kleinman's*, approximation, which is valid in the case where the field frequencies and their combinations are in the transparency windows of the matter. In this case, the polarization follows the field instantaneously, and for $m = 2$,

$$P_\alpha^{(2)}(t) = \bar{\chi}_{\alpha\beta\gamma} E_\beta(t) E_\gamma(t), \tag{6.9}$$

where $\bar{\chi}$ is some real constant tensor. In particular,

$$P_x^{(2)}(t) = \bar{\chi}_{xxx} E_x^2(t) + \bar{\chi}_{xxy} E_x(t) E_y(t) + \bar{\chi}_{xyx} E_y(t) E_x(t) + \bar{\chi}_{xyy} E_y^2(t) + \ldots \tag{6.10}$$

From this example, one can see that only the sum $\bar{\chi}_{\alpha\beta\gamma} + \bar{\chi}_{\alpha\gamma\beta}$ has a physical meaning; these components cannot be measured separately, and therefore the $\bar{\chi}$ tensor is symmetric in the last two indices while the $\bar{\chi}^{(m)}$ tensor is symmetric in all indices except the first one.[b] Assuming the field in (6.9) to be biharmonic, we obtain

$$P_\alpha^{(2)}(t) = (1/2)\bar{\chi}_{\alpha\beta\gamma} \mathrm{Re}\{\sum_{n=1,2} [E_{n\beta} E_{n\gamma}^* + E_{n\beta} E_{n\gamma} \exp(-2i\omega_n t)] + 2E_{1\beta} E_{2\gamma} \exp[-i(\omega_1 + \omega_2)t] + 2E_{1\beta} E_{2\gamma}^* \exp[-i(\omega_1 - \omega_2)t]\}. \tag{6.11}$$

The first term here describes optical rectification, the second one, harmonic generation, the third and the fourth ones correspond to the generation of sum and difference frequencies.

On the other hand, from the definition of the susceptibility (6.3) it follows that

$$P_\alpha^{(2)}(t) = \sum_{n=1,2} \chi_{\alpha\beta\gamma}(\omega_n, -\omega_n) E_{n\beta} E_{n\gamma}^* + \mathrm{Re}\{\sum_{n=1,2} \chi_{\alpha\beta\gamma}(\omega_n, \omega_n) E_{n\beta} E_{n\gamma} \exp(-2i\omega_n t) + \chi_{\alpha\beta\gamma}(\omega_1, \omega_2) E_{1\beta} E_{2\gamma} \exp[-i(\omega_1 + \omega_2)t] + \chi_{\alpha\beta\gamma}(\omega_1, -\omega_2) E_{1\beta} E_{2\gamma}^* \exp[-i(\omega_1 - \omega_2)t]\}. \tag{6.12}$$

(Here we have omitted the order index of the susceptibility, which we will sometimes do in what follows.)

Comparing (6.11) and (6.12), we see that in the dispersion-free approximation,

$$\chi(\omega, \omega') = 2\chi(\omega, \omega) = 2\chi(\omega, -\omega) = \bar{\chi}, \tag{6.13}$$

where $\omega \neq \omega' \neq 0$. Similarly, one can show that $\chi(\omega, 0) = 2\bar{\chi}$. The first equality in (6.13) is obviously still valid in the presence of dispersion, provided that ω and ω' are sufficiently close. Thus, *every component of the* $\bar{\chi}(\omega, \omega')$ *tensor, considered*

[b]Below, we show from the energy considerations that the $\bar{\chi}^{(m)}$ tensors are fully symmetric.

as a function of two variables, ω and ω', has a peculiarity on the lines $\omega' = \pm\omega$ where its values are twice as small as at the neighbouring points.

Higher-order susceptibilities have similar peculiarities at coincident frequency arguments. The corresponding coefficients can be found by repeating the derivation of (6.13).

6.1.3 °*Permutative symmetry*

Definitions (6.3), (6.4) lead to the invariance of the $\chi^{(m)}$ tensors to permutations of their frequency arguments together with the corresponding Cartesian indices,

$$\chi_{\alpha\beta\gamma}(\omega_1, \omega_2) \equiv \chi_{\alpha\gamma\beta}(\omega_2, \omega_1), \tag{6.14}$$

$$\chi_{\alpha\beta\gamma\delta}(\omega_1, \omega_2, \omega_3) \equiv \chi_{\alpha\beta\delta\gamma}(\omega_1, \omega_3, \omega_2) \equiv \dots \tag{6.15}$$

Indeed, (6.3) can be rewritten in other equivalent forms,

$$P^{(2)}_{0\alpha} = \chi_{\alpha\beta\gamma}(\omega_2, \omega_1)E_{2\beta}E_{1\gamma} = \chi_{\alpha\gamma\beta}(\omega_2, \omega_1)E_{2\gamma}E_{1\beta}.$$

Comparison of the latter with (6.3) yields (6.14).

It follows from (6.14), (6.15) that *tensors describing the generation of harmonics are symmetric with respect to all indices except the first one.*

Another general property of susceptibilities follows from the fact that both the polarization and the field are real values, which requires that the amplitudes should turn into their complex conjugates as the frequencies change their signs. Changing all frequency signs in (6.4) and doing complex conjugation, we obtain

$$\boldsymbol{P}_0 = \chi^*(-\omega_1, \dots, -\omega_m)\boldsymbol{E}_1 \dots \boldsymbol{E}_m.$$

Comparing this with (6.4), we find (see (4.5))

$$\chi^*(-\omega_1, \dots, -\omega_m) = \chi(\omega_1, \dots, \omega_m). \tag{6.16}$$

6.1.4 °*Transparent matter*

Note that spatio-frequency permutative relations (6.14), (6.15) do not concern the first index. Using $\chi^{(2)}$ as an example, let us show that its permutation is possible in the non-resonant case where all frequencies are away from the resonances of the matter. For the sake of symmetry, we introduce the third argument into the notation of the susceptibility,

$$\chi(\omega_1, \omega_2) \equiv \chi(-\omega_0; \omega_1, \omega_2) \equiv \chi^{\bar{0}12} = \chi^{0\bar{1}\bar{2}*}, \tag{6.17}$$

where the minus sign by the combination frequency $\omega_0 = \omega_1 + \omega_2$ provides that the sum of all three arguments of the susceptibility is zero. In the last equality, we took into account relation (6.16). Note that sometimes, other notation can be used as well,

$$\chi(-\omega_0; \omega_1, \omega_2) \equiv \chi(\omega_0 = \omega_1 + \omega_2).$$

Let us find the power absorbed in a unit volume of the matter due to the quadratic nonlinearity in the case of a three-frequency field. According to (4.13),

$$\mathcal{P}^{(2)} = \overline{\boldsymbol{E} \cdot \dot{\boldsymbol{P}}^{(2)}} = (1/2) \sum_{n=0}^{2} \omega_n \mathrm{Im} \boldsymbol{E}_n^* \cdot \boldsymbol{P}_n^{(2)} \equiv \mathcal{P}_0 + \mathcal{P}_1 + \mathcal{P}_2. \qquad (6.18)$$

Hence, according to definition (6.3), the partial powers are

$$\begin{aligned} \mathcal{P}_0 &= (1/2)\omega_0 \mathrm{Im} \chi_{\alpha\beta\gamma}^{\bar{0}12} E_{0\alpha} E_{1\beta} E_{2\gamma}, \\ \mathcal{P}_1 &= (1/2)\omega_1 \mathrm{Im} \chi_{\beta\alpha\gamma}^{\bar{1}0\bar{2}} E_{1\beta}^* E_{0\alpha} E_{2\gamma}^*, \\ \mathcal{P}_2 &= (1/2)\omega_2 \mathrm{Im} \chi_{\gamma\beta\alpha}^{\overline{21}0} E_{2\gamma}^* E_{1\beta}^* E_{0\alpha}. \end{aligned} \qquad (6.19)$$

Note that the signs and the values of the powers $\mathcal{P}_n$ depend on the field phases. With the help of (6.16), we find from (6.19) that

$$\mathcal{P}^{(2)} = (1/2)\mathrm{Im}[\omega_1(\chi_{\alpha\beta\gamma}^{\bar{0}12} - \chi_{\beta\alpha\gamma}^{1\bar{0}2}) + \omega_2(\chi_{\alpha\beta\gamma}^{\bar{0}12} - \chi_{\gamma\beta\alpha}^{21\bar{0}})] E_{0\alpha}^* E_{1\beta} E_{2\gamma}. \qquad (6.20)$$

If all three frequencies are far from resonances, then the absorption is absent and the matter only redistributes the energy between the three frequency components of the field. According to (6.19), the share of the nth component scales as ω_n.

In a transparent matter, $\mathcal{P} = 0$, and since the complex amplitudes E_{na} are arbitrary, the expression in the square brackets of (6.20) should turn to zero. Provided that the dispersion is weak, the frequencies can be also considered as arbitrary, so that each of the coefficients by ω_1 and ω_2 is zero as well. Hence, in addition to the automatic symmetry relations (6.14), (6.15), which do not concern the first index, a medium that is transparent at all three frequencies imposes the following bounds:

$$\chi_{\alpha\beta\gamma}^{\bar{0}12} = \chi_{\beta\alpha\gamma}^{1\bar{0}2}) = \chi_{\gamma\beta\alpha}^{21\bar{0}}. \qquad (6.21)$$

In the general case of an *arbitrary-order non-resonant susceptibility* $\chi^{(m)}$, *there is complete permutative symmetry with respect to all indices*. A possibility to have the permutation of the first indices leads to the Manley-Rowe relations (Sec. 6.3): $\mathcal{P}_0/\omega_0 = \mathcal{P}_1/\omega_1 = \mathcal{P}_2/\omega_2$. Resonance nonlinear susceptibilities have symmetries more limited than given by (6.21) (Sec. 6.3). For instance, the Raman susceptibility satisfies the relation

$$\chi_{xxxx}^{\overline{12}12} = \chi_{xxxx}^{\overline{21}12*}.$$

In the dispersion-free approximation, $\chi^{(m)}$ have no frequency dependence at all, and the susceptibility tensors are therefore symmetric in all indices.

As an example of using Eq. (6.21), consider the case $\omega_2 = -\omega_1$. Then, with an account for (6.13),

$$4\chi_{\alpha\beta\gamma}(0 = \omega_1 - \omega_1) = \chi_{\gamma\beta\alpha}(\omega_1 = \omega_1 + 0). \tag{6.22}$$

The presence of the factor 4 here can be verified by repeating the derivation of (6.13) in the presence of a constant field. The susceptibility on the left-hand side describes *optical rectification*, while the one on the right-hand side is responsible for the *linear electrooptic effect* (the *Pockels effect*), i.e., variation of the refractive index $n(\omega_1)$ at the frequency ω_1 scaling as the static field E_0. Indeed, polarization at the frequency ω_1, with an account for linear and quadratic susceptibilities, is

$$P_{1\alpha} = [\chi_{\alpha\beta}(\omega_1) + \chi_{\alpha\beta\gamma}(\omega_1 = \omega_1 + 0)E_{0\gamma}]E_{1\beta} \equiv (\chi_{\alpha\beta} + \Delta\chi_{\alpha\beta})E_{1\beta}. \tag{6.23}$$

The susceptibility increase $\Delta\chi$ will manifest itself in an anisotropic variation of $\Delta n(\omega_1)$. Thus, (6.22) sets a quantitative relation between two different phenomena, the Pockels effect, which has been known for a long time, and the optical rectification, which has only been discovered after the invent of lasers. Another example of such a pair of related phenomena is given by *Faraday's direct and inverse effects*. (The inverse Faraday effect is the emerging of static magnetization scaling as the intensity of a circularly polarized light wave.)

Thus, according to (6.14), (6.15), and (6.21), *nonlinear susceptibility tensors $\chi^{(m)}$ of a transparent material are invariant to all $(m + 1)!$ permutations of their space-frequency arguments.*

The absence of dissipation in the transparency windows enables one to define the energy v of the polarization of the matter (Sec. 4.1). If dispersion, i.e., delay of the response, is completely neglected, then, by analogy with (4.28),

$$v^{(2)}(t) = -(1/3)\bar{\chi}_{\alpha\beta\gamma}\boldsymbol{E}_\alpha\boldsymbol{E}_\beta\boldsymbol{E}_\gamma, \tag{6.24}$$

where $\boldsymbol{E} \equiv \boldsymbol{E}(t)$. In the thermodynamic approach, this energy should be added to the free energy density of the matter F and to other thermodynamic potentials (Sec. 4.1). Then, $\boldsymbol{P}, \boldsymbol{\chi}^{(1)}$, and $\boldsymbol{\chi}^{(2)}$ are determined, respectively, by the first, second, and third derivatives of $F(\boldsymbol{E})$ at the point $\boldsymbol{E} = 0$, see (4.32), (4.33).

Let the field contain three harmonics,

$$\boldsymbol{E}(t) = (1/2)\sum_n \boldsymbol{E}_n \exp(-i\omega_n t), \tag{6.25}$$

where

$$n = \pm 0, \pm 1, \pm 2,\ \omega_0 = \omega_1 + \omega_2,\ \omega_1 \neq \omega_2,\ \omega_n \neq 0,$$

$$\omega_{-n} \equiv -\omega_n,\ \boldsymbol{E}_{-n} \equiv \boldsymbol{E}_n^*,$$

then the time-averaged polarization energy is

$$v^{(2)} \equiv \overline{v^{(2)}(t)} = -(3!/3 \cdot 2^3)\bar{\chi}_{\alpha\beta\gamma}E^*_{0\alpha}E_{1\beta}E_{2\gamma} + \text{c.c.} \tag{6.26}$$

Dispersion should be taken into account by replacing $\bar{\chi}$ with $\chi(\omega, \omega')$, which is possible due to the symmetry (6.16), (6.21). As a result,

$$v^{(2)}(\{E_n\}) = -(1/4)(\chi^{\bar{0}12}_{\alpha\beta\gamma}E^*_{0\alpha}E_{1\beta}E_{2\gamma} + \chi^{0\overline{12}}_{\alpha\beta\gamma}E_{0\alpha}E^*_{1\beta}E^*_{2\gamma}). \tag{6.27}$$

This relation allows one to calculate the nonlinear polarization, as well as $\chi^{(2)}$, in terms of $v^{(2)}$ (see (4.36)),

$$P^{(2)}_{n\alpha} = -4\partial v^{(2)}/\partial E^*_{n\alpha}. \tag{6.28}$$

6.1.5 *The role of the material symmetry*

In different reference frames, the components of the field $\boldsymbol{E}$ and polarization $\boldsymbol{P}$ vectors, as well as the tensors $\chi^{(m)}$ relating them, are of course different. Under the rotation of a Cartesian frame of reference, the susceptibility transforms as

$$\tilde{\chi}^{(m)}_{\alpha'\beta'\ldots} = \sum_{\alpha,\beta,\ldots} a_{\alpha'\alpha}a_{\beta'\beta}\ldots\chi^{(m)}_{\alpha\beta\ldots}, \tag{6.29}$$

with the tilde denoting the components in the new frame of reference. Here, $\boldsymbol{a}$ is the matrix describing the transformation from the old components to the new ones, and we assume that *rotation* also includes the sign changes in some or all coordinates, i.e., *inversion* and *mirror reflection*. For instance, in the case of inversion, $a_{\alpha'\alpha} = -\delta_{\alpha'\alpha}$, and

$$\tilde{\chi}^{(m)_{\alpha\beta\ldots}} = (-1)^{m+1}\chi^{(m)}_{\alpha\beta\ldots}, \tag{6.30}$$

where $m + 1$ is the rank of the tensor. Thus, *as a result of the reference frame inversion, the components of odd-rank tensors (in particular, vectors and quadratic susceptibility tensors) change their signs,* $\tilde{E}_\alpha = -E_\alpha, \tilde{\chi}^{(2)}_{\alpha\beta\gamma} = -\chi^{(2)}_{\alpha\beta\gamma}, \ldots$.

For tensors describing the physical properties of a material, there is a special frame of reference where the tensor has its simplest form, with the maximal number of zero or equal components. In crystals, this 'natural' frame of reference coincides with the crystallographic one. For instance, real tensors of rank two are diagonal in the natural frame of reference, $\chi^{(1)}_{\alpha\beta} = \chi_\alpha\delta_{\alpha\beta}$.

Any unbounded medium, either amorphous or crystalline, has a certain symmetry of particles' positions averaged over their thermal motion. Formally, this symmetry of the medium is determined by a set (*group*) of a certain number of *symmetry elements*. In particular, the elements of a point[c] group of symmetry are

[c]The term 'point' is due to the fact that rotations leave one point (the origin) fixed, unlike translational transformations of the coordinates.

all rotations $\boldsymbol{a}$ of the reference frame, including mirror reflections and inversion, which leave the structure of the medium unchanged. For instance, many crystals, as well as optically non-active liquids and gases, are invariant with respect to inversion. Such media are called *centrally symmetric*.

Similarly, any macroscopic property of a medium, which is characterized by a certain tensor, has its own group of symmetry elements. The symmetry elements of a tensor are all rotations $\boldsymbol{a}^{(i)}$ that act according to rule (6.29) and do not change the components of the tensor. For instance, according to (6.30), all even-rank tensors are invariant with respect to inversion.

It seems obvious that the symmetry of a macroscopic property of a medium cannot be lower than the symmetry of its structure (the von Neumann principle, Ref. [Nye (1957)]). In other words, *the symmetry group of a property should include all symmetry groups of the structure*, i.e., the latter is a *subgroup* of the symmetry group of the property. Hence, if $\boldsymbol{a}$ is an element of a point symmetry group of a medium, then the tilde in (6.29) can be omitted. Then, (6.29) becomes a relation between different components of the $\boldsymbol{\chi}^{(m)}$ tensor. By substituting into (6.30) all symmetry elements $\boldsymbol{a}^{(i)}$ of the medium one by one, we obtain a homogeneous system of equations for $\chi^{(m)}_{\alpha\beta\gamma}$. In isotropic media and in crystals, such equations greatly reduce the number of susceptibility nonzero components, as well as make many components equal, sometimes up to the sign.

The most bright example follows from (6.30) in the case of centrally symmetric media. According to the von Neumann principle, all tensors describing the physical properties of such media should be also centrally symmetric, i.e., the tilde in (6.30) can be omitted. Hence, for even m, it should be $\chi^{(m)}_{\alpha\beta\ldots} = -\chi^{(m)}_{\alpha\beta\ldots}$, which leads to $\boldsymbol{\chi}^{(m)} = 0$. Thus, *in centrally symmetric media, all even-order susceptibilities are equal to zero.*

Note that this conclusion is not valid in the case of susceptibilities describing magnetic effects. This is because the magnetic field and the magnetization are pseudovectors (axial vectors, as they do not change their signs under the inversion of the coordinates). As a result, the corresponding susceptibilities are pseudotensors and do not transform according to (6.30) under inversion. In particular, the Faraday effect, described by the relation $\boldsymbol{P}_1 = \boldsymbol{\eta}(\omega_1 = \omega_1 + 0)\boldsymbol{E}_1\boldsymbol{H}_0$, is also possible in centrally symmetric media.

6.2 Models of optical anharmonicity

Depending on the features of the matter, its state, the frequencies of the incident fields, and other experimental conditions, various mechanisms can contribute in the observable effects. Below we will consider several particular classical models of optical anharmonicity and then present the general quantum scheme of calculating nonlinear susceptibilities.

6.2.1 *Anharmonicity of a free electron*

Let an electron (or any charged particle) be in the field of a plane monochromatic wave E polarized linearly along the x axis and propagating along the z axis. With an account for the magnetic part of the Lorentz force, non-relativistic equations of motion for the electron have the form ($e < 0$)

$$\ddot{X} + 2\gamma\dot{X} = \frac{e}{m}\left(E_x - \frac{1}{c}\dot{Z}H_y\right) = \frac{e}{m}\left(1 - \frac{1}{c}\dot{Z}\right)E,$$

$$\ddot{Z} + 2\gamma\dot{Z} = \frac{e}{mc}\dot{X}E, \tag{6.31}$$

where γ is a phenomenological damping constant providing that after the periodic field is switched off, the motion proceeds with a stationary amplitude, and

$$E \equiv E_x(Z,t) = \mathrm{Re}E_1 \exp[ikZ(t) - i\omega t] = H_y(Z,t) \tag{6.32}$$

is the field at the particle location. Damping can be caused by collisions, as well as by the reaction of the radiation, i.e., radiation losses (*radiation friction*). Suppose that the transverse displacement Z of the electron is small, then in (6.32) one can assume $Z = 0$ (dipole approximation, i.e., zero-order approximation in kZ).

Let us search the stationary solution to (6.31) in the form of a Fourier expansion using the method of successive approximations in the field amplitude,

$$\boldsymbol{R} = \boldsymbol{R}^{(1)} + \boldsymbol{R}^{(2)} + \cdots = \mathrm{Re}(\boldsymbol{R}_1 e^{-i\omega t} + \boldsymbol{R}_2 e^{-2\omega t} + \ldots), \tag{6.33}$$

where $\boldsymbol{R} \equiv \{X, Y, Z\}$. In the first approximation, one can neglect the effect of the magnetic field, so that the response of the electron is linear,

$$X_1^{(1)} = -\frac{e/m}{\omega^2 + 2i\omega\gamma}E_1. \tag{6.34}$$

Hence, the linear polarisability tensor of an electron, which defines the relation between the amplitudes of the dipole moment $\boldsymbol{d}_1 = e\boldsymbol{R}_1$ and the field E_1, is

$$\alpha_{\alpha\beta}(\omega) = -\frac{e^2}{m\omega(\omega + 2i\gamma)}\delta_{\alpha\beta}. \tag{6.35}$$

In our model, there is no neutralizing positive charge, and the dipole moment is defined with respect to the origin, $\boldsymbol{R} = 0$. After multiplying α by the density N of electrons, we find the linear susceptibility $\chi^{(1)}$ of cold (free of Doppler effect) plasma, see (4.52).

Thus, at $\omega \gg \gamma$, the linear polarisability of an electron is

$$\alpha = -\frac{e^2}{m\omega^2} = -r_e \lambda\!\!\!^{-2}, \tag{6.36}$$

where $r_e \equiv e^2/mc^2 \approx 3 \cdot 10^{-13}$ cm is the classical radius of the electron. Note that r_e is related to other typical length parameters through the *fine structure constant* $e^2/\hbar c^2 \approx 1/137$,

$$\lambda\!\!\!^{-}_0/2 = 137a_0 = 137^2\lambda\!\!\!^{-}_c = 137^3 r_e, \tag{6.37}$$

where $\lambda_0 \equiv 2\pi\lambda\!\!\!^{-} \equiv 1/R \approx 10^{-5}$ cm is the wavelength corresponding to the ionization potential of the hydrogen atom 13.6 eV, $R \equiv me^4/4\pi\hbar^3 c$ is the Rydberg constant, $a_0 \equiv \hbar^2/me^2 \approx 5 \cdot 10^{-9}$ is Bohr's radius and $\lambda\!\!\!^{-}_c \equiv \hbar/mc \approx 4 \cdot 10^{-11}$ cm is the Compton wavelength. Let $\lambda = \lambda_0$, then

$$-\alpha = 4a_0^3 \approx 6 \cdot 10^{-25}\text{cm}^3, \tag{6.38}$$

i.e., *polarisability of a free electron in the UV range is, on the order of magnitude, equal to the hydrogen atom volume, i.e., to the polarisability of a bound electron in the absence of a resonance.*

In order to find the second approximation, one should replace $\boldsymbol{R}$ by $\boldsymbol{R}^{(2)}$ in the left-hand sides of (6.31) and by $\boldsymbol{R}^{(1)}$ in the right-hand sides. Then the Lorentz force will have components with both zero and double frequency,

$$F_z^{(2)} = (e/c)\dot{X}^{(1)}E = (1/2)k\text{Im}\,\alpha(|E_1|^2 + E_1^2 e^{-2i\omega t}). \tag{6.39}$$

The double-frequency force causes longitudinal oscillations of the electrons with the frequency 2ω. Their amplitude, according to (6.31) and (6.39), is

$$Z_2^{(2)} = \frac{e^2 E_1^2}{8im^2 c\omega(\omega + i\gamma)(\omega + 2i\gamma)} \equiv \frac{1}{e}\beta_{zxx}E_1^2. \tag{6.40}$$

In the last equation, we have introduced the quadratic polarisability tensor $\boldsymbol{\beta}$ of a free particle, which provides the relation between the amplitudes of the field and the dipole moment $eZ_2^{(2)}$ at the double frequency. Thus, the quadratic polarisability of a free electron at $\omega \gg \gamma$ is

$$|\beta| \approx e^3/m^2 c\omega^3 = (e\lambda\!\!\!^{-}/mc^2)\alpha \equiv \alpha/E_{NL}^{free}. \tag{6.41}$$

Here, E_{NL} is a typical parameter equal to the field amplitude at which the linear and quadratic responses are equal, $Z^{(2)} = X^{(1)}$. At $\lambda = \lambda_0$,

$$E_{NL}^{free} = e/r_e\lambda\!\!\!^{-}_0 \approx 10^9\text{G}, \tag{6.42}$$

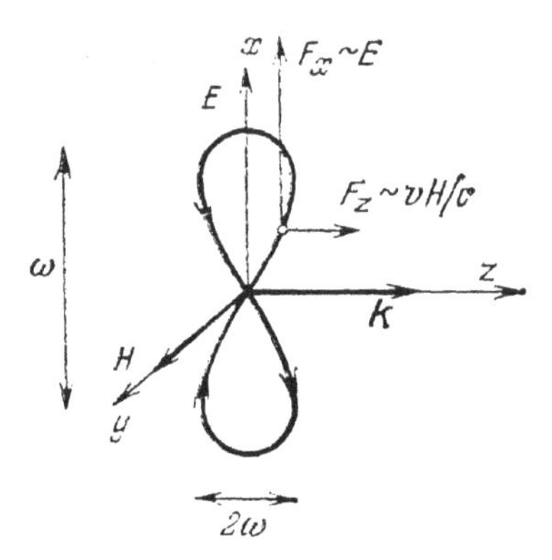

Fig. 6.2 Due to a plane monochromatic wave, an electron moves along a Lissajous pattern.

which corresponds to the intensity 10^{20} W/cm^2.[d] Below, we will show that the anharmonicity of a bound electron is two orders of magnitude as large as that of a free electron even in the absence of a resonance,

$$E_{NL}^{bound}/E_{NL}^{free} \approx \omega_0/137\omega,$$

where

$$E_{NL}^{bound} \approx e/2a_0^2 \approx 10^7\text{G}. \tag{6.43}$$

Multiplying β by the density of the electrons, we get the quadratic susceptibility of the plasma,

$$|\chi^{(2)}| = e^3N/m^2c\omega^3 = |\chi^{(1)}|/E_{NL}^{free}. \tag{6.44}$$

Thus, *one of the fundamental sources of the matter anharmonicity is the Lorentz force*. Note that the denominator of (6.44) contains the speed of light, which is typical for magnetic effects.

According to (6.41), the ratio $Z^{(2)}/X^{(1)}$ has an order of E_1/E_{NL}, while $kZ^{(2)} \sim (E_1/E_{NL})^2$. This justifies for using the dipole approximation in the calculation of β at $E_1 \ll E_{NL}$.

If the static force of light pressure is not taken into account, then, according to (6.34) and (6.40), an electron in a magnetic field moves in the xz plane along a figure of eight trajectory (Fig. 6.2). Stimulated oscillations of the electron along the field, $X^{(1)}(t)$, are accompanied by dipole emission in all directions, except the exact x one. This is what is called *Thomson scattering* or, taking into account recoil, *Compton scattering*. At the same time, longitudinal oscillations of

[d] In the Gaussian system of units, E and H have the same dimensionality; therefore, E can be measured in Gausses (1G = 300 V/cm).

the electron, $Z^{(2)}(t)$, lead to dipole emission at the frequency of the incident field second harmonic, i.e., the effect of *frequency doubling*. The second-harmonic emission is maximal in the transverse plane, xy, and absent along the primary field propagation direction. This structure of the free-electron quadratic polarisability tensor prevents the second-harmonic amplitudes in a macroscopic sample, such as plasma or a semiconductor, from adding up coherently.

Thus, electrons in a plasma, metal, or semiconductor provide, in addition to Thomson scattering, also non-coherent scattering with the double frequency and the intensity scaling as $\beta^2 N$. Bound electrons in atoms or molecules also have quadratic polarizability, which leads to a non-coherent scattering at the double frequency, termed *hyper-Rayleigh scattering*. From the quantum-mechanical viewpoint, it is interpreted as the absorption of two incident photons and the emission of a double-energy photon (Fig. 6.1(a)). If the phase velocities of the incident wave and its second harmonic are equal, $n(\omega) = n(2\omega)$ (the so-called phase matching condition), then, provided that the structure of the $\chi^{(2)}$ tensor is appropriate, weak nearly isotropic hyper-Rayleigh scattering is accompanied by a much more intense longitudinal emission, scaling as $\beta^2 N^2 = \chi^{(2)2}$ (Sec. 6.5).

6.2.2 °*Light pressure*

The constant component of the Lorentz force (6.39) determines the static light pressure force F_0 acting on the electron due to the traveling wave. According to (6.39), F_0 scales as the imaginary part of the electron linear polarisability, α'', i.e., the power of the scattered light, $\mathcal{P} = \omega\alpha''|E_1|^2/2$ (see (4.15)), or the interaction cross-section,

$$\sigma \equiv \mathcal{P}/I = 4\pi k\alpha'', \tag{6.45}$$

where $I = c|E_1|^2/8\pi$ is the intensity of the plane wave. Therefore, the force of light pressure can be represented in the form

$$F_0 = k\alpha''|E_1|^2/2 = \mathcal{P}/c = \sigma I/c. \tag{6.46}$$

This force accelerates the electron; however, collisions in plasma will lead to a constant speed of electron drift, $\dot{Z}_0 = F_0/m\tilde{\gamma}$, where $\tilde{\gamma} = 1/\tau$ and τ is the time between collisions. (Recall that γ is the oscillation damping rate, which can exceed $\tilde{\gamma}$). The constant current emerging along the beam, scaling as $|E_1|^2$, can be interpreted in terms of optical rectification. (In dielectrics, the term 'rectification', or '*dc-effect*', means the appearance of a static field $E_0 \sim |E_1|^2$.)

Let $\omega \gg \gamma$, then, according to (6.35), $\alpha'' = 2\gamma e^2/m\omega^3$. Let us estimate F_0 in the case where the damping of electron oscillations is only due to radiation losses (*radiation friction*). According to (6.31), the friction force, in the first order, is $-2\gamma m\dot{X}^{(1)}$. Multiplying it by the velocity $\dot{X}^{(1)}$ and averaging over a single period, we find the power of losses, $\mathcal{P} = \gamma m\omega^2|X_1^{(1)}|^2$. Setting equality between this expression and the power of dipole emission (5.19), we obtain

$$2\gamma/\omega = \alpha''/\alpha' = 2r_e/3\lambda\!\!\!^{-}. \tag{6.47}$$

The same result follows from (5.24) with a unity oscillator strength (4.62). Now, (6.46) takes the form

$$F_0 = r_e^2|E_1|^2/3. \tag{6.48}$$

Comparing (6.46) and (6.48), we find the Thomson cross-section of scattering by a free electron,

$$\sigma_T = 8\pi r_e^2/3. \tag{6.49}$$

From the photon viewpoint, light pressure is due to the fact that an electron acquires the momenta of absorbed photons, which are then symmetrically re-emitted in all directions via Thomson's (or Compton's) scattering. Let us stress that we have only considered the average value of the force, which has quantum fluctuations [Minogin (1986)].

We have found the force of light pressure acting on a free electron in the case of a traveling wave. Similar analysis can be carried out for a more complicated spatial configuration of light. For instance, an electron will be displaced in the field of a standing wave, which is the *Kapitsa-Dirac effect*. It is important that in a non-homogeneous field, the Lorentz force has a nonzero value averaged over the period even at $\alpha'' = 0$. In this case, F_0 is determined by α', which is much greater than α'', so that the latter can be assumed to be zero. This force is caused by the exchange between different plane waves and is called the *stimulated* one (in contrast to the *spontaneous* force (6.46)).

Displacement of an electron $\Delta\boldsymbol{R}(t)$ due to a monochromatic field is $\alpha\boldsymbol{E}(\boldsymbol{R}_0, t)/e$, where $\alpha = -me^2/\omega^2$ is the polarisability and $\boldsymbol{R}_0$ is the non-perturbed coordinate of the electron. Hence, we find the averaged Lorentz force,

$$\boldsymbol{F}_0 = \alpha\dot{\boldsymbol{E}} \times \boldsymbol{H}/c = k\mathrm{Im}\,\alpha\boldsymbol{E}_\omega \times \boldsymbol{H}_\omega^*/2, \tag{6.50}$$

where the last equality is also valid for the case of a complex polarisability. Note that in a plane traveling wave, $\dot{E}H \sim \sin\omega t\cos\omega t \to 0$, and that in the general case the force F_0 does not scale as the averaged Poynting vector, $\boldsymbol{S}_0 = c\mathrm{Re}\,\boldsymbol{E}_\omega \times \boldsymbol{H}_\omega^*/8\pi$.

Consider first the case of two plane waves,

$$\boldsymbol{E}_\omega = \sum_{n=1}^{2} \boldsymbol{E}_n \exp(i\boldsymbol{k}_n \cdot \boldsymbol{R}_0 + i\varphi_n),$$

$$\boldsymbol{H}_n = \hat{\boldsymbol{k}}_n \times \boldsymbol{E}_n, \;\; \boldsymbol{k}_n = \hat{\boldsymbol{k}}_n \omega/c.$$

At $\alpha'' = 0$, (6.50) contains only 'cross' components,

$$\begin{aligned}\boldsymbol{F}_0 &= (1/2)\alpha k \mathrm{Im}\,(\boldsymbol{E}_1 \times \boldsymbol{H}_2^* - \boldsymbol{E}_2^* \times \boldsymbol{H}_1)e^{i\psi} \\ &= (1/2)\alpha \mathrm{Im}[-\Delta\boldsymbol{k}(\boldsymbol{E}_1 \cdot \boldsymbol{E}_2^*) + \boldsymbol{E}_1(\boldsymbol{E}_2^* \cdot \boldsymbol{k}_1) - \boldsymbol{E}_2^*(\boldsymbol{E}_1 \cdot \boldsymbol{k}_2)]e^{i\psi}, \quad (6.51)\end{aligned}$$

where $\psi \equiv \Delta\boldsymbol{k} \times \boldsymbol{R}_0 + \varphi$, $\Delta\boldsymbol{k} \equiv \boldsymbol{k}_1 - \boldsymbol{k}_2$, $\varphi \equiv \varphi_1 - \varphi_2$.

It is easy to see that the part of $\boldsymbol{F}_0$ scaling as $\Delta\boldsymbol{k}$ can be represented in the 'gradient' form with the effective potential $-\alpha\overline{E^2}/2$ (from (4.28) it is clear why the factor $1/2$ appears),

$$\boldsymbol{F}_M = \alpha\nabla|E_\omega|^2/4. \quad (6.52)$$

Note that this part of $\boldsymbol{F}_0$, called the *Miller force*, disappears, according to (6.51), in the case where the waves are orthogonally polarized. For counter-propagating waves ($\boldsymbol{k}_2 = -\boldsymbol{k}_1$) with the same linear polarization, $\boldsymbol{F}_0 = \boldsymbol{F}_M$:

$$\boldsymbol{F}_0 = -\alpha\boldsymbol{k}_1 E_1 E_2 \sin(2\boldsymbol{k}_1 \cdot \boldsymbol{R}_0 + \varphi). \quad (6.53)$$

Due to this force, charged particles tend to bunch in the nodes of a standing wave.

In the general case, a field consists of a continuum of plane waves, and the light pressure force can be found by integrating (6.51) in $\boldsymbol{k}_1$ and $\boldsymbol{k}_2$. Note that in this case, the gradient part (6.52) maintains its form. It should be stressed that in the presence of the electron recoil, the interacting waves have different frequencies. The corresponding phenomenon is called the *stimulated Compton effect*.

Consider further the pressure of light acting on bound electrons in an atom or a molecule, i.e., on neutral polarisable particles. As the starting point we take the effective potential $\mathcal{V} = -\boldsymbol{d} \cdot \boldsymbol{E}(\boldsymbol{r})$, where $\boldsymbol{d}$ is the induced dipole moment and $\boldsymbol{r}$ is the coordinate of the particle center of mass. Hence (see(4.39)),

$$\boldsymbol{F} = \nabla\boldsymbol{d}(t) \cdot \boldsymbol{E}(\boldsymbol{r}, t), \quad (6.54)$$

or $F_\alpha = d_\beta \partial E_\beta/\partial x_\alpha$. Assuming

$$E_x = (1/2)E_1 e^{i(kz-\omega t)} + \text{c.c.} \equiv E^{(+)} + E^{(-)},$$

$$d_x = \alpha(\omega)E^{(+)} + \text{c.c.}$$

and selecting the constant component, we find (6.46) once again.

Consider first resonance pressure. Let, as before, the damping be only due to the emission by the particle, i.e., the resonance fluorescence (Sec. 5.2), then the width of the resonance is minimal (the natural width), $2\gamma_{rad} = A$. Then, according to (2.55), $\sigma = 2\pi ƛ^2$, and it follows from (6.45), (6.46) that $\alpha'' = ƛ^3/2$ and

$$F_0 = ƛ^2|E_1|^2/4. \tag{6.55}$$

Thus, *the resonance light pressure on a bound electron is* $ƛ^2/r_e^2 = 137^6$ *times as great as on a free one*, provided that the damping is only caused by radiation. This huge difference is due to the high ratio of the resonance and Thomson's cross sections.

Estimate (6.55) relates to the case where only the lower level is populated. In the general case,

$$\alpha'' = (1/2)ƛ^3\Delta = ƛ^3\Delta^{(0)}/2(1 + 2W_0T_1), \tag{6.56}$$

where Δ and $\Delta^{(0)}$ are relative population differences with an account for saturation and without it, respectively (Sec. 4.3), W_0 is the transition probability and T_1 is the time of longitudinal relaxation. In the case of radiation relaxation, $T_1 = 1/A$. According to (6.56), *in the presence of population inversion the light pressure force is directed oppositely to the light beam*, which in the photon language can be explained in terms of the recoil of photons emitted via stimulated transitions forward.

At strong saturation, the power $\mathcal{P}$ absorbed by the atom, according to (4.102), is $\hbar\omega\Delta^{(0)}/2T_1$, so that (6.46) leads to

$$F_0 = \hbar k\Delta^{(0)}/2T_1, \tag{6.57}$$

the force scales as the momentum of the incident photon times the number of photons scattered per unit time. Let $\lambda = 1\mu$ and $A = 10^6$ s^{-1}, then $F_0 = 3 \cdot 10^{-17}$ dyn, and at $m = 3 \cdot 10^{-23}$ g the acceleration is as high as 10^6 cm/s^2. The intensity of saturating light is in this case much greater than 10^{-4} W/cm^2.

Resonance pressure of laser light provides quite unusual applications. With its help, one can accelerate, displace, and focus beams of neutral molecules, separate isotopes, 'trap' molecules within a small space domain, and reduce their thermal velocities [Minogin (1986)].

6.2.3 *Striction anharmonicity*

Now, let all frequencies of the field be in the transparency range of the matter. Then the dispersion can be neglected and the polarisability of a particle can be considered as a real constant. In this case, the force takes the form

$$\boldsymbol{F} = \alpha\nabla\overline{E^2}/2 = \alpha\nabla\boldsymbol{E}^{(+)} \cdot \boldsymbol{E}^{(-)}. \tag{6.58}$$

Here, the bar denotes averaging over high-frequency components. Indeed, we are only interested in the static part of the force acting on a molecule as a whole, or at least in the part that is varying slowly compared with the frequencies of the field and the molecule. For notation $E^{(\pm)}$, see Sec. 7.2.

The gradient force (6.58) corresponds to the effective potential of the molecule $\mathcal{V} = -\alpha\overline{E^2}/2$.

The additional energy density of the matter and the force density in an optical field will be N times as high (see (4.35)),

$$v = -\chi\overline{E^2}/2 = -(n^2 - 1)\overline{E^2}/8\pi, \tag{6.59}$$

$$f = \chi\nabla\overline{E^2}/2, \tag{6.60}$$

where $\chi \equiv \chi^{(1)} \approx \alpha N$, N is the density of molecules, which is assumed to be uniform, and $n = \sqrt{\epsilon}$ is the refractive index.

Here, we have neglected the interaction between molecules, which is only possible for a sufficiently small N. One can show that in the general case (see Ref. [Landau (1982)]), one should make a substitution into (6.60),

$$\chi \to \rho\left(\frac{\partial\chi}{\partial\rho}\right)_T = \frac{\rho}{4\pi}\left(\frac{\partial\epsilon}{\partial\rho}\right)_T, \tag{6.61}$$

where ρ is the density of the matter. For instance, from the Clausius-Mossotti relation, which can be easily obtained from (4.48),

$$\frac{\epsilon - 1}{\epsilon + 2} = \frac{4\pi\alpha}{3m}\rho, \tag{6.62}$$

it follows that

$$\frac{1}{\rho}\left(\frac{\partial\rho}{\partial\epsilon}\right)_T = \left(\frac{\partial\ln\rho}{\partial\epsilon}\right)_T = \frac{3}{(\epsilon - 1)(\epsilon + 2)},$$

so that in the case of a dense medium, (6.60) should be multiplied by the Lorentz correction $(\epsilon + 2)/3$.

In a traveling plane wave, $\nabla\overline{E^2}$ has no constant component, hence $F = 0$. (Note that we are considering a range where $\alpha'' = 0$ and the *spontaneous* force (6.46) related to the dissipation is absent.) However, in a standing plane wave, $E = 2E_1\cos(kz)\cos(\omega t)$, and from (6.58), it follows that (compare with (6.53))

$$F_z = -\alpha k E_1^2 \sin(2kz). \tag{6.63}$$

Thus, *at $\alpha > 0$ particles gather in the anti-nodes of a wave*. The force (6.63), scaling as the polarisability, is called the *stimulated* force.

In limited beams of light, there is a static transverse gradient of the field square, and at $\alpha > 0$ the particles tend to move towards the beam axis. Under stationary conditions, the force density (6.60) should be compensated for due to the increase in the pressure, Δp, and the density of particles, ΔN, in the central part of the beam,

$$\Delta p = -v = \chi \overline{E^2}/2, \tag{6.64}$$

$$\Delta N/N = \Delta\rho/\rho = \beta_T \Delta p, \tag{6.65}$$

where β_T is the isothermic compressability of the medium. These equations describe *electrostriction in a light field.*

The increased density of particles in the light beam will cause a change in the susceptibility of matter,

$$\Delta\chi = \alpha\Delta N = \beta_T \chi^2 |E_1|^2/4. \tag{6.66}$$

At the same time, by definition,

$$P_1^{(3)} = \chi^{(3)}(\omega = \omega - \omega + \omega)|E_1|^2 E_1 = \Delta\chi E_1,$$

so that the electrostriction contribution to the cubic susceptibility is

$$\chi^{(3)} = \beta_T \chi^{(1)^2}/4. \tag{6.67}$$

Hence, we find the typical nonlinear parameter

$$E_{NL}^2 \equiv \chi^{(1)}/\chi^{(3)} = 4/\beta_T \chi^{(1)}. \tag{6.68}$$

In liquids, $n \approx 1.5$ ($\chi \approx 0.1$), and $\beta_T \approx 10^{-10}$ dyn. (Recall that $\beta \approx 1/\rho v^2$, where v is the speed of sound.) Hence, $\chi^{(3)} \approx 10^{-13}$ cm^2/erg, $E_{NL} \approx 10^6$ G.

The gradient force (6.58) and the corresponding pressure 6.64 are very important for applications: they enable one to generate strong ultra-sound waves using bi-harmonic laser fields. The same force causes *stimulated Mandelshtam-Brillouin scattering* (see below). Striction nonlinearity (6.67) is one of the reasons for the *self-focusing of light*. Another important mechanism, *the optical Kerr effect*, i.e., orientation of anisotropic molecules of a liquid in a linearly polarized light beam, will be considered further.[e]

6.2.4 *Anharmonic oscillator*

In classical Lorentz's dispersion theory (Sec. 4.2), electrons in atoms are assumed to be harmonic oscillators. It is natural to consider the optical nonlinearity of matter in terms of the anharmonic oscillator model. Let a particle be in the potential

$$\mathcal{V}(x) = m\omega_0^2 x^2/2 - m\eta x^3/3 - eEx, \tag{6.69}$$

[e]Editors' note: The same mechanism is used in the atom optics, namely, in making atomic beams scattered by an optical lattice.

where e is the particle charge and η is a small parameter defining how much the shape of the potential well differs from a parabolic one. (For simplicity, we consider a one-dimensional model.) From (6.69), we find the equation of motion

$$\hat{D}x \equiv \ddot{x} + 2\gamma\dot{x} + \omega_0^2 x = eE/m + \eta x^2. \tag{6.70}$$

Let the external field E be a bi-harmonic one. We search the stationary solution as a perturbative series,

$$x(t) = \sum_m x^{(m)}(t), \; x^{(m)} \sim \eta^{m-1}E^m, \tag{6.71}$$

where $m = 1, 2, 4, 8, \ldots$. Substituting (6.71) into (6.70) and setting equalities between terms of the same order of magnitude, we find the solution in the form of iterations,

$$\hat{D}x^{(1)} = eE/m, \; \hat{D}x^{(2m)} = \eta x^{(m)2}. \tag{6.72}$$

In the first order,

$$\begin{aligned} x^{(1)} &= \mathrm{Re}(x_1 e^{-i\omega_1 t} + x_2 e^{-i\omega_2 t}), \\ x_n &= \alpha(\omega_n)E_n/e, \; \alpha(\omega) = e^2/mD(\omega), \end{aligned} \tag{6.73}$$

where

$$D(\omega) \equiv \omega_0^2 - \omega^2 - 2i\gamma\omega = D^*(-\omega).$$

In the second order, according to (6.72), there are components of $x(t)$ with frequencies $0, 2\omega_1, 2\omega_2, \omega_1 \pm \omega_2$. Consider the response x_3 at the frequency $\omega_3 \equiv \omega_1 + \omega_2$. It follows from (6.72) at $m = 1$ that

$$x_3 = \eta x_1 x_2/D(\omega_3) \equiv \beta(\omega_3 = \omega_2 + \omega_1)E_2E_1/e,$$

where

$$\beta(\omega_3 = \omega_2 + \omega_1) = \eta e^3/m^2 D(\omega_3)D(\omega_2)D(\omega_1). \tag{6.74}$$

At $\omega_1 = \pm\omega_2$, β contains an additional factor of $1/2$ (see (6.13)). According to (6.14), (6.15), (6.16), polarisability β is invariant to the permutation of the last two arguments, as well as to a simultaneous change in the signs of all three frequencies and the imaginary unity. However, according to (6.74), the first argument can be interchanged with the second or the third ones only in the absence of a resonance at one of the frequencies, with $|\omega_n - \omega_0| \gg \gamma$ (compare with (6.21)). From the polarisability β of a single oscillator one can pass to the susceptibility $\chi^{(2)}$ of the medium by multiplying the polarisability by the density N of the particles.

Thus, *the model of an anharmonic oscillator predicts a dramatic increase in the quadratic polarisability β (by a factor of $Q \equiv \omega_0/2\gamma$) in the vicinity of* *resonances, where one of its three arguments is close to the eigenfrequency* ω_0. In this case, β becomes a complex value. From the comparison of (6.72) and (6.73), it follows that β scales as the product of the linear susceptibilities at the corresponding frequencies,

$$\beta(\omega_3 = \omega_2 + \omega_1) \sim \eta\alpha(\omega_3)\alpha(\omega_2)\alpha(\omega_1). \tag{6.75}$$

Note that the quadratic susceptibilities of many dielectric crystals in the transparency range (between the lattice and electron eigenfrequencies) satisfy the relation

$$\chi^{(2)} \sim n(\omega_3)n(\omega_2)n(\omega_1), \tag{6.76}$$

with the same scaling factor for different crystals (*Miller's empirical rule*).

Our model does not take into account the difference between the *actual (local)* field E_{loc} and the *macroscopic* field E, which is averaged over the atomic inhomogeneities. According to Lorentz, in a cubic crystal $E_{loc}/E = (\epsilon + 2)/3$. (This correction is only valid for dielectrics, while in metals and semiconductors, $E_{loc} = E$.) Polarization P of the medium, caused by a given external polarization P_{ext}, is also $(\epsilon+2)/3$ times as large as P_{ext}. As a result, the quadratic susceptibility $\chi^{(2)}$, defined with respect to macroscopic parameters, is

$$\chi^{(2)} = \frac{\epsilon(\omega_3)+2}{3}\frac{\epsilon(\omega_2)+2}{3}\frac{\epsilon(\omega_1)+2}{3}\chi^{(2)}_{loc}, \tag{6.77}$$

where $\chi^{(2)}_{loc}$ is defined for local parameters. In non-cubic crystals, this correction is a tensor. Equations (6.75)–(6.77) indicate that *there is a close connection between the linear and nonlinear properties of a medium.*

Let a displacement $x = a_0$ correspond to the case where the linear part of the force in (6.70), $eE_0 = m\omega_0^2 a_0$, is equal to the nonlinear one, $\eta m a_0^2$. (Here, E_0 is a typical field keeping the charge near the equilibrium position.) Then, $\eta = \omega_0^2/a_0$, and for $\gamma, \omega_n \ll \omega_0$ we have an estimate

$$E_{NL} \equiv \alpha/\beta \sim m\omega_0^4/\eta e \sim m\omega_0^2 a_0/e = E_0. \tag{6.78}$$

Thus, the ratio of the quadratic polarization and the linear one is on the order of E_1/E_0 (compare with (6.42)). For a hydrogen atom, a_0 should be understood as Bohr's radius, $\hbar^2/me^2$, and ω_0, as the ionization edge frequency, $e^2/2\hbar a_0$. Then, $E_0 = e/2a_0^2 \approx 10^7$ G.

The cubic potential (6.69), according to (6.72), creates only even harmonics, $2\omega, 4\omega, 8\omega, \ldots$. For the formation of odd harmonics, it is necessary that the potential has a term $\sim x^4$. Note that even at $E = 0$, the potential (6.69) is not centrally symmetric: it changes its sign under the coordinate inversion, $x \to -x$. It is useful to consider a three-dimensional model using the potential [Akhmanov (1964)]

$$\mathcal{V}(\boldsymbol{r}) = m\omega_{0\alpha}^2 x_\alpha x_\alpha/2 - \eta_{\alpha\beta\gamma} m x_\alpha x_\beta x_\gamma. \tag{6.79}$$

6.2.5 Raman anharmonicity

Spontaneous Raman effect, or *spontaneous Raman scattering* (SpRS)[f] was discovered by Raman and, independently, by Mandelshtam and Landsberg, in 1928, much ahead of the advent of lasers. The corresponding stimulated effect (StRS) was first observed only in 1962.

In SpRS, monochromatic light (*the pump*) incident on the matter polarizes molecules with an optical frequency ω_1. As a result, the molecules acquire the dipole moment

$$d(t) = \alpha_1 E_1 \cos\omega_1 t, \tag{6.80}$$

where, for simplicity, we assume the linear polarisability of a molecule $\alpha_1 \equiv \alpha(\omega_1)$ to be a real scalar. Emission of molecular dipoles (6.80) leads to the *Rayleigh scattering*.

Let us now take into account the intra-molecular oscillations of the nuclei with the eigenfrequency $\Omega_0 \ll \omega_1$, which are excited due to collisions. Oscillations of the nuclei near the equilibrium positions, $Q(t)$, slowly modulate the electron 'cloud' surrounding them. In this case, all electronic parameters of the molecule are modulated as well, including its optical polarisability, $\alpha(t) = (\partial\alpha/\partial Q)Q(t)$. This picture is based on the so-called *adiabatic approximation*, which implies that the electron eigenfrequencies ω_0 much exceed Ω_0 (usually, $\omega_0/2\pi c \sim 10^5$ cm^{-1}, $\Omega_0/2\pi c \lesssim 10^3$ cm^{-1}). With the oscillations of the nuclei taken into account, (6.80) takes the form of an amplitude-modulated oscillation,

$$d(t) = \left(\alpha_1 + \frac{\partial\alpha}{\partial Q} Q_0 \cos\Omega_0 t\right) E_1 \cos\omega_1 t, \tag{6.81}$$

where Q_0 is the amplitude of the nuclei oscillations.

As a result, the radiation scattered by the dipoles contains, in addition to the '*carrier*' (Rayleigh's) frequency ω_1, two sidebands: the *Stokes* one, $\omega_1 - \Omega_0$, and the *anti-Stokes* one, $\omega_1 + \Omega_0$. In the case of a multi-atom molecule, the Raman spectrum contains its normal oscillations (some oscillations do not influence α due to symmetry). With an account for the anisotropy of α and the rotations of molecules, the induced dipole moment $d(t)$ will be also modulated by the typical rotational frequencies.

This modulation approach is based on a single nonlinear parameter, $\partial\alpha/\partial Q$, introduced by Placzek, and provides an explicit description of the spontaneous effect. (Here, the term 'spontaneous' relates to the field components with

[f]Editors' note: in the original text, the Russian term is used, which is '*spontaneous combination scattering*'.

the frequencies $\omega_1 \pm \Omega_0$, which are absent in the incident field and appear 'spontaneously'.)

In order to describe *stimulated* nonlinear effects caused by the parametric coupling between nuclei and electrons, it is useful to consider the model of two coupled oscillators, with the potential

$$\mathcal{V}(x, Q) = m\omega_0^2 x^2/2 + M\Omega_0^2 Q^2/2 - \eta x^2 Q - eEx. \tag{6.82}$$

Here, x and ω_0 are the coordinate and the eigenfrequency of the electron, Q and Ω_0 are the corresponding values for the nuclei, η is the coupling parameter scaling as $\partial\alpha/\partial Q$. It follows from (6.82) that

$$\ddot{x} + 2\gamma\dot{x} + \omega_0^2 x = eE/m + 2\eta Qx/m, \tag{6.83}$$

$$\ddot{Q} + 2\Gamma\dot{Q} + \Omega_0^2 Q = \eta x^2/M. \tag{6.84}$$

This model was proposed for the description of StRS by Platonenko and Khokhlov in 1964. According to (6.84), the force acting on the the nuclei scales as the square of the electron displacement; therefore, *the nuclei oscillations will be enhanced in the case where the difference between the two field frequencies is close to* Ω_0. Let the incident field be a biharmonic one, and $\omega_1 - \omega_2 \equiv \Omega \sim \Omega_0$.

In the linear approximation in the field, $Q^{(1)} = 0$, and

$$x_n^{(1)} = eE_n/mD_n, \; D_n \equiv \omega_0^2 - \omega_n^2 - 2i\gamma\omega_n, \; n = 1, 2. \tag{6.85}$$

In the expression for the force, ηx^2, we only leave the resonance terms with the frequency Ω scaling as $x_1^{(1)} x_2^{(1)*}$, then the amplitude of stimulated molecular oscillations with the frequency Ω is

$$Q_\Omega = \frac{\eta x_1^{(1)} x_2^{(1)*}}{2M(\Omega_0^2 - \Omega^2 - 2i\Gamma\Omega)} = \frac{(\eta/M)(e/m)^2}{2D_0 D_1 D_2^*} E_1 E_2^*. \tag{6.86}$$

Thus, *an optical biharmonic field with a proper frequency difference 'excites' the intra-molecular nuclei oscillations through the electron shell.* These oscillations, coherent with the incident light, add to the equilibrium thermal ones and cause additional incoherent scattering at the anti-Stokes frequency, $\omega_3 = \omega_1 + \Omega = 2\omega_1 - \omega_2$, and at the second Stokes frequency, $\omega_4 \equiv \omega_2 - \Omega = 2\omega_2 - \omega_1$. In addition, cubic polarization is induced at the initial field frequencies, leading to the amplification of the field E_2 with the lower frequency and attenuation of the field E_1 with the higher frequency. It is namely the effect of *Raman amplification* that causes stimulated Raman scattering.

By substituting (6.85) and (6.86) into (6.83), (6.84), we find

$$
\begin{aligned}
x_1^{(3)} &= \frac{\eta x_2^{(1)} Q_\Omega}{mD_1} = \frac{\eta^2 e^3/2Mm^4}{D_0 D_1^2 D_2 D_2^*}|E_2|^2 E_1, \\
x_2^{(3)} &= \frac{\eta x_1^{(1)} Q_\Omega^*}{mD_2} = \frac{\eta^2 e^3/2Mm^4}{D_0 D_1 D_1^* D_2^2}|E_1|^2 E_2, \\
x_3^{(3)} &= \frac{\eta x_1^{(1)} Q_\Omega}{mD_3} = \frac{\eta^2 e^3/2Mm^4}{D_0 D_1^2 D_2^* D_3} E_1^2 E_2^*.
\end{aligned}
\tag{6.87}
$$

After multiplying $x_n^{(3)}$ by eN, we find the cubic polarisability. As a rule, in experiment the frequencies of the field ω_n ($n = 1, 2, 3$) are much less than the electron transition frequencies ω_0; therefore, $D_n \approx \omega_0^{-2}$ (*non-resonance* RS). Within this approximation, (6.87) leads to

$$
\begin{aligned}
\chi^{(3)}(\omega_1 = \omega_1 - \omega_2 + \omega_2) &= \chi^{(3)}(\omega_2 = \omega_2 - \omega_1 + \omega_1)^* \\
&= \chi^{(3)}(\omega_3 = 2\omega_1 - \omega_2) \\
&= C/[\Omega_0^2 - (\omega_1 - \omega_2)^2 - 2i\Gamma(\omega_1 - \omega_2)],
\end{aligned}
\tag{6.88}
$$

where $C \equiv \eta^2 e^4 N/Mm^4\omega_0^8$

These nonlinear susceptibilities describe, respectively, *Raman absorption* (since $\chi^{(3)}(\omega_1)'' > 0$ at $\omega_1 > \omega_2$), *Raman amplification* ($\chi^{(3)}(\omega_2)'' < 0$) and *coherent anti-stokes Raman scattering* (CARS) with the intensity scaling as $|\chi^{(3)}(\omega_3)|^2 I_1 I_2$. Besides, it follows from (6.88) that in the presence of a monochromatic wave with a fixed frequency ω_L within the transparency range, another monochromatic wave, with a variable frequency ω, will have resonance dispersion in two regions, $\omega_L \pm \Omega_0$ (Fig. 6.3). The width 2Γ of these resonances is determined by the decay rate of molecular oscillations, and this *Raman (induced) dispersion* is anomalous in the Stokes range.

Let us find the relation between the nonlinear parameters η and $\partial\alpha/\partial Q$. We substitute $Q = Q_0 \cos \Omega_0 t$ into Eq. (6.83). In the first order in η, it leads to the relation $x_2 = \eta e E_1 Q_0/m^2 D_1 D_2$. Comparing it with (6.81), we get

$$
\frac{\partial\alpha}{\partial Q} = \frac{2e^2}{m^2 D_1 D_2} \approx \frac{2e^2}{m^2\omega_0^4}\eta. \tag{6.89}
$$

By means of the two-oscillator model, one can also describe the spontaneous Raman scattering. To this end, the right-hand side of (6.84) should be replaced by the *stochastic Langevin force* $f(t)$ causing thermal (and quantum) fluctuations of $Q(t)$. This force is delta-correlated, and one can find its spectral density by setting

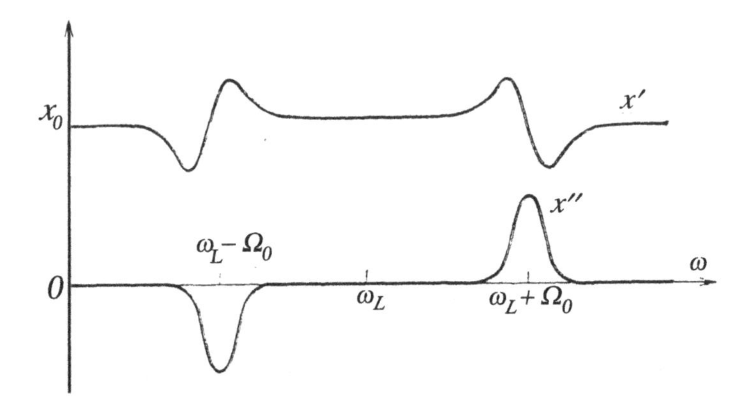

Fig. 6.3 Raman susceptibility. Due to the effect of the pump with the frequency ω, the susceptibility of matter acquires additional resonances at frequencies $\omega_L \pm \Omega_0$, where Ω_0 are the eigenfrequencies of the molecules. It is important that the Stokes resonance has negative losses (bottom left) and anomalous dispersion of the refractive index (top left).

equality between the fluctuation energy Q and the equilibrium energy of the oscillator. Another method of describing SpRS is based on the Raman analogue of the FDT (Sec. 7.7), stating that the polarization fluctuations of matter at frequency ω are determined by the imaginary part of the cubic susceptibility [Fain (1972); Klyshko (1980)],

$$\langle P^*(\omega)P(\omega')\rangle = (\hbar/\pi)\delta(\omega - \omega')\mathcal{N}(-\Omega)\chi^{(3)}(\omega = \omega - \omega_L + \omega_L)''|E_L|^2, \quad (6.90)$$

where

$$\mathcal{N}(\Omega) \equiv [\exp(\hbar\Omega/\kappa T) - 1]^{-1} = -\mathcal{N}(-\Omega) - 1, \; \Omega \equiv \omega_L - \omega. \quad (6.91)$$

Here, at $\Omega < 0$ (the anti-Stokes range), $\mathcal{N}$ has the meaning of the equilibrium number of phonons $\mathcal{N}_0$, while at $\Omega > 0$ (the Stokes range), $\mathcal{N} = -(\mathcal{N}_0 + 1)$. In the last expression, the unity describes the quantum fluctuations of the nucleus coordinate Q, which cause the Stokes scattering even at $T = 0$, when the anti-Stokes scattering is absent. The elementary process corresponding to the Stokes scattering is a two-photon one. It includes the annihilation of a pump photon and the creation of a Stokes photon and a phonon (Fig. 6.1(g)).

Certainly, the two-oscillator model, similarly to the anharmonic-oscillator one, is only qualitative. A quantitative calculation of the susceptibility, even the linear one, is very complicated and requires the knowledge of the wave functions and eigenfrequencies of the system (see below).

Inelastic scattering can be also due to the excitation of other degrees of freedom of the medium, for instance, electronic ones. In this case, the radiation frequency changes by a quantity equal to the frequency of some electron transition in an atom or a molecule, $\omega_1 - \omega_2 = \omega_{mn} \equiv (\mathcal{E}_m - \mathcal{E}_n)/\hbar$. If the incident field contains two frequencies such that $\omega_1 + \omega_2 = \omega_{mn} > 0$ and the molecule is at the ground level n, then two photons can be absorbed simultaneously. In the inverse process, the excited molecule emits two photons, spontaneously or via stimulated emission.

In macroscopic matter, light interacts not only with local internal oscillations of the particles, but also with the *collective excitations*, for instance, with acoustic, temperature, spin, plasma waves, and with the oscillations of the molecules' orientations.

The equilibrium chaotic part of these waves modulate the refractive index (see (6.66)), and the corresponding sidebands $\omega_2 = \omega_1 \pm \Omega$ appear in the spectrum of the scattered light. The scattering can be viewed as resulting from the diffraction of the incident light by a traveling grating formed by waves of pressure, temperature, and so on. From the quantum viewpoint, a photon $\hbar\omega_1$ of the incident light (the pump) gets scattered with a simultaneous birth or annihilation of a matter excitation quantum $\hbar\Omega$ (a phonon, a magnon, a plasmon, an exciton, a polariton etc.). For the scattering by propagating excitations, it is typical that the modulation frequency depends on the observation direction, i.e., on the angle of scattering, $\Omega = \Omega(\vartheta)$. This dependence follows from the diffraction Bragg's condition or, in other words, the *phase-matching* condition (the momentum conservation law), $\boldsymbol{k}_1 - \boldsymbol{k}_2 \pm \boldsymbol{q} = 0$, and the dispersion dependence for the scattering wave, $\boldsymbol{q} = \boldsymbol{q}(\Omega)$, where $\boldsymbol{q}$ is the wavevector of the matter excitation. The effect of phase matching on the Raman anharmonicity can be formally taken into account by assuming the cubic susceptibility to depend not only on the frequencies but also on the wave vectors (*spatial dispersion*).

For describing stimulated scattering by acoustic waves (MBS, *Mandelshtam-Brillouin scattering*) and other collective excitations, one should take into account the non-equilibrium coherent part of these excitations caused by the biharmonic pump. It is clear from (6.64) how sound can be excited by light: due to electrostriction, there appears a source of alternating pressure in the matter, with the difference frequency Ω,

$$\Delta p(\boldsymbol{r}, t) = (1/2)\chi^{(1)} \left[\sum_n (E_n/2) \exp(-i\omega_n t) + \text{c.c.} \right]^2$$

$$= (1/8)\chi^{(1)} E_1 E_2^* \exp(-i\Omega t) + \ldots \qquad (6.92)$$

This source creates density waves $\Delta\rho$, propagating with the speed of sound v. If the pump waves are plane ones, then $\Delta\rho \sim \exp[i(\boldsymbol{k}_1 - \boldsymbol{k}_2) \cdot \boldsymbol{r}]$, and the stimulated sound wave will have the maximal intensity at $|\boldsymbol{k}_1 - \boldsymbol{k}_2| = q = \Omega/v$. This phase-matching condition 'chooses' from the continuous spectrum of acoustic excitations, spanning a range from zero to approximately 10^{11} Hz, one (or two, considering the difference between v for transverse and longitudinal waves in amorphous solids) discrete component, with the frequency

$$\Omega = v|\boldsymbol{k}_1 - \boldsymbol{k}_2| \approx 2vk\sin(\vartheta/2). \tag{6.93}$$

The width of this spectral component is determined by the sound absorption coefficient.

6.2.6 *Temperature anharmonicity*

It follows from (6.93) that at $\vartheta \neq 0$, light scattering by acoustic waves, i.e., waves of pressure, Δp, and density, $\Delta\rho$, is inelastic, $\Omega \sim v \neq 0$. (To be precise, the maximum of the scattering corresponds to $\Omega \neq 0$.) According to (6.93), the scattering can be centered around $\Omega = 0$ only if it is due to non-propagating excitations, for which $v = 0$ or which decay sufficiently fast. Such scattering can be caused by temperature fluctuations, ΔT, or entropy fluctuations, $\Delta S \sim \Delta T$, as well as by concentration fluctuations, ΔC, in mixtures and solutions. The values $x \equiv p, T, C, \ldots$ (or $\rho, S, \ldots$) are thermodynamic parameters describing the macroscopic state of a medium, and their oscillations, both thermal ('spontaneous') and stimulated (coherent) ones, break the optical homogeneity of the medium ($\Delta n = (\partial n/\partial x)\Delta x$) and lead to the scattering of light, spontaneous or stimulated. All these types of scattering occur with a frequency shift that is small compared with the case of scattering by molecular oscillations and are called the *Rayleigh, or molecular, scattering* [Fabelinsky (1965)]. The last term emphasizes the difference from scattering by macroscopic inhomogeneities, such as dust particles and other objects.

The integral intensity of spontaneous scattering by the parameter x scales as the mean square $\overline{\Delta x^2}$, and it can be calculated thermodynamically. At the same time, the spectral distribution od the intensity is determined by the kinetic equations describing the evolution of the field $x(\boldsymbol{r}, t)$. For instance, for $x = T$ the kinetics is given by the diffusion equation,

$$\dot{T} - a\nabla^2 T = \mathcal{P}/c_p\rho, \tag{6.94}$$

where a is the temperature conductance, $\mathcal{P}$ is the power of external heat sources per unit volume, c_p is the specific heat capacity at constant pressure. (Strictly speaking, one should take into account not only ΔT but also the simultaneous variation of pressure, Δp, due to heating; here, for simplicity, we ignore the relation between temperature and pressure waves.)

At $\mathcal{P} = 0$, (6.94) describes spontaneous temperature scattering, and its solution can be represented as a sum of plane waves exponentially decaying in time,

$$\Delta T(\boldsymbol{r}, t) = \sum_q T_q e^{i\boldsymbol{q}\cdot\boldsymbol{r} - \gamma t}, \tag{6.95}$$

with $\gamma \equiv aq^2$. Similarly to the case of scattering by acoustic waves, the scattering of light with the wavevector $\boldsymbol{k}_1$ in the direction $\boldsymbol{k}_2$ is caused by the 'temperature grating' with $\boldsymbol{q} = \pm(\boldsymbol{k}_1 - \boldsymbol{k}_2)$. However, according to (6.95), this grating is not moving, and diffraction by it results in an *elastic* (unshifted) line in the spectrum of the scattered light, with the bandwidth

$$\Delta\omega = 2\gamma = 8ak^2 \sin^2(\vartheta/2). \tag{6.96}$$

In liquids, $\Delta\omega \sim 10^8$ s^{-1} at $\vartheta = 90°$. The inverse linewidth of the temperature (entropy) scattering, $\tau_T \equiv 1/\gamma = \lambda\!\!\!^{-2}/a$, has the meaning of the typical time of temperature diffusion by a distance of one wavelength $\lambda\!\!\!^- \equiv |\boldsymbol{k}_1 - \boldsymbol{k}_2|^{-1}$, i.e., the relaxation time of the temperature grating.

The mechanism of the stimulated temperature scattering (StTR) and the corresponding anharmonicity is evident in the presence of some absorption (StTR-2). Indeed, in the case of a biharmonic field, the external force in (6.94) has a variable component,

$$\mathcal{P}(\boldsymbol{r}, t) = \omega\chi''\overline{E^2(\boldsymbol{r}, t)} = \omega\chi''\mathrm{Re}\boldsymbol{E}_1 \cdot \boldsymbol{E}_2 e^{i(\boldsymbol{q}\cdot\boldsymbol{r} - \Omega t)}, \tag{6.97}$$

which creates a temperature wave,

$$\Delta T = \mathrm{Re} T_\Omega e^{i(\boldsymbol{q}\cdot\boldsymbol{r} - \Omega t)}. \tag{6.98}$$

Its amplitude can be found by substituting (6.97) and (6.98) into (6.94),

$$T_\Omega = \frac{\omega\chi''/c_p\rho}{\gamma - i\Omega} \boldsymbol{E}_1 \cdot \boldsymbol{E}_2^*. \tag{6.99}$$

Here, $\omega \approx \omega_1 \approx \omega_2 \gg |\Omega|$ and $\chi'' \equiv \chi''(\omega)$.

The traveling coherent wave of temperature (6.98) modulates the susceptibility χ (mainly due to the density decrease caused by temperature expansion),[g]

$$\Delta\chi = \left(\frac{\partial\chi}{\partial T}\right)_p \Delta T \approx -\chi\frac{\Delta T}{T}, \tag{6.100}$$

[g]In an ideal gas, $N = p/\kappa T$, therefore, $\chi = \alpha p/\kappa T$ and, if the dependence of α on T is neglected, $(\partial\chi/\partial T)_p = -\chi/T$.

therefore, the wave of the susceptibility has the amplitude

$$\chi_\Omega = \frac{\omega\chi''/c_p\rho}{\gamma - i\Omega}\left(\frac{\partial\chi}{\partial T}\right)_p \boldsymbol{E}_1 \cdot \boldsymbol{E}_2^*. \tag{6.101}$$

As a result, cubic polarization emerges, with the frequencies $\omega_1, \omega_2, \omega_3 \equiv 2\omega_1 - \omega_2, \omega_4 \equiv 2\omega_2 - \omega_2$,

$$\begin{aligned}\boldsymbol{P}^{(3)}(t) &= \Delta\chi(t)\boldsymbol{E}(t) \\ &= \mathrm{Re}\chi_\Omega(\boldsymbol{E}_2 e^{-i\omega_1 t} + \boldsymbol{E}_1^* e^{i\omega_2 t} + \boldsymbol{E}_1 e^{-i\omega_3 t} + \boldsymbol{E}_2^* e^{i\omega_4 t})/2.\end{aligned} \tag{6.102}$$

Let $\boldsymbol{E}_1, \boldsymbol{E}_2$ be parallel to the x axis, then, from the definition of the cubic susceptibility we find that

$$\chi_{xxxx}^{\bar{1}21\bar{2}} = \chi_{xxxx}^{2\overline{12}1} = \chi_{xxxx}^{\bar{3}11\bar{2}} = \chi_{xxxx}^{4\overline{22}1} = \frac{\omega\chi''}{2\rho c_p[aq^2 + i(\omega_2 - \omega_1)]}\left(\frac{\partial\chi}{\partial T}\right)_p. \tag{6.103}$$

Note that StRS-2 has an interesting feature: here, *the anti-Stokes components are amplified* (compare with (6.88)), i.e., the energy of the field is transferred, in the course of propagation, from the low-frequency components into the high-frequency ones, since $(\frac{\partial\chi}{\partial T})_p < 0$.

In order to estimate the temperature anharmonicity, assume that

$$\left(\frac{\partial\chi}{\partial T}\right)_p = -\frac{\chi}{T}, \ c_p\rho = \frac{5}{2}\kappa N, \ \omega\chi'' \equiv \frac{n^2}{4\pi\tau_E}, \ aq^2 \equiv \frac{1}{\tau_T}, \tag{6.104}$$

where $\tau_E = n/\alpha c$ is the relaxation time of the field. Then, at $|\omega_1 - \omega_2|\tau_T \ll 1$,

$$E_{NL}^2 \equiv \frac{\chi^{(1)}}{\chi^{(3)}} \approx \frac{40\pi U\tau_E}{3n^2\tau_T}, \tag{6.105}$$

where $U = 3\kappa TN/2$ is the internal energy density. Thus, if $\tau_T = \tau_E$, which at $\tau_E = 10^{-8}$ s corresponds to $\alpha = 0.003$ cm^{-1}, then the nonlinear polarization becomes equal to the linear one when the field energy $n^2E_1^2/8\pi$ equals the thermal energy of the matter. Note that if one passes from the temperature conductance a to the heat conductance $\lambda = a\rho c_p$, then E_{NL}^2 can be represented as $8\pi\lambda q^2 T\tau_E$.

6.2.7 *Electrocaloric anharmonicity*

Temperature anharmonicity can take place even in a completely transparent material, due to the *optical electrocaloric* effect. The corresponding stimulated scattering is called StTS-1.

Consider the simplest model describing the effect of the electric field on the temperature of a non-absorbing material. After the field (continuous or alternating) is switched on, the energy levels of the molecules get shifted due to the Stark effect, and their populations do not correspond any more to the temperature of the thermostat. (The role of the thermostat is usually played by the translational and rotational degrees of freedom of the molecules, or vibrations of the crystal lattice.) During the relaxation time T_1, populations are re-distributed, which is accompanied by the change of the thermostat energy.[h] As a result, the thermostat temperature is changed. Note that an analogous *magnetocalorical effect* is used for obtaining ultra-low temperatures (*adiabatic demagnetization*). A more rigid explanation follows from the definition of the temperature for a closed system (microcanonical ensemble),

$$1/T \equiv \partial S/\partial U \equiv \kappa\partial(\ln g)/\partial U, \tag{6.106}$$

where S is the entropy, U the internal energy and $g(U)$ the density of energy states. The latter depends on the configuration of the levels and hence changes after the field is turned on.

Let us estimate the contribution of the electrocaloric effect in the temperature anharmonicity. According to (4.35), when a dielectric is polarized, its thermodynamic potentials get an increase $v = -\chi|E_1|^2/4$ per unit volume. Let us choose T and ρ as independent parameters, then the entropy is determined in terms of the temperature derivative of the Gibbs' potential $\Phi(T, p)$; therefore, the variation of S due to the polarization is

$$\Delta S = -\left(\frac{\partial \Delta\Phi}{\partial T}\right)_p = \frac{1}{4}\left(\frac{\partial \chi}{\partial T}\right)_p |E_1|^2. \tag{6.107}$$

After multiplying ΔS by T, we obtain the increase of the heat ΔQ, and after multiplying it by $-T/c_p\rho$, we find the temperature increase,

$$\Delta T = -\frac{T}{4c_p\rho}\left(\frac{\partial \chi}{\partial T}\right)_p |E_1|^2. \tag{6.108}$$

Comparing this with (6.94), we see that the role of the absorbed power in a transparent material is played by

$$\mathcal{P}_{equiv} = -\frac{d\Delta Q}{dt} = -\frac{T}{4}\left(\frac{\partial \chi}{\partial T}\right)_p \frac{d|E_1|^2}{dt}. \tag{6.109}$$

In the case of a biharmonic field, we obtain

$$\mathcal{P}_{equiv}(\boldsymbol{r}, t) = -\frac{\Omega T}{2}\left(\frac{\partial \chi}{\partial T}\right)_p \operatorname{Im} \boldsymbol{E}_1 \cdot \boldsymbol{E}_2^* e^{i(\boldsymbol{q}\cdot\boldsymbol{r}-\Omega t)}. \tag{6.110}$$

---

[h]Here, the finite heat capacity of the external degrees of freedom is taken into account.

Comparing (6.110) and (6.97), we find the ratio of the electrocaloric anharmonicity and the anharmonicity (6.103) due to the dissipation,

$$\frac{\chi_{ec}^{(3)}}{\chi_{ab}^{(3)}} \sim \frac{\Omega T}{2\omega\chi''}\left(\frac{\partial\chi}{\partial T}\right)_p = \frac{2\pi\Omega T}{\alpha c n}\left(\frac{\partial\chi}{\partial T}\right)_p . \tag{6.111}$$

Hence, it follows that the absorption coefficient equivalent to the electrocaloric effect is not high,

$$\alpha_{equiv} \equiv \frac{2\pi\Omega T}{cn}\left|\frac{\partial\chi}{\partial T}_p\right| \sim 10^{-4}\ \mathrm{cm}^{-1}, \tag{6.112}$$

where we assumed $\Omega = 10^8\ \mathrm{s}^{-1}$, $T = 300$ K, $(\partial\chi/\partial T)_p = -10^{-4}\ \mathrm{K}^{-1}$.

Finally, note that due to the relation between the density and temperature waves, electrostriction also contributes into the temperature anharmonicity (see [Apanasevich (1977)]).

6.2.8 *Orientation anharmonicity*

As we have already mentioned, rotation of anisotropic molecules in a gas also modulates the scattered light, which leads, due to the quantization of the rotational motion, to the appearance of discrete sidebands near the Rayleigh (non-shifted) and Raman lines in the spectrum of spontaneous scattering. However, if the density of the particles is high, a molecule cannot make a full rotation during the orientation relaxation time τ; therefore, in liquids the rotational lines overlap and the Rayleigh line acquires a broad 'pedestal' spanning tens of inverse centimeters, the so-called *Rayleigh wing*. Light scattering by orientation fluctuations of molecules is also called *anisotropic (depolarized) scattering*. In the framework of the macroscopic description, one can assume that anisotropic scattering is caused by the symmetry breaking of the medium (which is otherwise isotropic), i.e., the scattering is due to the symmetry fluctuations. A schematic shape of the scattered spectrum, with an account for the matter excitations considered above, is shown in Fig. 6.4.

Interaction of light with the orientation motion of molecules is another source of optical anharmonicity. This type of anharmonicity manifests itself in the *Kerr effect*, discovered as early as in the 19th century, in which $\Delta n \sim E_0^2$, the *optical Kerr effect* and *self-focusing*, with $\Delta n \sim |E_1|^2$, and in *stimulated Rayleigh-wing scattering*.

Let us estimate the contribution of the orientation anharmonicity in the cubic susceptibility. Consider first a non-polar molecule. In a field $\boldsymbol{E}(t)$ it acquires induced dipole moment $\boldsymbol{d}(t) \approx \alpha(\omega) \cdot \boldsymbol{E}(t)$ (we neglect the absorption) and the

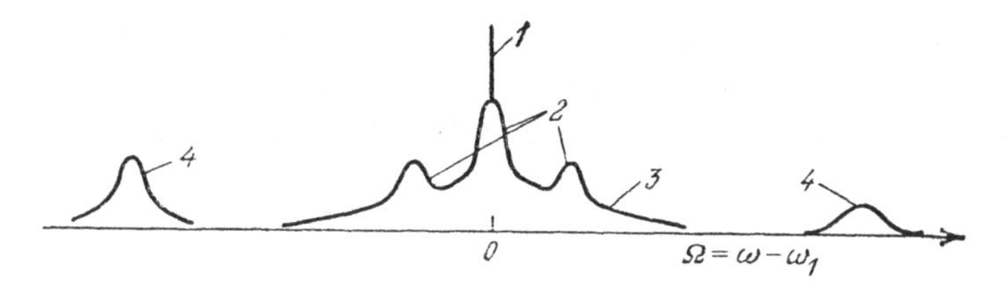

Fig. 6.4 Main types of scattering, the corresponding nonlinearities and typical frequencies: 1, temperature (entropy) scattering by temperature fluctuations is related to the electrocaloric anharmonicity ($\Delta\Omega \sim 10^8$ s^{-1}); 2, Mandelshtam-Brillouin scattering ($\Delta\Omega \sim 10^{10}$ s^{-1}) by pressure fluctuations is related to striction anharmonicity; 3, the Rayleigh wing is caused by anisotropy fluctuations and is related to the orientation anharmonicity ($\Delta\Omega \sim 10^{11}$ s^{-1}); 4, Raman scattering by the internal vibrations of molecules ($\Delta\Omega \sim 10^{14}$ s^{-1}). The first three types are called *molecular*, or *Rayleigh, scattering*.

time-averaged energy (see (4.35))

$$\mathcal{V} = -\mathrm{Re}\boldsymbol{E}^{(-)} \cdot \alpha \cdot \boldsymbol{E}^{(+)}. \tag{6.113}$$

Hence, *an anisotropic molecule tends to turn with respect to the field to maximize its polarisability*. However, in equilibrium matter, rotation of molecules by the field is hindered by their interaction with the neighbors: relaxation processes restore the equilibrium state with chaotic orientations of the molecules, which the field tends to order. Competition between the field and the thermal motion results in a dynamic equilibrium, with the degree of orientation being on the order of $\mathcal{V}/\kappa T$. In this case, the liquid becomes birefringent, similarly to a uniaxial crystal with the axis parallel to $\boldsymbol{E}$ (if the field is linearly polarized). This phenomenon is called the *optical Kerr effect*.

Let the polarisability anisotropy be $\Delta\alpha(\omega)$. For instance, for a linear molecule, $\Delta\alpha = \alpha_{\|} - \alpha_{\perp}$. Then the susceptibility variation $\Delta\chi(\omega_2)$ at a frequency ω_2 induced by the field at a frequency ω_1 will be equal, on the order of magnitude, to $\Delta\alpha(\omega_2)N$ times the degree of orientation,

$$\Delta\chi \approx \Delta\alpha(\omega_1)\Delta\alpha(\omega_2)N|E_1|^2/\kappa T. \tag{6.114}$$

Thus, the cubic susceptibility for non-polar molecules can be estimated as

$$\chi^{(3)} \approx \Delta\alpha(\omega_1)\Delta\alpha(\omega_2)N/\kappa T. \tag{6.115}$$

Assuming that the anisotropy is high, $\Delta\alpha \approx \alpha = \chi^{(1)}/N$, we get

$$E_{NL}^2 \equiv \chi^{(1)}/\chi^{(3)} \approx \kappa T/\alpha. \tag{6.116}$$

Let $\alpha \approx a_0^3 \approx 10^{-24}$ cm^3 and $T = 300$ K, then $E_{NL} = 2 \cdot 10^5$ G, and if $\chi^{(1)} = 0.1$, then $\chi^{(3)} = 10^{-12}$ cm^3/erg (compare with (6.42), (6.43), (6.68)).

If a molecule has a constant dipole moment $\boldsymbol{d}_0$ and the orienting field is a static one or it varies slowly compared with the orientation relaxation time ($\tau \sim 10^{-12}$ s), then the effective energy is $\mathcal{V} = -\boldsymbol{d}_0 \cdot \boldsymbol{E}$. (The induced moment can be neglected in this case.) The degree of orientation will then scale as $(\mathcal{V}/\kappa T)^2$ since linear electrooptic effect is forbidden in liquids (Sec. 6.1). As a result,

$$\chi^{(3)}(\omega, -0, 0) \approx \Delta\alpha(\omega) N (d_0/\kappa T)^2 \sim 10^{-10}\text{cm}^3/\text{erg}, \tag{6.117}$$

$$E_{NL} \approx \kappa T/d_0 \sim 10^4\text{G}. \tag{6.118}$$

(Here, we assumed $d_0 = 1$ D.) Thus, the Kerr effect in polar liquids is much stronger than in non-polar ones.

Suppose now that the orienting field is an optical biharmonic one, with $\Omega \equiv \omega_1 - \omega_2 < 1/\tau$, then $\mathcal{V}$ and, correspondingly, the degree of orientation $\mathcal{V}/\kappa T$ will contain an alternating component with the frequency Ω. As a result, the polarization at the probe field frequency ω_3 (which, in principle, can coincide with ω_1 or ω_2) will be modulated, i.e., the matter will emit coherent field with the frequencies $\omega_3 \pm \Omega$. Thus, orientation anharmonicity provides a resonance contribution with the width $2/\tau$ to the cubic susceptibility $\chi^{(3)}(\omega_3, -\omega_2, \omega_1)$ at $(\omega_1 - \omega_2) \leq 1/\tau$. If the orientation relaxation is taken into account, this contribution becomes complex. Its imaginary part corresponds to the amplification of the ω_2 field, with the maximum at $\omega_2 = \omega_1 - 1/\tau$, and leads to the stimulated Rayleigh-wing scattering.

Orientation anharmonicity can be quantitatively described in terms of the molecular orientation distribution function, which is stationary at $\Omega\tau \ll 1$:

$$P(\boldsymbol{\theta}) = Ce^{-\mathcal{V}/\kappa T} = C\left[1 - \frac{\mathcal{V}}{\kappa T} + \frac{1}{2}\left(\frac{\mathcal{V}}{\kappa T}\right)^2 - \ldots\right], \tag{6.119}$$

where C is a normalization factor (depending, of course, on the temperature and the field), the energy $\mathcal{V}(\boldsymbol{\theta})$ is defined in (6.113) and $\boldsymbol{\theta}$ is a set of three angles determining the orientation of the molecule in a laboratory frame of reference (the Euler angles).

In the case of non-polar molecules, it is sufficient to do the expansion up to the linear in $1/\kappa T$ term,

$$C = P^{(0)}(1 + \langle\mathcal{V}\rangle^{(0)}/\kappa T), \;\; P(\theta) = P^{(0)}\left[1 - (\mathcal{V} - \langle\mathcal{V}\rangle^{(0)})/\kappa T\right], \tag{6.120}$$

where

$$\langle\mathcal{V}\rangle^{(0)} \equiv \int d^3\theta P^{(0)}\mathcal{V}, \;\; P^{(0)} \equiv 1/\int d^3\theta = 1/8\pi^2. \tag{6.121}$$

The angle-averaged induced dipole moment has the form

$$\langle \boldsymbol{d} \rangle = \langle \alpha(\theta) \rangle \cdot \boldsymbol{E} + \langle \boldsymbol{\beta}(\theta) \rangle : \boldsymbol{EE} + \ldots \tag{6.122}$$

The angular brackets denote averaging with the perturbed distribution function $P(\boldsymbol{\theta})$. If we neglect the intra-molecular anharmonicity ($\beta = 0$), then $\langle \boldsymbol{d} \rangle = \boldsymbol{\alpha} \cdot \boldsymbol{E}$, where

$$\begin{aligned} \boldsymbol{\alpha} \equiv \langle \alpha \boldsymbol{\theta} \rangle &\equiv \int d^3\theta P(\boldsymbol{\theta}) \alpha(\boldsymbol{\theta}) \\ &= \boldsymbol{\alpha}^{(0)} - (\langle \alpha(\boldsymbol{\theta}) \mathcal{V}(\boldsymbol{\theta}) \rangle^{(0)} - \boldsymbol{\alpha}^{(0)} \langle \mathcal{V} \rangle^{(0)}) / \kappa T, \end{aligned} \tag{6.123}$$

$$\alpha^{(0)}_{\alpha\beta} \equiv \int d^3\theta P^{(0)} \alpha_{\alpha\beta}(\boldsymbol{\theta}) = \delta_{\alpha\beta}(\alpha_{xx} + \alpha_{yy} + \alpha_{zz})/3. \tag{6.124}$$

Here, $\boldsymbol{\alpha}^{(0)}$ is the linear polarisability averaged over the equilibrium distribution function. Hence, we find the correction to polarisability,

$$\boldsymbol{\alpha} - \boldsymbol{\alpha}^{(0)} = \int d^3\theta [\alpha(\boldsymbol{\theta}) - \boldsymbol{\alpha}^{(0)}] \operatorname{Re} \boldsymbol{E}^{(-)} \cdot \alpha(\boldsymbol{\theta}) \cdot \boldsymbol{E}^{(+)} / 8\pi^2 \kappa T. \tag{6.125}$$

Here, $\boldsymbol{\alpha}(\boldsymbol{\theta})$ is the linear polarisability tensor in the laboratory frame of reference for a molecule with a given orientation $\boldsymbol{\theta}$.

At $\Omega\tau \gtrsim 1$ it is necessary to take into account the variation of the distribution function in time, $P = P(\boldsymbol{\theta}, t)$, which is described by the kinetic equation [Apanasevich (1977)] or a Focker-Planck-type equation. Within the exponential-relaxation approximation, these equations yield a usual dispersion dependence with a pole at $\omega_1 = \omega_2$,

$$\chi^{(3)}(\omega_2 = \omega_2 - \omega_1 + \omega_1) \sim 1/[1 + i(\omega_1 - \omega_2)\tau]. \tag{6.126}$$

Hence, the low-frequency (Stokes) component of the field is amplified, $\chi^{(3)''} < 0$, and this amplification is maximal at $|\Omega| = 1/\tau$.

6.2.9 °*Quantum theory of nonlinear polarization*

Nonlinear polarisabilities $\beta, \gamma, \ldots$ of molecules and the susceptibility of matter $\chi^{(2)} \approx N\beta$, $\chi^{(3)} \approx N\gamma, \ldots$ can be calculated similarly to the linear polarisability (Sec. 4.2), using the density-matrix equation with the phenomenological damping constants. However, it is easier to use the general formula (3.75) for the response $\langle f(t) \rangle$ of a quantum system to an external perturbation with the energy $\mathcal{V}(t)$; the damping should be added from general considerations at the final stage of the calculation.

We are interested in the stationary response of the system to a periodic perturbation; therefore, the lower integration limits in (3.75) should be chosen as $-\infty$.

The upper limits can be chosen as $+\infty$ provided that the causality is taken into account by introducing the step functions $\theta(t - t_1), \ldots, \theta(t_{k-1} - t_k)$ into the integrand. Let us introduce the damping, which is necessary for achieving a stationary regime and practically inevitable in any system (see Fig. 6.5),

$$\theta(t) = e^{-\epsilon t}\ (t > 0),\ \ \theta(t) = 0\ (t < 0), \tag{6.127}$$

where ϵ is some positive constant, which will be further replaced by ν_{mn}.

Assuming in (3.75) $f = d_\alpha$ and $\mathcal{V} = -\boldsymbol{d} \cdot \boldsymbol{E}$, for the induced dipole moment of order k we obtain (see also Ref. [Fain (1972)])

$$\begin{aligned}\langle d_\alpha(t)\rangle^{(k)} &= (i/\hbar)^k \int dt_1 \ldots \int dt_k\, \theta(t - t_1) \ldots \theta(t_{k-1} - t_k) \\ &\times \mathrm{Tr}\,\{\rho[\ldots[d'_\alpha(t), d'_{\alpha_1}(t_1)], \ldots, d'_{\alpha_k}(t_k)]\} E_{\alpha_1}(t_1) \ldots E_{\alpha_k}(t_k).\end{aligned} \tag{6.128}$$

Here, ρ is the equilibrium density operator and the operators $d'_\alpha(t)$ are considered in the interaction picture. The integrand in (6.128) depends on $k + 1$ time arguments, only k of them being independent, as one can easily verify. This tensor function is called the *response function* of the system or its *Green's function*. One also says that it is a *causal* Green's function and, at $k > 1$, a *multi-time* one.

Let the field have a discrete spectrum,

$$\boldsymbol{E}(t) = \sum_p \boldsymbol{E}_p \exp(-i\omega_p t)/2,\ \ p = \pm 1, \pm 2, \ldots, \tag{6.129}$$

then the integration in (6.128) is elementary. For instance, for $k = 1$ we find

$$\begin{aligned}\langle d_\alpha(t)\rangle^{(1)} &= \frac{i}{2\hbar} \int_{-\infty}^{t} dt_1 E_{p\beta} \exp[-i\omega_p t_1 + \epsilon(t_1 - t)] \\ &\times \rho_{nn} \left[d_{nm}^{(\alpha)} d_{mn}^{(\beta)} \exp(i\omega_{nm} t + i\omega_{mn} t_1) - d_{nm}^{(\beta)} d_{mn}^{(\alpha)} \exp(i\omega_{nm} t_1 + i\omega_{mn} t) \right] \\ &= \frac{\rho_{nn}}{2\hbar} \left(\frac{d_{nm}^{(\alpha)} d_{mn}^{(\beta)}}{\omega_{mn} - \omega_p - i\epsilon} - \frac{d_{nm}^{(\beta)} d_{mn}^{(\alpha)}}{\omega_{nm} - \omega_p - i\epsilon} \right) E_{p\beta} \exp(-i\omega_p t).\end{aligned} \tag{6.130}$$

Here, summation over the state indices m, n and over the Cartesian indices $\alpha, \beta = x, y, z$ (which are sometimes written as superscripts, to make the notation more compact), as well as over the frequency index p, is implied. If the damping is taken into account by changing ϵ to the damping rate γ_{mn} of the density matrix non-diagonal element, then the linear susceptibility $\chi_{\alpha\beta}^{(1)}$ defined by (6.130) will coincide with (4.57).

Similarly, at $k = 2$, (6.128) yields

$$\langle d_\alpha \rangle^{(2)} = -\frac{1}{4\hbar^2}\int_{-\infty}^{t} dt_1 \int_{-\infty}^{t_1} dt_2 E_{p\beta}E_{q\gamma}\exp[-i(\omega_p t_1 + \omega_q t_2) + \epsilon(t_2 - t)]$$

$$\times \rho_{nn}\left[d_{nm}^{(\alpha)} d_{ml}^{(\beta)} d_{ln}^{(\gamma)} \exp[(i\omega_{nm}t + i\omega_{ml}t_1 + i\omega_{ln}t_2)] + \ldots\right]$$

$$= \left[\frac{\rho_{nn} d_{nm}^{(\alpha)} d_{ml}^{(\beta)} d_{ln}^{(\gamma)}}{4\hbar^2(\omega_{mn} - \omega_p - \omega_q - i\epsilon)(\omega_{ln} - \omega_q - i\epsilon)} + \ldots\right]$$

$$\times E_{p\beta}E_{q\gamma}\exp[-i(\omega_p + \omega_q)t]. \tag{6.131}$$

This expression contains only the contribution of the first term of the double commutator $[[\boldsymbol{d}(t), \boldsymbol{d}(t_1)], \boldsymbol{d}(t_2)]$, since the other three differ only in signs and in the permutations of the state indices l, m, n.

Consider sum-frequency generation, $\omega_1 + \omega_2 = \omega_0$. At $\omega_1 \neq \omega_2$, the double sum over frequencies contains two terms oscillating with the frequency ω_0: the one with $q = 1, p = 2$ and the one with $q = 2, p = 1$. Therefore, the ω_0 component of the dipole moment can be represented as a sum of two terms differing by a permutation of indices $1, \gamma$ and $2, \beta$,

$$d_{0\alpha}^{(2)} = E_{2\beta}E_{1\gamma}\prod_{\beta\gamma}^{21}\frac{\rho_{nn}}{2\hbar^2}\left(\frac{d_{nm}^{(\alpha)} d_{ml}^{(\beta)} d_{ln}^{(\gamma)}}{D_{mn}^{(0)} D_{ln}^{(1)}} + \ldots\right), \tag{6.132}$$

where

$$D_{mn}^{(p)} \equiv \omega_{mn} - \omega_p - i\gamma_{mn},$$

$\prod$ is the operator of summing over various permutations, and ϵ has been replaced by γ_{mn}. Note that the dispersion function $1/D$ is the Fourier transform of the step function $\theta(t)$ (Fig. 6.5).

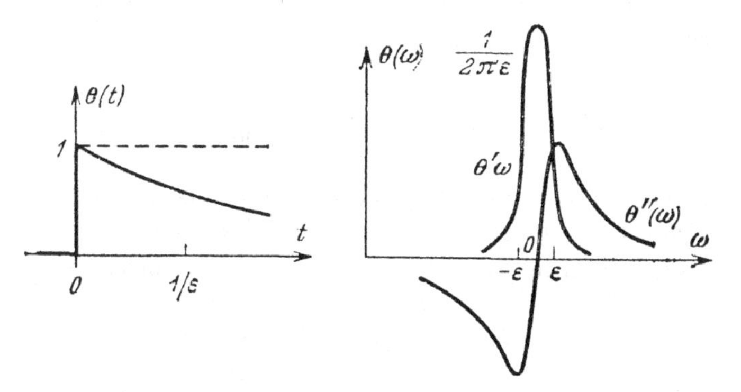

Fig. 6.5 The Heaviside step function multiplied by $e^{-\epsilon t}$ and its Fourier transform are used for taking into account the effects of causality and damping on the response of a system.

Thus, the quadratic susceptibility of matter consisting of N similarly oriented non-interacting molecules can be written in terms of the unperturbed level populations $N_n = \rho_{nn}N$, frequencies ω_{mn}, and the transition moments $\boldsymbol{d}_{mn}$ as

$$\chi^{\bar{0}12}_{\alpha\beta\gamma} = \frac{1}{2\hbar^2}\prod_{\beta\gamma}^{21}\sum_{lmn} N_n \left(\frac{d^{(\alpha)}_{nm} d^{(\beta)}_{ml} d^{(\gamma)}_{ln}}{D^{(0)}_{mn} D^{(1)}_{ln}} + \frac{d^{(\gamma)}_{nm} d^{(\beta)}_{ml} d^{(\alpha)}_{ln}}{D^{(0)}_{nl} D^{(1)}_{nm}} \right.$$

$$\left. - \frac{d^{(\beta)}_{nm} d^{(\alpha)}_{ml} d^{(\gamma)}_{ln}}{D^{(0)}_{lm} D^{(1)}_{ln}} - \frac{d^{(\gamma)}_{nm} d^{(\alpha)}_{ml} d^{(\beta)}_{ln}}{D^{(0)}_{lm} D^{(1)}_{nm}} \right), \tag{6.133}$$

where the superscripts 0 and $\bar{0}$ relate to the frequencies $\pm\omega_0$.

It is easy to see that expression (6.133) has all symmetry features described in Sec. 6.1. The $\prod$ operator provides the spatio-frequency symmetry with respect to the last two indices (6.14), (6.15). The property (6.16) follows from the relations $D^{(p)}_{mn} = -D^{(\bar{p})*}_{nm}$ and $\boldsymbol{d}^*_{mn} = \boldsymbol{d}_{nm}$, so that changing the signs of all frequencies and the imaginary units only interchanges the terms in (6.133): the first one is swapped with the second one and the third one, with the fourth one.

The fact that in centrally symmetric media $\chi^{(2)} = 0$ also follows from (6.133). The eigenstates of a system with a center of symmetry possess a certain parity: $\varphi_n(-\boldsymbol{r}) = \pm\varphi_n(\boldsymbol{r})$; therefore, $\boldsymbol{d}_{mn} = 0$ if φ_m and φ_n have the same parity. Hence, at least one of the three moments relating the states l, m, n is equal to zero.

The symmetry property in the case of transparent matter, (6.21), also follows from (6.133) at $\gamma_{mn} = 0$. In this case, the last terms in (6.133) can be combined in pairs,

$$\frac{1}{D^{(0)}_{lm} D^{(1)}_{ln}} + \frac{1}{D^{(0)}_{lm} D^{(2)}_{nm}} = \frac{1}{D^{(1)}_{ln} D^{(2)}_{nm}}.$$

As a result, out of 8 terms only $6 = 3!$ are left, which differ by permutations of index pairs $(\alpha, \bar{0}), (\beta, 2), (\gamma, 1)$,

$$\chi^{\bar{0}21}_{\alpha\beta\gamma} = \frac{1}{2\hbar^2}\sum_{lmn} N_n \prod_{\alpha\beta\gamma}^{\bar{0}21} \frac{d^{(\alpha)}_{nm} d^{(\beta)}_{ml} d^{(\gamma)}_{ln}}{(\omega_{mn} - \omega_0)(\omega_{ln} - \omega_1)}. \tag{6.134}$$

Susceptibility of the next order is calculated similarly. In each new order of the perturbation theory, factors of the form $d_{ln}/\hbar D^{(p)}_{ln}$ are added. Hence, we get an estimate for the optical anharmonicity in the transparency range,

$$E_{NL} \equiv \frac{\chi^{(k)}}{\chi^{(k+1)}} \approx \frac{\hbar\omega_0}{ea_0} \equiv E_0, \tag{6.135}$$

where a_0, ω_0, E_0 are the typical size, frequency, and internal field of the molecule. For the hydrogen atom, $E_0 = 13.6\text{V}/a_0 \approx 10^7$ G. Assuming $\chi^{(1)} = 0.1$, we obtain

$\chi^{(2)} = 10^{-8}\ \mathrm{G}^{-1}, \chi^{(3)} = 10^{-15}\ \mathrm{G}^{-2}$. Even this crude estimate gives a correct idea about the orders of magnitude of the susceptibilities. At resonance, $\chi^{(k)}$ increases.

6.2.10 °*Probability of multi-photon transitions*

If we are only interested in energies, then the nonlinear interaction between radiation and matter can be described in terms of probabilities or cross sections of multi-photon transitions, as it was done in Chapter 2 for the case of linear single-photon interaction. Then, as a rule, the field does not have to be quantized, i.e., one can use the semi-classical theory, but the results of calculations can be interpreted in the photon language.

As an example, let us find the probability of a two-photon elementary process describing Raman scattering and two-photon (induced) absorption or emission. Substituting the transition amplitude in the first-order approximation (2.24) into (2.21), for the amplitude of the two-photon transition from level a to level b we obtain

$$c_{ba}^{(2)}(t) = (i\hbar)^{-2} \int_{t_0}^{t} dt_2 \mathcal{V}'_{bn}(t_2) \int_{0}^{t_2} dt_1 \mathcal{V}'_{na}(t_1). \tag{6.136}$$

Here, the n index, in which summation is implied, numerates all intermediate (*virtual*) non-perturbed states through which the transition can occur. This expression reflects an important feature of quantum dynamics: all possible virtual states participate in a transition, even those seemingly violating the conservation laws, with the only restriction, following from the causality principle: $t > t_2 > t_1 > t_0$.

Let us substitute into (6.136) the dipole perturbation $\mathcal{V}' = -\boldsymbol{d}' \cdot \boldsymbol{E}$ and the biharmonic field (6.129), assuming that $t_0, t = \pm\infty$,

$$\begin{aligned} c_{ba}^{(2)} &= -(2\hbar)^{-2} \sum_{nqp} \boldsymbol{d}_{bn} \cdot \boldsymbol{E}_q \boldsymbol{d}_{na} \cdot \boldsymbol{E}_p \\ &\quad \times \int_{-\infty}^{\infty} dt_2 \int_{-\infty}^{t_2} dt_1 \exp[i(\omega_{bn} - \omega_q)t_2 + i(\omega_{na} - \omega_p)t_1] \\ &= -\sum_{nqp} \frac{\boldsymbol{d}_{bn} \cdot \boldsymbol{E}_q \boldsymbol{d}_{na} \cdot \boldsymbol{E}_p}{4\hbar^2 i(\omega_{na} - \omega_p)} \int_{-\infty}^{\infty} dt_2 \exp[i(\omega_{ba} - \omega_q - \omega_p)t_2]. \end{aligned} \tag{6.137}$$

The lower integration limit in the integral over t_1 makes no contribution due to the adiabatic start of the perturbation or due to the damping (see (6.127) at $\epsilon \to +0$). The integral left in (6.137) is one of the delta-function representations: $\int dt_2 \cdots = 2\pi\delta(\omega_{ba} - \omega_q - \omega_p)$. Thus, in the second order of the perturbation theory, the field excites a molecule only provided that the algebraic sum of the two field frequencies coincides with the frequency of the transition $a \to b$. This condition generalizes Bohr's postulate for a single-photon resonance.

Let two field frequencies ω_1 and ω_2 satisfy the condition of the 'combination' resonance, $\omega_1 + \omega_2 \approx \omega_{ba}$, then (6.137) yields

$$c_{ba}^{(2)} = i(2\pi/\hbar)(\boldsymbol{E}_1 \cdot \boldsymbol{M}^{12} \cdot \boldsymbol{E}_2)\delta(\omega_{ba} - \omega_1 - \omega_2),$$

$$M_{\alpha\beta}^{12} = M_{\beta\alpha}^{21} \equiv \frac{1}{4\hbar}\sum_n \left(\frac{d_{bn}^{(\alpha)} d_{na}^{(\beta)}}{\omega_{na} - \omega_2} + \frac{d_{bn}^{(\beta)} d_{na}^{(\alpha)}}{\omega_{na} - \omega_1}\right). \tag{6.138}$$

If $\omega_{ba} > 0$, i.e., the initial state of the molecule is the lower one, then (6.138) gives the probability amplitude for the molecule to absorb two photons (at $\omega_1, \omega_2 > 0$) or to absorb one photon ($\omega_1 > 0$) and emit another photon, the Stokes one ($\omega_2 < 0$). In the last case, (6.138) is the amplitude of a Stokes Raman process. Similarly, at $\omega_{ba} < 0$, (6.138) describes two-photon emission or anti-Stokes scattering. One can say that the two terms in (6.138) differ in the sequence the ω_1 and ω_2 photons are absorbed (or emitted).

Note that there are four types of two-photon emission: stimulated, spontaneous-stimulated, stimulated-spontaneous and purely spontaneous (Fig. 6.1(f)). They correspond to the four terms in the expression $(N_1+1)(N_2+1) = N_1 N_2 + N_1 + N_2 + 1$, which follows from (6.138) after replacing E_p with operators. (Here, N_p are the initial photon numbers.)

It follows from (6.138) that the maximal contribution to the transition amplitude is provided by virtual states with the minimal energy deficit $\hbar(\omega_{na} - \omega_p)$. Note that the various transition 'paths' (contributions of various virtual states) may differ in signs and cancel each other (*quantum interference of states*).

In order to find the transition probability from (6.138), it is necessary to define the square of the delta function:

$$|\delta(\omega)|^2 = \delta(\omega) \lim_{T\to\infty} \int_{-T/2}^{T/2} dt e^{i\omega t}/2\pi = \delta(\omega)T/2\pi. \tag{6.139}$$

From (6.138), (6.139), we obtain the probability of a two-photon transition per unit time, i.e., the transition rate (compare with the derivation of (2.36)),

$$W_{ba}^{(2)} = 2\pi |K_{ba}^{(2)}|^2 \delta(\omega_{ba} - \omega_2 - \omega_1)/\hbar^2, \tag{6.140}$$

$$K_{ba}^{(2)} \equiv \boldsymbol{E}_1 \cdot \boldsymbol{M}^{12} \cdot \boldsymbol{E}_2. \tag{6.141}$$

According to (6.140), the two-photon transition $a \to b$ is possible if the virtual transitions $a \to n$ and $n \to b$ are allowed. In a centrally symmetric medium, the a and b levels should have the same parity; in this case, the single-photon transition between them will be forbidden (the *alternative prohibition*). Thus, *two-photon spectroscopy and, in particular, Raman spectroscopy enable the study of levels that are not accessible for linear spectroscopy.*

For taking into account the finite width of the resonance, one should replace $\delta(\omega)$ by the normalized form factor $g(\omega)$ (see (2.37)). The resulting expression coincides with the probability of a single-photon transition (2.37), with the only exception that the Rabi frequency $|\boldsymbol{d}_{ba} \cdot \boldsymbol{E}_1|/\hbar \equiv \Omega$ is replaced by $2K_{ba}^{(2)}$. Probabilities of k-photon transitions have a similar structure: each additional photon ω_p participating in the transition adds to $K_{ba}^{(k)}$ a factor of the form $\boldsymbol{d}_{mn} \cdot \boldsymbol{E}_p/\hbar(\omega_{na} - \omega_p - \cdots - \omega_1) \sim \Omega/\omega_0 \sim E_p/E_{NL}$ (compare with (6.135)).

As in the case of single-photon transitions (Sec. 2.3), the transition rate $W^{(2)}$ determines the cross-section $\sigma^{(2)}$ and the absorption (amplification) coefficient $\alpha^{(2)}$, as well as the imaginary part of the cubic susceptibility $\chi^{(3)}(\omega_2, -\omega_1, \omega_1)$. The cross section of a stimulated two-photon transition will be defined as the transition rate in the case of unity densities of the photon fluxes,

$$\sigma_{stim}^{(2)} \equiv \frac{W^{(2)}}{F_1 F_2}, \quad F_p \equiv \frac{I_p}{\hbar\omega_p} = \frac{c}{8\pi\hbar\omega_p}|E_p|^2. \tag{6.142}$$

For taking into account the finite transition linewidth, we replace the delta function in (6.140) by the normalized form factor g (Sec. 2.2),

$$\sigma_{stim}^{(2)} = 128\pi^3 k_1 k_2 |\boldsymbol{e}_1 \cdot \boldsymbol{M}^{12} \cdot \boldsymbol{e}_2|^2 g(\omega_{ba} - \omega_2 - \omega_1), \tag{6.143}$$

where $k_n \equiv \omega_n/c$.

The two-photon absorption coefficient for the field with the frequency ω_2 in the presence of the second field, with the frequency ω_1, is (compare with (1.4))

$$\alpha^{(2)}(\omega_2) = \pm\sigma_{stim}^{(2)} \Delta N F_1, \tag{6.144}$$

where $\Delta N \equiv N_a - N_b$ is the population difference per unit volume.

The power absorbed from the field ω_2 by a unit volume of matter is

$$\mathcal{P}_2 = \omega_2 \mathrm{Im} \boldsymbol{P}_2^{(3)} \cdot \boldsymbol{E}_2^*/2 = \omega_2 \mathrm{Im} \chi_{\alpha\beta\gamma\delta}^{\overline{2}121} E_{2\alpha}^* E_{1\beta}^* E_{2\gamma} E_{1\delta}/2. \tag{6.145}$$

On the other hand, $\mathcal{P}_2 = \alpha^{(2)} I_2$. Comparing (6.145) and (6.144), for the case of a real transition dipole moment $\boldsymbol{d}_{mn}$ we find that

$$\mathrm{Im}\chi_{\alpha\beta\gamma\delta}^{\overline{2}121} = \pm 4\pi\hbar^{-1} \Delta N M_{\alpha\beta}^{21} M_{\delta\gamma}^{21} g(\omega_{ba} - \omega_1 - \omega_2). \tag{6.146}$$

Finally, the complete expression for the cubic nonlinearity follows from (6.146) after replacing g by $1/\pi(\omega_{ba} - \omega_1 - \omega_2 - i\gamma)$. It is not difficult to verify that a straightforward calculation of $\chi^{(3)}$ from (6.128) in a single-resonance approximation yields the same result.

Equations (6.140)–(6.143) describe purely stimulated two-photon transitions. In the case of spontaneous-stimulated transitions (leading to the spontaneous Raman scattering), the role of $|E_2|^2$ is played by the quantum fluctuations of the field, which provide an equivalent photon flux density of $F_{2vac} = c/L^3$ in each field mode in the vicinity of the frequency ω_2, with L^3 being the quantization volume (Sec. 7.3). After summing over all modes in the vicinity of the frequency $\omega_{ba} - \omega_1$, we find

$$\sigma^{(2)}_{sp.st} \equiv W^{(2)}_{sp.st}/F_1 = (c/L^3)\sum_{k_2}\sigma^{(2)}_{stim} = \int d\omega_2 \int_{4\pi} d\Omega_2 k_2^2 \sum_{\nu_2}\sigma^{(2)}_{stim}/(2\pi)^3$$
$$\equiv \int_{4\pi} d\Omega_2 \sum_{\nu_2}(d\sigma^{(2)}_2/d\Omega)_{sp.st}, \tag{6.147}$$

where ν_2 is the polarization index. After substituting here (6.143), we obtain the differential cross-section per unit solid angle and a single polarization type of the field ω_2,

$$(d\sigma^{(2)}_2/d\Omega)_{sp.st} = 16k_1k_2^3|\boldsymbol{e}_1 \cdot \boldsymbol{M}^{12} \cdot \boldsymbol{e}_2|^2. \tag{6.148}$$

Further, substituting F_1 by c/L^3 and summing over all modes ω_1 of the field, from 0 up to the transition frequency ω_{ab}, we find the rate of a purely spontaneous transition,

$$W^{(2)}_{sp} = (c/L^3)\sum_{k_1}\sigma^{(2)}_{sp.st} = (2/\pi^3)\int_0^{\omega_{ba}} d\omega_1 \int_{4\pi} d\Omega_1 d\Omega_2 \sum_{\nu_1\nu_2} k_1^3k_2^3|\boldsymbol{e}_1 \cdot \boldsymbol{M}^{12} \cdot \boldsymbol{e}_2|^2. \tag{6.149}$$

For a rough estimate of the two-photon decay probability in the optical range, assume that all frequencies in (6.149) are equal to $\omega_0 = e^2/2\hbar a_0 = \alpha c/2a_0$, and the polarisability M of the transition is equal to a_0^3, with $\alpha = 1/137$. Neglecting all other numerical factors, we obtain

$$W^{(2)}_{sp} \sim \omega_0(a/2)^6 \approx 50\,\mathrm{s}^{-1}.$$

An accurate calculation for the $2s \to 1s$ in a hydrogen atom yields 8 s^{-1}.

Thus, $W^{(2)}_{sp}$ is $137^3 \sim 10^6$ times as small as $W^{(1)}_{sp}$. Despite the small probability, two-photon decay can be easily observed using two PMTs and a coincidence circuit. Note that *spontaneous (as well as thermal) two-photon emission, in contrast to the single-photon one, has a continuous spectrum*, which has no relation to the discrete spectrum of an unperturbed atom. For the emission from a heated body, the statistics of two-photon radiation also differs from the one of single-photon radiation, which is caused by the fact that photons are emitted in pairs. Thus, the anharmonicity of matter leads, in principle, to the deviation of the thermal radiation statistics from a Gaussian one [Klyshko (1980)].

6.2.11 *Conclusions*

Thus, a large number of mechanisms contributes to the optical anharmonicity of macroscopic matter. Quadratic susceptibility $\chi^{(2)}$, as a rule, is related to the nonlinearity of bonded electrons. It differs from zero only in piezoelectric crystals and is on the order of $10^{-7} - 10^{-9}$ G^{-1} provided that all frequencies belong to the optical transparency window. The cubic susceptibility $\chi^{(3)}$ in condensed transparent matter is also caused by the electronic nonlinearity ($\chi^{(3)} \sim 10^{-15}$ G^{-2}) if all frequencies belong to the optical range. However, if the difference Ω of two frequencies coincides with the frequency of a molecular vibration, then $\chi^{(3)}$ becomes as high as $10^{-12} - 10^{-13}$ due to the combined electron-nuclear (Placzek or Raman) nonlinearity. At $\Omega \sim 0$, the main contribution in the case of solids is that of electrostriction ($\chi^{(3)} \sim 10^{-13}$); in liquids, the orientation (Kerr's) nonlinearity is added ($\chi^{(3)} \sim 10^{-12}$). Electrocaloric anharmonicity, usually, results in $\chi^{(3)} \lesssim 10^{-13}$. Extremely strong nonlinear optical effects can be observed in liquid crystals and in plasma. Note that the nonlinear electrodynamics of plasma is well described by the Landau-Vlasov kinetic equations (see [Silin (1961)]).

6.3 Macroscopic nonlinear optics

Thus, using classical or quantum macroscopic models, we have found polarization $\boldsymbol{P}$ of matter in a given field $\boldsymbol{E}$, i.e., we have excluded the variables of the medium. By substituting $\boldsymbol{D} = \boldsymbol{E} + 4\pi\boldsymbol{P}$ into macroscopic Maxwell's equations, we obtain a closed system of equations for $\boldsymbol{E}, \boldsymbol{H}$ describing the emission and propagation of electromagnetic field in matter with an account for the matter nonlinearity. Manifestations of nonlinearity in the optical range are extremely diverse and depend on the properties of both the medium and the initial field, such as the amplitude, the spatial and temporal spectra. The most important parameter is of course the ratio E/E_{NL}, which, as a rule, is much less than the unity.

6.3.1 *Initial relations*

Below, we consider the main types of stationary effects. The field can be represented as a sum of independent spectral components,

$$\begin{gathered} \boldsymbol{E}(\boldsymbol{r}, t) = (1/2) \sum_n \boldsymbol{E}_n(\boldsymbol{r}) \exp(-i\omega_n t), \\ \boldsymbol{E}_{-n} = \boldsymbol{E}_n^*, \ \omega_{-n} = -\omega_n, \ n = \pm 1, \pm 2, \dots . \end{gathered} \tag{6.150}$$

The amplitudes of monochromatic waves $\boldsymbol{E}_n(\boldsymbol{r})$ satisfy the system of Helmholtz wave equations that are related to each other due to the anharmonicity of the medium, which is considered to be non-magnetic,

$$c^2\nabla \times \nabla \times \boldsymbol{E}_n - \omega_n^2 \epsilon_n \cdot \boldsymbol{E}_n = 4\pi\omega_n^2 \boldsymbol{P}_n^{NL}(\boldsymbol{r}). \quad (6.151)$$

These relations can be easily obtained from Maxwell's equations (4.9)–(4.12). Here, the linear part of the polarization is included into ϵ_n. The amplitude of the nonlinear polarization at the frequency $-\omega_n$, which is the source of the macroscopic field $E_{n_1}^*$, is defined through the nonlinear polarisability (Sec. 6.1),

$$\boldsymbol{P}_{-n_1}^{NL}(\boldsymbol{r}) = \sum_{m=3}^{\infty} \boldsymbol{\chi}^{(m-1)}(\omega_{n_1}; \omega_{n_2}, \dots, \omega_{n_m}) \vdots \boldsymbol{E}_{n_2}(\boldsymbol{r}) \dots \boldsymbol{E}_{n_m}(\boldsymbol{r}), \quad (6.152)$$

where

$$\sum_{i=1}^{m} \omega_{n_i} = 0. \quad (6.153)$$

For describing non-coherent (noise) fields, such as thermal or Raman radiation, one should add Langevin random sources in the right-hand side of (6.151). According to the fluctuation-dissipation theorem (FDT), their spectral density scales as $\chi^{(1)''}$ or $\chi^{(3)''}$, respectively (see (6.90)). Another method of calculating the noise radiation from a nonlinear medium is based on Kirchhoff-type equations, which directly express the noise intensity in terms of the medium temperature and the solutions to the dynamic equations (6.150) (consider, for instance, (6.225), (6.307), (7.6)).

6.3.2 *Classification of nonlinear effects*

It follows from Maxwell's equations (see (4.13)) that the specific power of radiation absorbed from the field by the matter at point $\boldsymbol{r}$ due to anharmonicity within a single period is

$$\mathcal{P}_{NL}(\boldsymbol{r}) = (i/4)\sum_{n} \omega_n \boldsymbol{E}_n \cdot \boldsymbol{P}_{-n} = (i/4)\sum_{m=3}^{\infty}\sum_{n_1} \omega_{n_1} \boldsymbol{\chi}^{(m-1)} \vdots \boldsymbol{E}_{n_1} \dots \boldsymbol{E}_{n_m}. \quad (6.154)$$

Each term $\mathcal{P}^{(m)}(\boldsymbol{r})$ in the last sum describes *m-frequency interaction*. If the field amplitudes are represented in the form

$$\boldsymbol{E}_n(\boldsymbol{r}) \equiv \sum_{\alpha} \hat{x}_\alpha |E_{n\alpha}(\boldsymbol{r})| \exp[i\varphi_{n\alpha}(\boldsymbol{r})], \quad (6.155)$$

where $\alpha = x, y, z$, then both the sign and the absolute value of $\mathcal{P}^{(m)}$ will depend on $\varphi^{(m)}(\boldsymbol{r}) \equiv \varphi_{n_1\alpha_1} + \dots + \varphi_{n_m\alpha_m}$, i.e, on the relation between the field spectral components. Such interactions are called *parametric*. It is clear from (6.154) that the parametric interaction of m harmonics provides an effect that is accumulated in space only under the condition $\varphi^{(m)}(\boldsymbol{r}) = \text{const}$, which is possible only for plane waves, such that $\varphi_{n\alpha}(\boldsymbol{r}) = \boldsymbol{k}_n \cdot \boldsymbol{r}$, $\boldsymbol{k}_{-n} \equiv -\boldsymbol{k}_n$. Then, the wave vectors of the interacting waves should form a closed polygon,

$$\Delta \boldsymbol{k}^{(m)} \equiv \sum_{i=1}^{m} \boldsymbol{k}_{n_i} = 0. \tag{6.156}$$

This condition, with an account for (6.153), is called the *spatial phase-matching condition.*

Often, it is only a single low-order term in (6.154) that is important ($m = 3$ or 4, i.e., three- or four-frequency interaction), with a certain combination of frequency signs s_n. This combination can be put into correspondence to an elementary multi-photon process involving m photons; then (6.153) and (6.156) can be interpreted as the conservation laws of the field energy, $\sum \hbar\omega_n$, and momentum, $\sum \hbar \boldsymbol{k}_n$. In a parametric process, the energy of matter does not change, i.e., the initial level $\mathcal{E}_a$ and the final level $\mathcal{E}_b$ coincide, $\omega_{ba} = 0$ (Fig. 6.1(a)).

Further, one can consider *resonance* parametric processes, for which one or more virtual levels coincides with the real ones. Then, non-parametric effects become also important (see below), i.e., there appears linear (or multi-photon) absorption (or emission), $\alpha \neq 0$, and the energy of the field is not conserved. However, provided that the absorption is low, the phase-matching condition (6.156) is still valid. Note that, for instance, at $m = 4$, single, double, and triple resonances are possible. The efficiency of resonance interactions strongly depends on the frequencies ω_n (even without taking phase matching into account), and the corresponding susceptibilities take complex values.

Outside of resonances, $\alpha \approx 0$, susceptibilities are real, and their dispersion is weak; therefore, the efficiency of non-resonance parametric processes has only indirect frequency dependence, through the phase-matching condition, which at fixed directions of $\boldsymbol{k}_n$ is satisfied only for a particular set of frequencies and at fixed frequencies, only for a certain set of directions. Therefore, the efficiency of parametric interactions manifests *joint frequency-angular dispersion.*

Let us return to the sum (6.154). Among the terms of even orders m, there are degenerate ones, containing subscript pairs $n_i = -n_j$, for which $\varphi^{(m)} \equiv 0$. The corresponding interactions, depending only on the *intensities* of the field harmonics, $|E_n|^2$, are called *non-parametric*. The phase-matching condition is satisfied for them automatically. The power absorbed at the frequency ω_1 due to the non-parametric interaction with an l-frequency field, according to (6.154), is

$$\mathcal{P}_1(\boldsymbol{r}) = (1/2)\omega_1 \mathrm{Im}\chi^{(2l-1)}(-\omega_1; \omega_1 \cdots - \omega_l, \omega_l) \vdots \boldsymbol{E}_1^* \boldsymbol{E}_1 \ldots \boldsymbol{E}_l^* \boldsymbol{E}_l. \qquad (6.157)$$

It has a constant sign for a given set of frequencies and slowly (compared to $e^{i\boldsymbol{k}\cdot\boldsymbol{r}}$) varies in space for any configuration of the field, since the phases of the harmonics do not influence Eq. (6.157).

In non-parametric elementary processes, such as, for instance, two-photon absorption or a Raman transition, the final energy of the matter differs from the initial one (Fig. 6.1), and therefore, the field energy is not preserved.[i] For a non-parametric process, in addition to the trivial (in this case) condition (6.153) there is a resonance at

$$\sum_{n=1}^{l} \omega_n = \omega_{ba}. \qquad (6.158)$$

In the presence of additional intermediate resonances, a non-parametric process is called a *cascaded* or *resonance* one. Such processes have complicated dispersion dependence on each field frequency ω_n separately. Examples are resonance Raman scattering and cascaded two-photon absorption in a three-level system (Fig. 6.1(e)).

In the case of non-resonance (non-cascaded) processes, $\alpha^{(1)} = 0$, and the efficiency of non-parametric interaction depends, according to (6.158), only on the sum of all field frequencies, $\sum \omega_n$. This case can be called a single-resonance one.

Note that in non-parametric processes, new spectral components appear only due to spontaneous or spontaneous-stimulated transitions, and in the absence of a feedback the radiation is noisy, *non-coherent*, even if the pump is coherent and has a fixed phase. An example is inelastic scattering of light. At the same time, parametric processes can lead to the generation of *coherent* fields with new frequencies (generation of harmonics).

The effects of nonlinear optics can be additionally classified according to the number of essential spectral components or according to the number of plane waves (modes). For instance, single-frequency non-parametric (degenerate) effects include *nonlinear absorption and dispersion* ($l\omega_1 = \omega_{ba}$) and *saturation* ($\omega_1 - \omega_1 + \cdots + \omega_1 = \omega_{ba}$). Examples of two-frequency non-parametric effects are *induced absorption and dispersion* ($\omega_1 + \omega_2 = \omega_{ba}$) and *Raman interaction* ($\omega_1 - \omega_2 = \omega_{ba}$), leading to the *stimulated Raman scattering* (StRs) and the *inverse Raman effect*, i.e., induced absorption at the anti-Stokes frequency.

[i]Note, however, that *induced dispersion*, accompanying multi-photon absorption, similarly to linear dispersion, corresponds to a virtual process with $\omega_{ba} = 0$.

The *second harmonic generation* is a degenerate case of three-frequency parametric interaction ($\omega_1 \pm \omega_2 \pm \omega_3 = 0$). Four-frequency parametric effects include third-harmonic generation and *coherent anti-Stokes Raman scattering* (CARS), which forms the basis of *active spectroscopy*. CARS is determined by the resonance part of the cubic susceptibility $\chi^{(3)}(\omega_4 = \omega_3 - \omega_2 + \omega_1)$ at $\omega_1 - \omega_2 = \omega_4 - \omega_3 \approx \omega_{ba}$. An example of a parametric single-frequency four-wave interaction is *optical phase conjugation* due to $\chi^{(3)}(\omega = \omega - \omega + \omega)$ at $\boldsymbol{k}_2 = -\boldsymbol{k}_1, \boldsymbol{k}_4 = -\boldsymbol{k}_3$.

All these effects will be considered in more detail below. Sometimes, several effects can be manifested simultaneously; for instance, StRS can be accompanied by self-focusing and CARS. However, often, by choosing the experimental conditions one can select a single effect. In our analysis, for simplicity we will assume that this is the case.

It is also convenient to distinguish between the effects according to other pairwise features: spontaneous – stimulated, stationary – non-stationary (Chapter 5).

It should be stressed that the terms 'spontaneous' and 'stimulated' have no rigid definitions in quantum electronics. In linear optics, spontaneous and stimulated emission is considered to be a non-stationary process involving a single molecule, while the joint stationary radiation from heated matter is called thermal radiation. At the same time, the term 'spontaneous scattering' is understood as scattering by thermal (or, at $\hbar\Omega/\kappa T \gg 1$, quantum) fluctuations of various parameters of the matter (Sec. 6.2) at small pump intensity I_1, although this process is explained in terms of spontaneous-stimulated two-photon transitions (Sec. 6.2). If I_1 is increased, the intensity of the scattered light grows first linearly and then, at $|\alpha^{(2)}|l > 1$, exponentially (for the Stokes component). As a result, the efficiency of frequency conversion can be as high as tens of per cent, and it is to this spontaneously emitted and amplified radiation that the term StRS is usually applied. The same term is sometimes used for the effect of external Stokes field amplification.

The term *coherent* is even more ambiguous, even if we restrict ourselves to nonlinear optics. It is used, on the one hand, for the effects of non-stationary resonance interaction, like SIP (Sec. 5.1), and on the other hand, for parametric stationary effects like second harmonic generation. In the first case, it means that the field and the matter have the same phase and in the second case, that different components of the field have the same phase.

6.3.3 *The role of linear and nonlinear dispersion*

The nonlinearity of the wave equations of hydrodynamics and gas dynamics is most apparent in the appearance of shock waves, i.e., in the transformation of sine acoustic waves into sawtooth ones. (A well-known example is wave breaking in shallow water.) In spectral language, this effect is explained by higher-harmonic generation, which enriches the spectrum of an excitation in the course of its propagation.

However, due to the refractive-index dispersion $n(\omega)$, light shock waves do not emerge: harmonics with the frequencies $\omega_1, 2\omega_1, \ldots$ propagate with different phase velocities $c/n(p\omega_1)$, so that the sign of the interaction energy and, correspondingly, the amplitudes of the harmonics have fast spatial oscillations (see (6.154)). As a result, the amplitudes of higher harmonics do not increase in the course of propagation ('non-accumulating interaction'), and another small parameter appears in the theory, the ratio of the coherence length $l_{coh} = \pi/\Delta k$ to the length of the medium l. Here, Δk is the wave mismatch (6.156), which is equal, in the case of collinear pth-harmonic generation, to $k_p - pk_1 = [n(p\omega_1) - n(\omega_1)]p\omega_1/c$.

It is only under special conditions that the phase velocities of two or three harmonics can be matched, using birefringence or anomalous dispersion. Thus, *the phase matching condition (6.156) restricts the number of efficiently interacting spectral components of the field in parametric interactions.* For non-parametric processes, the restriction is due to the resonance condition (6.158), i.e., the dispersion of $\chi^{(3)}, \chi^{(5)}, \ldots$.

The efficiency of nonlinear processes is usually increased by using pulsed pumping with Q-switched ($\tau \sim 10$ ns) or mode locked ($\tau \sim 10$ ps)[j] lasers. Clearly, for the interaction of several light pulses with different central frequencies to accumulate in space, it is also necessary that their group velocities coincide,

$$u_p = d\omega/dk_p = c(n_p + \omega_p dn_p/d\omega)^{-1}$$

(the *group matching* condition); otherwise, the pulses will separate in the course of propagation.

6.3.4 °*One-dimensional approximation*

Further, as a rule, we will apply the *one-dimensional approximation*, which is most useful in nonlinear (as well as linear) optics. Also known as plane-waves approximation, it allows one to pass from partial-derivatives equations (6.151) to usual equations. It reflects correctly the main features of many effects involving pump beams not too much convergent or divergent and samples with the length l not too large. The divergence can be often taken into account in the final formulas,

[j]Editors' note: At present, mode locking provides pulse durations as small as tens of femtoseconds.

at least qualitatively, by summing up the contributions from all essential plane waves. We will go beyond this approximation only in the description of self-focusing in the end of Sec. 6.4.

An arbitrary field can be represented as a four-dimensional Fourier integral,

$$\boldsymbol{E}(\boldsymbol{r},t) = (L/2\pi)^3 \int d^3k d\omega \boldsymbol{E}(\boldsymbol{k},\omega) \exp[i(\boldsymbol{k}\cdot\boldsymbol{r} - \omega t)], \tag{6.159}$$

and similarly for $\boldsymbol{H}$. The factor $(L/2\pi)^3$ is added from dimensionality considerations, see (7.100). This is the so-called *$\boldsymbol{k},\omega$-representation*. Here, $\boldsymbol{k}$ and ω are independent real variables. In the case of a homogeneous transparent linear medium without internal sources, $\boldsymbol{k}$ and ω are not independent any more (Sec. 4.2). Then,

$$\boldsymbol{E}(\boldsymbol{k},\omega) = (1/2) \sum_{\nu=1,2} \sum_{s=\pm} \boldsymbol{e}_\nu(\boldsymbol{k}) E_\nu^{(s)}(s\boldsymbol{k}) \delta(\omega - s\omega_\nu(\boldsymbol{k})), \tag{6.160}$$

$$E_\nu^{(+)}(\boldsymbol{k}) = E_\nu^{(-)}(\boldsymbol{k})^*. \tag{6.161}$$

We assume the medium to be non-gyrotropic; therefore, $\boldsymbol{\epsilon}$ and $\boldsymbol{e}_\nu$ are real. Substituting (6.160) into (6.159), we find

$$\boldsymbol{E}(\boldsymbol{r},t) = \frac{1}{2}\left(\frac{L}{2\pi}\right)^3 \int d^3k \sum_{\nu s} \boldsymbol{e}_\nu(\boldsymbol{k}) E_\nu^{(s)}(\boldsymbol{k}) \exp[i\psi_s(\boldsymbol{k})], \tag{6.162}$$

where

$$\psi_s(\boldsymbol{k}) \equiv s[\boldsymbol{k}\cdot\boldsymbol{r} - \omega(\boldsymbol{k})t], \;\; s = \pm, \; \nu = 1,2.$$

For instance, in the case of a plane monochromatic polarized wave,

$$E_\nu^{(s)}(\boldsymbol{k}) = (E_1\delta_{s^+} + E_1^*\delta_{s^-})\delta_{\nu\mu}\delta^{(3)}(\boldsymbol{k} - \boldsymbol{k}_1).$$

Unit polarization vectors $e_\nu(\boldsymbol{k})$ and the dispersion law $\omega_\nu(\boldsymbol{k})$ are determined by the $\boldsymbol{\epsilon}(\omega)$ tensor,

$$c^2\boldsymbol{k}\times[\boldsymbol{k}\times\boldsymbol{e}_\nu] + \omega_\nu^2\boldsymbol{\epsilon}(\omega_\nu)\cdot\boldsymbol{e}_\nu = 0. \tag{6.163}$$

Hence, if $\boldsymbol{\epsilon}$ is real,

$$\omega_\nu(-\boldsymbol{k}) = \omega_\nu(\boldsymbol{k}), \;\; \boldsymbol{e}_\nu(\boldsymbol{k}) = \boldsymbol{e}_\nu(-\boldsymbol{k}) = \boldsymbol{e}_\nu^*(\boldsymbol{k}).$$

The wave vector $\boldsymbol{k}$ and the type of polarization ν define a plane monochromatic wave, or a *mode*. It is convenient to make the set of modes countable using a cube with periodicity L^3 (Sec. 7.3); then the Fourier integral turns into a series, in which a mode is labeled by a single subscript $k \equiv \{\boldsymbol{k},\nu,s\}$ (which also includes the sign s of the frequency),

$$\boldsymbol{E}(\boldsymbol{r},t) = (1/2)\sum_k \boldsymbol{e}_k E_k \exp(i\psi_k), \;\; E_k \equiv E_\nu^{(s)}(\boldsymbol{k}). \tag{6.164}$$

In a transparent linear medium without sources, the mode amplitudes E_k are independent complex numbers defined by the boundary conditions.

Consider now a nonlinear absorbing plane layer ($0 < z < l$) in the medium, which has the same ϵ', and a stationary field incident on it from both sides. Then, since the model has no dependence on x, y, t, the amplitudes E_k are only functions of z. They are called the *slowly varying amplitudes* (SVA). Actually, the transition from the real field $\boldsymbol{E}(\boldsymbol{r}, t)$ to the z-dependent mode amplitudes $E_k(z)$ is a three-dimensional Fourier transform in the variables x, y, t (a non-complete transformation, unlike (6.159)) or the transition to the *$\omega, \boldsymbol{k}_\perp, z$-representation.*

Note that a mode can be defined by fixing, instead of $\boldsymbol{k}$, the frequency, $\omega = \omega_\nu(\boldsymbol{k})$, $\boldsymbol{k}_\perp$, and the sign σ_k of the longitudinal component k_z. Then, k_z is determined by the dispersion law (for an isotropic medium, $k_z = \sigma_k(\epsilon'\omega^2/c^2 - k_\perp^2)^{1/2}$). In experiment, mode expansion is achieved by placing a spectral device into the far-field zone of the emitter; in this case, the spherical angles ϑ_k, φ_k of $\boldsymbol{k}$ are also fixed (taking into account refraction). Thus, the subscript k stands for one of the following sets of values:

$$k \equiv \{\boldsymbol{k}, \nu, s\} = \{\omega, \boldsymbol{k}_\perp, \sigma, \nu, s\} = \{\omega, \vartheta, \varphi, \nu, s\}. \tag{6.165}$$

As we will show in what follows, in the case of sufficiently small nonlinearity and absorption, Fourier transformation of the Helmholtz equations (6.151) w.r.t. x, y leads to a system of ordinary equations for SVA (for $\sigma_k > 0$),

$$\left(\frac{d}{dz} + \frac{\alpha_k}{2}\right) E_k(z) = \frac{2\pi i\omega_k}{c\overline{n_k}} P_k^{NL}(z) \exp(-ik_z z). \tag{6.166}$$

Here, the following notation was introduced:

$$\alpha_k \equiv \omega_k \boldsymbol{e}_k \cdot \boldsymbol{\epsilon}'' \cdot \boldsymbol{e}_k / c\overline{n_k}, \quad \overline{n_k} \equiv n_k \cos\theta_k \cos\rho_k \approx ck_z/\omega_k, \tag{6.167}$$

where θ_k is the angle between the ray (Poynting) vector $\boldsymbol{s}_k$ of the mode and the z axis, ρ_k is the anisotropy angle, i.e., the angle between $\boldsymbol{k}$ and $\boldsymbol{s}_k$, and P_k^{NL} is the Fourier component of the nonlinear polarization with the frequency $\omega_k > 0$ and the transverse wave vector $\boldsymbol{k}_\perp \equiv \{k_x, k_y\}$, which is parallel to the $\boldsymbol{e}_k$ vector. It can be written as

$$\begin{aligned} P_k^{NL} &\equiv (4\pi|u_{kz}|/L) P^{NL}(\omega_k, \boldsymbol{q}, z) \\ &= (2|u_{kz}|/L^3) \int dx\, dy\, dt\; \boldsymbol{e}_k \cdot \boldsymbol{P}^{NL}(\boldsymbol{r}, t) \exp[i(\omega_k t - k_x x - k_y y)], \end{aligned} \tag{6.168}$$

$$\boldsymbol{q} \equiv \{k_x, k_y\}, \quad \boldsymbol{r} \equiv \{x, y, z\}, \quad u_{kz} \equiv \partial\omega_\nu(\boldsymbol{k})/\partial k_z. \tag{6.169}$$

According to (6.166), linear absorbtion results in an exponential dependence of the mode amplitude on z, while nonlinearity leads to a coupling (mixing) of modes. Let us stress that in a one-dimensional model, a mode $\boldsymbol{k}$ can be only excited by a component of the source field having the same frequency and transverse wavevector. The source field can have any dependence on z but the most efficient component is the one that has the propagation constant K_z close to k_z, due to the phase matching condition (compare with (4.23)).

The boundary conditions for system (6.166) are given by the amplitudes $E_k(z_0)$ of the plane waves incident from the left and from the right: $z_0 = 0$ for $\sigma = +$ and $z_0 = l$ for $\sigma = -$. The solution to (6.166) yields the rule of mode transformation by the nonlinear layer, i.e., the *scattering matrix* of the layer. Certainly, for describing real experiments the model should include reflection and refraction at the boundaries (with an account for the nonlinearity, see Ref. [Bloembergen (1965)]), as well as consider the limited interaction cross-section in x and y. In this case, even in the linear approximation one gets the 'diffraction' coupling between the modes having close directions. This effect is more conveniently described in the $\omega, \boldsymbol{r}$ representation (6.151) (see the self-focusing section in 6.3).

Let us find, with the help of (6.166), the variation rate of the energy flux longitudinal component carried by the mode k, in $\hbar\omega_k$ units:

$$\begin{aligned} F_{kz} &\equiv I_k \cos\theta_k/\hbar\omega_k = c[\boldsymbol{E}_k \times \boldsymbol{H}_k^*]\cdot\hat{z}/16\pi\hbar\omega_k + \text{c.c.} \\ &= c\overline{n_k}|E_k(z)|^2/8\pi\hbar\omega_k, \end{aligned} \tag{6.170}$$

where $\hat{z}$ is a unit vector along the z axis. For calculating F_{kz}, we multiply (6.166) by E_k^* and sum with the complex conjugated expression,

$$(d/dz + \alpha_k)F_{kz} = -\text{Im}P_k^{NL}E_k^*\exp(-ik_z z)/2\hbar. \tag{6.171}$$

The right-hand side has, according to (6.154), a simple meaning: this is the energy (in $\hbar\omega_k$ units) absorbed by a unit volume of the matter per unit time due to the nonlinear polarization, i.e., this is the rate of the photon density decrease, $-\dot{N}_k$, in mode k. In other words, (6.171) is the *transfer equation* (or continuity equation) for the photons of the macroscopic field,

$$\nabla \boldsymbol{F}_k + \dot{N}_k = 0, \;\; \boldsymbol{F}_k = \boldsymbol{u}_k N_k. \tag{6.172}$$

The nonlinear polarization P_k^{NL} in the right-hand side of (6.166) or (6.171) is given by equations (6.152) and (6.158) through the hierarchy of susceptibilities $\chi^{(m)}$, $m \geqslant 2$. In the case of an m-frequency interaction, equation (6.166) takes the form

$$(d/dz + \alpha_k/2)E_1^* = (2\pi s_1\omega_1/ic\overline{n_1})\chi^{(m-1)}\vdots\boldsymbol{e}_1\boldsymbol{E}_2\ldots\boldsymbol{E}_m\exp(i\Delta k_z), \tag{6.173}$$

where

$$\Delta k \equiv \sum_{i=1}^{m} s_i k_{iz}, \quad \boldsymbol{E}_i \equiv \boldsymbol{e}_{\nu_i} E_{\nu_i}^{(s_i)}(\boldsymbol{k}_i), \tag{6.174}$$

and integration w.r.t. x, y, t in (6.168) yields the following restrictions for the interacting modes and the sign indices:

$$\Delta \boldsymbol{k}_\perp \equiv \sum_{i=1}^{m} s_i \boldsymbol{k}_{i\perp} = 0, \tag{6.175}$$

$$\Delta\omega \equiv \sum_{i=1}^{m} s_i \omega_i = 0. \tag{6.176}$$

In addition, accumulation of the interaction along z requires that the longitudinal wave mismatches are sufficiently small,

$$\Delta k \equiv \sum_{i=1}^{m} s_i k_{iz} \lesssim 1/l. \tag{6.177}$$

Let us re-normalize the mode amplitudes so that their squares are equal to the longitudinal photon flux density. (To simplify the notation, we assume $F_i \equiv F_{k_i z} > 0$.)

$$a_i \equiv (c\overline{n_i}/8\pi\hbar\omega_i)^{1/2} E_i, \tag{6.178}$$

$$|a_i|^2 = F_i. \tag{6.179}$$

As a result, (6.171) and (6.173) take the form

$$\left(\frac{d}{dz} + \frac{\alpha_1}{2}\right) a_1^* = \frac{s_1}{2i} \beta_{1\dots m} a_2 \dots a_m \exp(i\Delta k z), \tag{6.180}$$

$$\left(\frac{d}{dz} + \alpha_1\right) F_1 = s_1 \mathrm{Im} \beta_{1\dots m} a_2 \dots a_m \exp(i\Delta k z), \tag{6.181}$$

where

$$\beta_{1\dots m} \equiv \frac{1}{2\hbar}\left(\frac{8\pi\hbar}{c}\right)^{m/2} \left(\frac{\omega_1 \dots \omega_m}{\overline{n_1} \dots \overline{n_m}}\right)^{1/2} \chi^{(m-1)} \vdots \boldsymbol{e}_1 \dots \boldsymbol{e}_m. \tag{6.182}$$

These equations will be analyzed for several typical cases in Secs. 6.4, 6.5.

6.3.5 *The Manley-Rowe relation and the permutation symmetry*

Suppose that only a single interaction between m-frequency field components $n_1, \ldots, n_m$ (see (6.154)) is essential, corresponding to an l-photon process. (For a parametric interaction, $l = m$, while for a non-parametric one, $l = m/2$.) Let this interaction be non-resonance for some pair of frequencies ω_1, ω_2. To be more exact, let the photons with these frequencies be created or annihilated only together, simultaneously, which is the case if the corresponding virtual level is far from any real one. Then,

$$s_1 \dot{N}_1 = s_2 \dot{N}_2, \tag{6.183}$$

where

$$s_n \equiv \text{sign}(n) \equiv \text{sign}(\omega_n). \tag{6.184}$$

These relations are similar to the Manley-Rowe equations used in the oscillation theory for describing nonlinear circuits [Migulin (1978)]. The signs of the frequencies are determined by the energy conservation law (6.153) or (6.158), $\dot{N}_n$ is the rate of the flux variation for photons with the frequency $|\omega_n|$. Apparently, $\hbar|\omega_n|\dot{N}_n + \mathcal{P}_n = 0$, where $\mathcal{P}_n = \omega_n \text{Im} E_n^* P_n/2$ is the power absorbed by matter at frequency ω_n.

From (6.154) and (6.183), it follows for the case of a parametric interaction that

$$\text{Im}(\chi^{12\ldots} \vdots \boldsymbol{E}_1 \boldsymbol{E}_2 \cdots - \chi^{21\ldots} \vdots \boldsymbol{E}_2 \boldsymbol{E}_1 \ldots) = 0, \tag{6.185}$$

where

$$\chi^{12\ldots} \equiv \chi(\omega_1; \omega_2, \ldots) = (\chi^{\bar{1}\bar{2}\ldots})^*. \tag{6.186}$$

Let us substitute (6.155) into (6.185),

$$\sum_{\alpha\beta\ldots} |E_{1\alpha} E_{2\beta} \ldots | \text{Im}\{(\chi^{12\ldots}_{\alpha\beta\ldots} - \chi^{21\ldots}_{\beta\alpha\ldots}) \exp[i(\varphi_{1\alpha} + \varphi_{2\beta} + \ldots)]\} = 0. \tag{6.187}$$

Here, all amplitudes and phases of the field harmonics can be varied independently; therefore, the expression in round brackets is identically equal to zero,

$$\chi^{12\ldots}_{\alpha\beta\ldots} = \chi^{21\ldots}_{\beta\alpha\ldots}. \tag{6.188}$$

This permutation relation is more general than (6.21) where the medium was assumed to be transparent at all frequencies.

In the case of a non-parametric interaction, the phases are irrelevant, and we obtain from (6.157) and (6.183) that

$$\text{Im}(\chi^{\bar{1}1\bar{2}2\ldots} \vdots \boldsymbol{E}_1^* \boldsymbol{E}_1 \boldsymbol{E}_2^* \boldsymbol{E}_2 \cdots - \chi^{\bar{2}2\bar{1}1\ldots} \vdots \boldsymbol{E}_2^* \boldsymbol{E}_2 \boldsymbol{E}_1^* \boldsymbol{E}_1 \ldots) = 0, \tag{6.189}$$

$$\sum_{\alpha\beta\ldots} |E_{1\alpha}E_{1\beta}E_{2\gamma}E_{2\delta}\ldots| \operatorname{Im}(\chi^{\bar{1}1\bar{2}2\ldots}_{\alpha\beta\gamma\delta\ldots} - \chi^{\bar{2}2\bar{1}1\ldots}_{\gamma\delta\alpha\beta\ldots}) = 0. \tag{6.190}$$

Hence, we obtain the invariance to 'block' permutations,

$$\chi^{\bar{1}1\bar{2}2\ldots}_{\alpha\beta\gamma\delta\ldots} = \chi^{\bar{2}2\bar{1}1\ldots}_{\gamma\delta\alpha\beta\ldots}. \tag{6.191}$$

We have already observed such symmetry for Raman transitions and, in general, two-photon transitions (Sec. 6.2). Assuming in (6.191) $\omega_1 \equiv \omega_L > 0$, $\omega_2 \equiv -\omega_S < 0$, we obtain, with the help of (6.16),

$$\chi^{\bar{L}LS\bar{S}}_{\alpha\beta\gamma\delta}(\omega_L = \omega_L + \omega_S - \omega_S) = \chi^{\bar{S}SL\bar{L}}_{\gamma\delta\alpha\beta}(\omega_S = \omega_S + \omega_L - \omega_L)^*. \tag{6.192}$$

Strictly speaking, (6.192) leads to the symmetry of only the imaginary parts of the susceptibilities; however, using relations of Kramers-Kronig type (4.8), one can show that the real and imaginary parts of $\chi^{(m)}$ have the same symmetry. Relation (6.191) also follows from (6.188), if one considers a resonance parametric process with the frequencies $\omega_3 = -\omega_1, \omega_4 = -\omega_2$:

$$\chi^{12\bar{1}\bar{2}\ldots}_{\alpha\beta\gamma\delta\ldots} = \chi^{21\bar{1}\bar{2}\ldots}_{\beta\alpha\gamma\delta\ldots}. \tag{6.193}$$

By performing complex conjugation and using partial symmetry (6.14), (6.15), and (6.16), we obtain (6.191).

Permutation symmetry of susceptibilities, or the Manley-Rowe relations, lead to several simple integrals of motion for SVA. For instance, in the case of non-resonance interaction, the coupling coefficients β (6.182) are invariant to the permutations of all indices; therefore, the right-hand sides of Eqs. (6.181) for the intensities differ only in signs. Hence, it follows for $\alpha_k = 0$ that

$$s_1 dF_1/dz = s_2 dF_2/dz = \cdots = s_m dF_m/dz = \operatorname{Im}(\beta_{1\ldots m} a_1 \ldots a_m e^{i\Delta kz}), \tag{6.194}$$

where the sign factors $s_i = \pm 1$ in the case $k_{iz} > 0$ are determined by the energy conservation in the corresponding elementary process. Hence, for systems (6.180), (6.181) we find $m-1$ integrals of motion, which allow the intensities in $m-1$ modes to be expressed in terms of the intensity in a single mode,

$$F_k(z) = s_1 s_k F_1(z) + C_k, \; C_1 = 0, \tag{6.195}$$

and the constants C_k are determined by the boundary values $F_k(0)$.

If all essential frequencies, including the combination ones, are in the transparency windows (i.e., parametric processes are non-resonance), then there is no dissipation of the field energy and there must exist one more integral of motion corresponding to the conservation of the total light energy flux. However, one can easily see that the total energy flux is expressed in terms of C_k constants. To this end, let us multiply relations (6.195) by $\hbar\omega_k$ and add them,

$$I_z(z) = \sum_{k=1}^{m} \hbar\omega_k F_k(z) = \sum_{k=1}^{m} \hbar\omega_k C_k. \tag{6.196}$$

In parametric interactions, there is yet another integral, which defines the phase difference

$$\varphi(z) \equiv \sum_{k=1}^{m} s_k \varphi_k(z) \tag{6.197}$$

of the complex amplitudes $a_k \equiv \sqrt{F_k}\exp(is_k\varphi_k)$ in terms of $F_k(z)$ and $\varphi(0)$,

$$s_n F_n \Delta k + 2\beta(F_1 \cdots F_m)^{1/2}\cos[\varphi(z) + \Delta k z] = C. \tag{6.198}$$

Here, n is any of the subscripts $1, \ldots, m$. By differentiating the left-hand part of (6.198), one can see, due to (6.180), (6.194), and the relation $\beta = \beta^*$, that it is independent of z.

6.3.6 °*Derivation of one-dimensional equations*

In order to derive equations (6.166) for SVA in the ω, q, z-representation, let us write the frequency components of the field $\boldsymbol{E}_n(\boldsymbol{r})$,

$$\boldsymbol{E}(\omega, \boldsymbol{r}) = (1/2)[\boldsymbol{E}_n(\boldsymbol{r})\delta(\omega - \omega_n) + \boldsymbol{E}_n^*(\boldsymbol{r})\delta(\omega + \omega_n)],$$

which enter the Helmholtz equations (6.151), in terms of the mode amplitudes $E_k(z) = 4\pi|u_{kz}|E(\omega, \boldsymbol{q}, z)/L$ defined in (6.164). Let $\omega > 0$ and the polarization type be fixed, then

$$\begin{aligned} 2E(\omega, \boldsymbol{r}) &= \int \frac{dt}{2\pi} e^{i\omega t} \int d^3k \left(\frac{L}{2\pi}\right)^3 E_k(z) \exp(i\boldsymbol{k}\cdot\boldsymbol{r} - i\omega(\boldsymbol{k})t) \\ &= \left(\frac{L}{2\pi}\right)^3 \int d^3k E_k(z) \exp(i\boldsymbol{k}\cdot\boldsymbol{r})\delta(\omega - \omega(\boldsymbol{k})) \\ &= \left(\frac{L}{2\pi}\right)^3 \int d^2q \sum_{\pm} |u_{kz}|^{-1} E_k(z) \exp[i\boldsymbol{q}\cdot\boldsymbol{\rho} \pm ik_z(\omega, \boldsymbol{q})z], \end{aligned} \tag{6.199}$$

where $\boldsymbol{q} \equiv \{k_x, k_y\}$, $\boldsymbol{\rho} \equiv \{x, y\}$, $u_{kz} \equiv \partial\omega(\boldsymbol{k})/\partial k_z$. The function $k_z(\omega, \boldsymbol{q}) \equiv k_z > 0$ is defined here indirectly, through the dispersion law, $\omega(\boldsymbol{q}, k_z) = \omega$. In deriving (6.199), we have used the property of the delta function

$$\delta[f(x)] = \sum_i a_i \delta(x - x_i), \;\; f(x_i) \equiv 0, \; a_i \equiv |df/dx|_{x=x_i}^{-1}. \tag{6.200}$$

Further, assume that only a single mode with a certain sign of k_z is excited efficiently, then

$$E(\omega, \boldsymbol{r}) = \frac{L^3}{16\pi^3} \int d^2q |u_{kz}|^{-1} E_k(z) \exp(i\boldsymbol{k} \cdot \boldsymbol{r}),$$
$$\boldsymbol{k} \equiv \{q_x, q_y, k_z(\omega, \boldsymbol{q})\}. \tag{6.201}$$

While substituting (6.201) into (6.151), let us take into account that usually both absorption and nonlinearity are low. Then, $E_k(z)$ vary very little at a distance of a wavelength, and the second derivatives can be neglected (for more detail on the SVA method, see [Akhmanov (1964)]),

$$|d^2 E_k/dz^2| \ll |k_z dE_k/dz|. \tag{6.202}$$

As a result,

$$\nabla \times \nabla \times \boldsymbol{e}E(z)e^{i\boldsymbol{k}\cdot\boldsymbol{r}} \approx e^{i\boldsymbol{k}\cdot\boldsymbol{r}}\{-\boldsymbol{k} \times (\boldsymbol{k} \times \boldsymbol{e}) + i\boldsymbol{k} \times (\nabla \times \boldsymbol{e}) + i\nabla \times (\boldsymbol{k} \times \boldsymbol{e})\}E(z), \tag{6.203}$$

where the k index is so far omitted. The first term here will be further canceled due to (6.163). Let us take the inner products of (6.151) and (6.203) with $\boldsymbol{e}$ and use the vector identity

$$\boldsymbol{e} \cdot \{\boldsymbol{k} \times (\nabla \times \boldsymbol{e}) + \nabla \times (\boldsymbol{k} \times \boldsymbol{e})\} = 2\{\boldsymbol{e} \times (\boldsymbol{e} \times \boldsymbol{k})\} \cdot \nabla, \tag{6.204}$$

then (6.151) takes the form

$$\left(\frac{iL^3}{16\pi^3}\right) \int d^2q |u_{kz}|^{-1} e^{i\boldsymbol{k}\cdot\boldsymbol{r}} \{2c^2[\boldsymbol{e} \times (\boldsymbol{e} \times \boldsymbol{k})] \cdot \nabla - \omega_k^2 \boldsymbol{e} \cdot \boldsymbol{\epsilon}'' \cdot \boldsymbol{e}\} E_k(z)$$
$$= 4\pi\omega_k^2 \boldsymbol{e} \cdot \boldsymbol{P}^{NL}(\omega_k, \boldsymbol{r}). \tag{6.205}$$

Here, the vector $\boldsymbol{e} \times (\boldsymbol{e} \times \boldsymbol{k}) \sim \boldsymbol{E}_k \times \boldsymbol{H}_k$ is parallel to the ray vector $\boldsymbol{s}_k$, i.e., to the direction of the energy flux at $\epsilon'' = P^{NL} = 0$, and $\boldsymbol{\nabla} = \hat{z}d/dz$. Let θ denote the angle between $\boldsymbol{s}_k$ and $\hat{z}$ and ρ be the angle between $\boldsymbol{s}_k$ and $\boldsymbol{k}$, which is usually small; then,

$$[\boldsymbol{e} \times (\boldsymbol{e} \times \boldsymbol{k})] \cdot \hat{z} = -k \cos\theta \cos\rho \equiv -\bar{n}\omega_k/c \approx -k_z. \tag{6.206}$$

Equation (6.166) is obtained by acting on (6.205) by the operator $\int d^2\rho \exp(-i\boldsymbol{q}' \cdot \boldsymbol{\rho})$.

6.4 Non-parametric interactions

Non-parametric (non-coherent) effects of nonlinear optics, i.e., effects like multi-photon absorption, are described by odd-order susceptibilities of the form $\chi^{(2m-1)}(\omega_1 = \omega_1 - \cdots - \omega_m + \omega_m)$. The rate of the mode k_1 amplitude variation scales then as the local amplitude $E_1(z)$ of the same mode, there is no phase matching factor $\exp(i\Delta kz)$, and one-dimensional SVA equations (6.166) can be represented in the form

$$\left(\frac{d}{dz} + \frac{\alpha_1}{2} - \frac{2\pi i\omega_1\chi_1}{c\overline{n_1}}|E_2|^2\cdots|E_m|^2\right)E_1 = 0, \tag{6.207}$$

$$\chi_1 \equiv \chi^{(2m-1)}(\omega_1 = \omega_1 - \cdots - \omega_m + \omega_m)\vdots \boldsymbol{e}_1^*\boldsymbol{e}_1\ldots\boldsymbol{e}_m^*\boldsymbol{e}_m. \tag{6.208}$$

Thus, the real part of χ_1 determines the local variation of the propagation constant (i.e., of the wavelength or the phase velocity) of mode k_1, which scales as the local intensities of other modes (*induced dispersion* effect), while the imaginary part of χ_1 determines additional absorption or amplification due to the energy of other modes (*nonlinear or induced absorption* effect).

Multiplying (6.207) by E_1^* or directly using (6.181), we find the system of equations for the intensities ($F_i \equiv F_{iz}$),

$$[d/dz + \alpha_1 + \alpha_1^{(m)}(z)]F_1(z) = 0, \tag{6.209}$$

where

$$\alpha_1^{(m)} \equiv \beta_1^{(m)}F_2\cdots F_m,\ \beta_1^{(m)} \equiv \frac{1}{2\hbar}\left(\frac{8\pi\hbar}{c}\right)^m\frac{\omega_1\cdots\omega_m}{\overline{n_1}\cdots\overline{n_m}}\chi_1''.$$

Recall that the signs of F_i and $\overline{n_i}$ coincide and are determined by the sign of the $\boldsymbol{k}_i$ projection onto the z axis; therefore, F_1 can increase in the course of propagation only if $\chi_1'' < 0$.

Below, we will consider the basic types of non-parametric effects: nonlinear absorption (including the saturation effect), Raman amplification and absorption, spontaneous and stimulated scattering, self-focusing and self-modulation, as well as their role in optics applications and in spectroscopy.

6.4.1 *Nonlinear absorption*

Consider the single-mode case. The imaginary part of the cubic susceptibility,

$$\chi^{(3)''} \equiv \mathrm{Im}\chi^{(3)}(\omega = \omega - \omega + \omega)\vdots \boldsymbol{e}^*\boldsymbol{e}\boldsymbol{e}^*\boldsymbol{e},$$

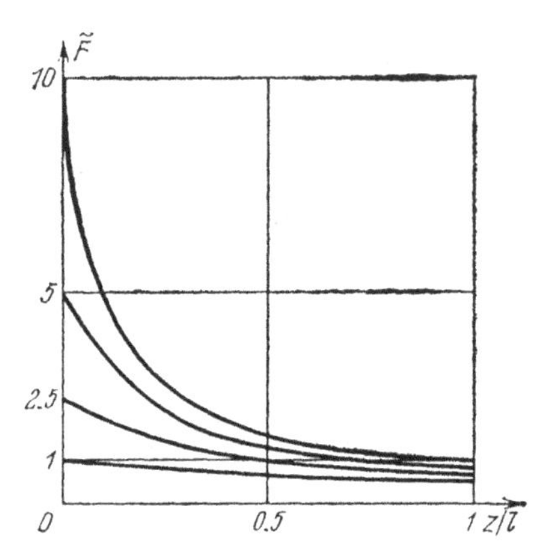

Fig. 6.6 Variation of the intensity of light in the course of propagation and the limitation effect under two-photon absorption: $\tilde{F}$ is the photon flux density times βl, where β is the two-photon absorption coefficient, and l is the thickness of the layer.

leads to the violation of the exponential Bouger's law for the intensity variation in matter. This effect can be related to two-photon absorption (see (6.146)), in which $\chi^{(3)\prime\prime} > 0$, and to the saturation of a single-photon resonance, where, according to (4.99),

$$\chi^{(3)\prime\prime} = -\chi^{(1)\prime\prime} d_{ab}^2 T_1 T_2/\hbar^2[1 + (\omega_0 - \omega)^2 T_2^2].$$

Let a plane monochromatic wave be incident on a transparent (in the linear approximation) medium orthogonally to the boundary. Then, according to (6.209),

$$dF/dz + \beta F^2 = 0, \; \beta \equiv 32\pi^2\hbar\omega^2\chi^{(3)\prime\prime}/c^2n^2 > 0. \tag{6.210}$$

The solution to this equation can be easily found,

$$F(z) = F(0)/[1 + \beta z F(0)] \to 1/\beta z. \tag{6.211}$$

The last equation takes place at $F(0) \gg 1/\beta z$ and describes the *limitation effect* (Fig. 6.6), in which the intensity $F(z)$ of the transmitted light does not depend on the incident light intensity $F(0)$.

If $F(0)$ has fluctuations (in particular, due to the photon structure of the field), the two-photon absorption will smear them, i.e., at the output of the layer photons will be distributed in time and space more uniformly than at the input (the *antibunching effect*, see Sec. 7.6).

For two-photon inter-band transitions in semiconductors like CdS, $\chi^{(3)\prime\prime} \approx 10^{-13}$ cm^3/erg at $\lambda_0 \sim 0.7\ \mu$ and $n \sim 2$. Then, $\beta/\hbar\omega \approx 1$ cm/GW, i.e., the limitation level at $l = 1$ cm for $I \equiv \hbar\omega F$ is on the order of 1 GW/cm^2. Note that at such intensities, the structure of the matter can be changed, there can be a phase transition and plasma can be created. Such *optical breakdown* in a transparent gas or condensed matter can start from multi-photon absorption or ionization.

Ionization of an inert-gas atom requires the absorption of about 10 photons from the neodimium laser ($\hbar\omega \sim 1$ eV). According to (6.209), the number of such transitions per unit volume and time is

$$\dot{N} = dF/dz = (8\pi/c)^{10}\chi^{(19)\prime\prime}I^{10}/2\hbar \sim (E_1/E_{NL})^{20}, \qquad (6.212)$$

with $E_{NL} \sim \hbar\omega_0/d_0$. Although this value is extremely small, the effect can be observed by focusing a Q-switched laser [Delone (1978)]. Note that the high power of the field in (6.212) increases the effect of its fluctuations. Indeed, according to (7.41), radiation with Gaussian statistics is $\langle I^{10}\rangle/\langle I\rangle^{10} = 10! \sim 3\cdot 10^6$ times as efficient as a non-fluctuating radiation with the same $\langle I\rangle$.

As it was shown by Keldysh (see Ref. [Elyutin (1982)], Sec. 29), at very large field values E_1 or small frequencies ω, so that the condition $d_0E_1 < \hbar\omega$ is violated, the dramatic power dependence of (6.212) on E_1 gets slower due to the factor $\exp(-\hbar\omega_0/d_0E_1)$, which is typical for the effect of *tunnel ionization* in a constant field.

Power dependence of the form (6.212) can be also violated due to the dissociation of multi-atom molecules in the field of IR-lasers under a single-photon resonance with one of the vibrational frequencies of the molecule, $\omega \sim \Omega_v$. In the case of a CO_2 laser ($\lambda \sim 10\,\mu, \hbar\omega \sim 0.1$ eV), dissociation requires several tens of photons. However, experiments with short pulses, where collision relaxation is too slow to be revealed, manifest a weak dependence on E_1. This effect is probably caused by the fast transfer of the absorbed energy to a large number of other vibrations ('intra-molecular relaxation'). Such effects are studied by a new field in quantum electronics, *IR photochemistry*, and used for the laser separation of isotopes.

In *photochrome materials* [Barachevsky (1977)], the limitation (darkening) effect is observed at much lower intensities, even in sunlight, where $I \sim 0.1$ W/cm^2, due to the 'photochemical' anharmonicity, which is very strong but inertial.

The lowest nonlinearity threshold is observed for the photographic process, where the photochemical reaction (creation of blackening centers, silver particles in AgBr microcrystallites) sometimes requires an energy density (exposure) as low as $It \sim 10^{-8}$ J/cm^2. It is noteworthy that for the creation of a single center, several photons are required ($m > 1$), i.e., this is a highly nonlinear process. At small exposures, the degree of blackening D of a developed negative scales as the number of centers; therefore, $D = \eta I^m$, where η is the film sensitivity, and we ignore the statistical variance of m. The relative intensity decrease for the probe light after passing through a negative with a small thickness l is

$$\Delta I/I = -D = -\alpha^{(m+1)}l, \tag{6.213}$$

where $\alpha^{(m+1)} \equiv \eta I^m/l$.

Various types of photochemical nonlinearity are used for holography and, in general, for information recording. More 'fast' optical nonlinearities form the basis of *dynamical holography* and the *optical phase conjugation* effect (Sec. 6.5) where recording and reading are simultaneous.

Thus, even a material that is transparent in the usual sense becomes absorbing or reflecting at sufficiently high intensities. The study of such effects can have very important applications in connection with the problem of laser thermonuclear synthesis.

Saturation of a single-photon resonance (Sec. 4.3), on the contrary, leads to the bleaching of matter (in the absence of the population inversion). In this case, the intensity fluctuations become more pronounced, which is used for mode locking in picosecond lasers. According to the two-level model, the absorption coefficient $\alpha(F)$ (or, at $\alpha < 0$, the amplification coefficient) at large intensity F has the form $\alpha_0/(1 + F/F_s)$, with the parameter F_s determined by the relaxation times T_1, T_2 and the transition moment d_{ab} (see (4.96)). Hence, (6.210) is replaced by the nonlinear transfer equation for a plane wave in the presence of saturation (i.e., at a single-photon resonance),

$$\frac{dF}{dz} + \frac{\alpha_0 F}{1 + F/F_s} = 0. \tag{6.214}$$

Note that here, in contrast to (6.210), contributions of an infinite number of odd-order susceptibilities are taken into account. The solution to (6.214) can be easily obtained in an indirect form,

$$-\alpha_0 z = \ln(F/F_0) + (F - F_0)/F_s, \tag{6.215}$$

where $F \equiv F(z)$, $F_0 \equiv F(0)$. The two terms in the right-hand side of (6.215) correspond to the exponential and linear $F(z)$ dependencies, respectively (Fig. 6.7). At strong saturation, the exponential Bouger's law becomes linear,

$$F \approx \left(1 - \frac{\alpha_0 z}{1 + F_0/F_s}\right) F_0, \tag{6.216}$$

and in the limit $F_0 \gg F_s$ the matter is completely bleached (compare with the nonstationary SIT effect described in Sec. 5.1).

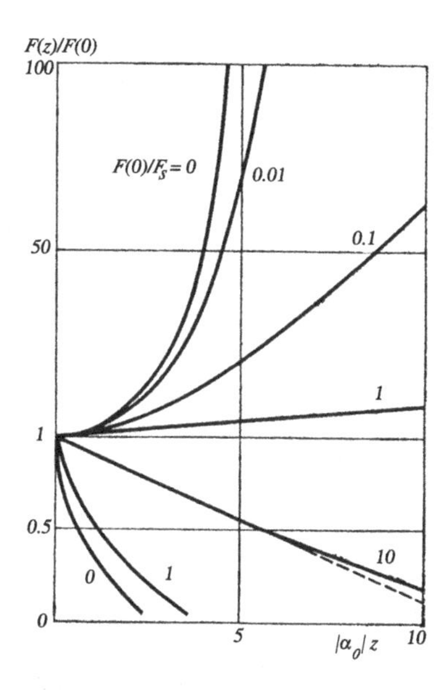

Fig. 6.7 Intensity F of a wave versus the distance z in an amplifying or absorbing medium in the presence of saturation. At strong saturation, the exponential variation of F is replaced by a linear one (the dashed line), α_0 is the amplification coefficient for a weak signal, F_s is the signal intensity at which α_0 is reduced twice.

6.4.2 *Doppler-free spectroscopy*

Induced absorption due to a two-photon transition and induced bleaching due to the saturation of a single-photon transition form the basis for two interesting spectroscopic methods that allow one to overcome Doppler's broadening of resonances masking the fine details of the spectra.

The scheme of a two-photon Doppler-free spectroscope is shown in Fig. 6.8. The gas under study is placed into a standing wave with the frequency ω, which is scanned in the vicinity of the transition half-frequency $\omega_0/2$. Due to the linear Doppler effect, a traveling wave gets a frequency shift $\omega' - \omega = -\boldsymbol{k} \cdot \boldsymbol{v}$, where ω is the field frequency in a laboratory frame of reference, ω' is the same in the reference frame of a molecule moving with the velocity $\boldsymbol{v}$, $\boldsymbol{k}$ is the field wave vector. In a standing wave, half of the photons have the wavevector $\boldsymbol{k}$ and half have the wavevector $-\boldsymbol{k}$; therefore, their Doppler shifts differ in signs. As a result, if a two-photon transition occurs through one photon absorbed from the 'forward' wave and the other photon, from the 'backward' one, the shifts are fully compensated, so that the observed resonance will look like a sharp peak with the natural or collision width, on a broad Doppler's pedestal.

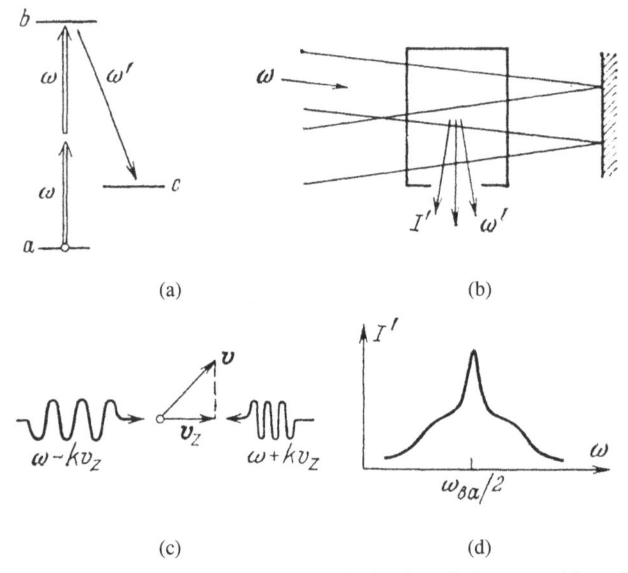

Fig. 6.8 Doppler-free two-photon spectroscopy: (a) the levels (ω is the scanned laser frequency and ω' is the frequency of the observed fluorescence); (b) the scheme of the spectroscope; (c) the idea of Doppler-shift compensation; (d) the signal as a function of the laser frequency.

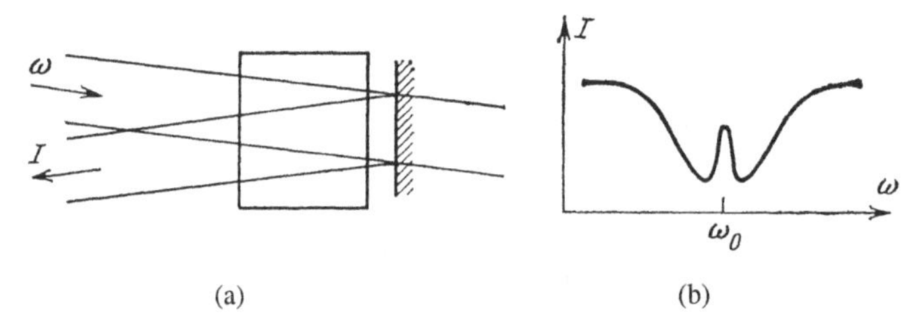

Fig. 6.9 Doppler-free saturation spectroscopy: (a) the scheme of the saturation spectroscope; (b) the observed signal as a function of the laser frequency.

Note that it is convenient to register the resonance indirectly, by observing single-photon fluorescence accompanying the transition of a molecule from the excited level to a third level. A disadvantage of the two-photon spectroscopy is the necessity to use a strong field, because of the small transition probability, which leads to a considerable shift of the resonance due to the optical Stark effect (see (5.48)).

The Doppler-free saturation method [Letokhov (1975)] also uses counter-propagating waves with the same frequency (Fig. 6.9). The forward wave has a larger intensity and causes strong saturation of molecules whose velocities have appropriate projections $v_z = (\omega - \omega_0)/k$ on the wave propagation direction. The backward probing wave at $\omega \neq \omega_0$ interacts with another group of molecules, which have the opposite sign of v_z, and therefore is strongly absorbed. However, at $\omega = \omega_0$, we have $v_z = 0$, and the resonance is bleached for the probe wave due to the saturation caused by the forward wave. Hence, the readings of the detector registering the power of the probe wave after the sample will grow at $\omega \approx \omega_0$. The width of the resonance will be determined by the natural, collision, or field-induced (4.101) broadening.

For more detailed description of nonlinear spectroscopy, see Refs. [Letokhov (1975); Akhmanov2 (1981); Bloembergen (1977); Steinfeld (1978); Walther (1976); Letokhov (1983)]. Let us mention other types of Doppler-free spectroscopy: the methods of *quantum beats* (Sec. 5.2), *molecular beams*, *buffer gas*. In the last method, an inert gas is added to reduce the free path of the molecules under study, so that the free path becomes much less than the wavelength. This leads to a narrow peak with the homogeneous broadening appearing in the center of a broad Doppler line.

6.4.3 *Raman amplification*

Consider now non-parametric interaction of two modes with different frequencies. Here, the most interesting effect is *Raman amplification*, which is closely related to the inelastic scattering of light. In the macroscopic description, the scattering is explained by a resonance of the cubic susceptibility $\chi^{(3)}(\omega_2, -\omega_1, \omega_1)$ at $\omega_1 - \omega_2 \equiv \Omega \sim \Omega_0 > 0$. Here, the combination frequency Ω is close to the frequency $\Omega_0 \equiv \omega_{ba}$ of some matter excitation. Usually, Ω_0 is one of the eigenfrequencies of molecular vibrations ($\Omega_0 \sim 10^2 - 10^3$ cm^{-1}) or the frequency of an acoustic wave ($\Omega_0 \sim 0 - 0.1$ cm^{-1}). In the first case, the scattering is called the *Raman* one, in the second one, the *Mandelshtam-Brillouin* one. In piezoelectric crystals, one also observes light scattering by polaritons, $\Omega_0 \lesssim 10^3$ cm^{-1} (Secs. 4.2, 6.5). In this case, similar to the case of acoustic-phonon scattering, $\chi^{(3)}$ and Ω_0 depend on the angle between $\boldsymbol{k}_1$ and $\boldsymbol{k}_2$. Therefore these effects are actually resonance ones, and they will be considered in more detail in Sec. 6.5.

It is important that the imaginary part of the cubic susceptibility is negative in the vicinity of the Stokes resonance. As a result, the energy is transferred from the high-frequency field components to the low-frequency ones. The spectrum becomes 'redder' and the matter is heated; as a result, the population difference $N_a - N_b$ for the transition with the frequency ω_{ba} reduces.

Let two polarized plane waves with the frequencies $\omega_{1,2} > 0$ be incident on a transparent medium, orthogonally to the boundary. From (6.209), we find the equations for the energy fluxes,

$$dF_1/dz + \beta F_1 F_2 = 0, \tag{6.217}$$

$$dF_2/dz - \beta F_1 F_2 = 0, \tag{6.218}$$

where

$$\beta \equiv -32\pi^2 \hbar k_1 k_2 \chi^{(3)\prime\prime}/n_1^2 n_2^2 = \sigma_{st}^{(2)}(N_a - N_b) > 0, \tag{6.219}$$

$$\begin{aligned}\chi^{(3)\prime\prime} &\equiv \mathrm{Im}\chi^{(3)}(\omega_2 = \omega_2 - \omega_1 + \omega_1)\vdots e_2^* e_2 e_1^* e_1 \\ &= -\mathrm{Im}\chi^{(3)}(\omega_1 = \omega_1 - \omega_2 + \omega_2)\vdots e_1^* e_1 e_2^* e_2.\end{aligned} \tag{6.220}$$

The last equation follows from (6.88) (see also (6.146), (6.192)); it provides conservation of the total number of photons in Raman two-photon transitions,

$$F_1(z) + F_2(z) = C. \tag{6.221}$$

By substituting (6.221) into (6.217), (6.218), we find

$$F_1 = \frac{C}{F_{10} + F_{20}e^{\beta Cz}} F_{10}, \tag{6.222}$$

$$F_2 = \frac{C}{F_{20} + F_{10}e^{-\beta Cz}} F_{20}, \tag{6.223}$$

where $F_{i0} \equiv F_i(0)$ and $C \equiv F_{10} + F_{20}$. At a sufficiently large distance, $F_1 \to 0$, $F_2 \to C$, i.e., all photons become Stokes ones (Fig. 6.10). The released energy is then spent on the excitation of matter; this phenomenon is used for the generation of ultrasound via stimulated Mandelshtam-Brillouin scattering. In practice, the energy conversion is not complete because of several effects we did not take into account here. In particular, these are spontaneous-stimulated transitions and the generation of new components with the frequencies $2\omega_1 - \omega_2$ and $2\omega_2 - \omega_1$, as well as waves with the same frequencies ω_1, ω_2 but with different polarization etc.

Let now $F_{10} \gg F_{20}$, then at sufficiently small z the variation of the high-frequency field can be neglected (the *undepleted-pump approximation*). In this case, (6.223) yields exponential Raman amplification,

$$\begin{aligned} F_2 &= F_{20}\exp(\alpha_2 z), \\ \alpha_2 \equiv \beta F_{10} &= -32\pi^2 \omega_2 \chi^{(3)\prime\prime} I_{10}/c^2 n_1 n_2 > 0. \end{aligned} \tag{6.224}$$

In practice, for strong Raman resonances with molecular vibrations in liquids, α_2 is as high as 1 cm^{-1} at $I \gtrsim 0.1$ GW/cm^2. Thus, *Raman anharmonicity enables light*

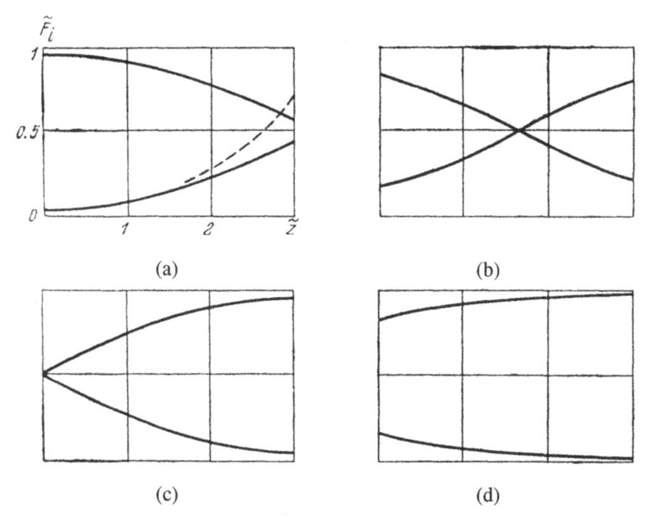

Fig. 6.10 Raman interaction of two waves at various initial intensities. The horizontal axis is the distance z in $1/\beta C$ units and the vertical one shows the intensities F_i in units $C = F_{10} + F_{20}$, i.e., the functions $\tilde{F}_1 = (1 + \alpha e^{\tilde{z}})^{-1}$ and $\tilde{F}_2 = 1 - \tilde{F}_1$, where $\alpha = F_{20}/F_{10}$ is the relative initial intensity of the Stokes wave. (a) $\alpha = 1/25$, the dashed line shows the asymptotics describing the exponential amplification of the Stokes wave in the undepleted-pump approximation; (b) $\alpha = 1/5$; (c) $\alpha = 1$; (d) $\alpha = 5$. The anti-Stokes wave, roughly, decays exponentially.

amplification without the population inversion. Raman amplification (or, for anti-Stokes frequencies, attenuation) is used in spectroscopy, as well as for shifting the frequency of lasers. If the feedback is added, with the help of a mirror, an amplifier becomes an oscillator (a *Raman laser*).

6.4.4 *Spontaneous and stimulated scattering*

Even without a feedback or an external signal at frequency ω_2, with $F_{10} \neq 0$ the matter will emit incoherent radiation at frequencies $\omega_1 \pm \Omega_0$. At $\alpha_2 l \gg 1$, the radiation at Stokes frequencies ω_2, i.e., the intrinsic noise of the Raman amplifier, can be comparable with the pump, $F_2 \sim F_1$. This effect is called *stimulated Raman scattering* (StRS). The emerging noise field with the central frequency $\omega_1 - \Omega_0$ plays the role of the pump for the second Stokes component, with the frequency $\omega_1 - 2\Omega_0$, which excites the third one, and so on. In addition, due to four-frequency parametric interactions like $2\omega_1 - \omega_2 \to \omega_3 \equiv \omega_1 + \Omega_0$, intense anti-Stokes components appear, and a quantitative analysis of the phenomenon becomes difficult.

Let us estimate the power of StRS in the undepleted-pump approximation, without taking into account higher-order Stokes and anti-Stokes components.

At $F_{20} = 0$, the Stokes field is generated due to spontaneous (to be precise, spontaneous-stimulated) transitions. According to the Kirchhoff law (Sec. 7.1), which is valid for two-photon transitions as well [Klyshko (1980)], the intensity of the noise at the frequency ω_2 at the output of the amplifier is (in photons per mode)

$$N = (\mathcal{N} + 1)(G - 1), \tag{6.225}$$

where

$$G \equiv \exp(\alpha_2 l), \ \mathcal{N} \equiv \frac{N_b}{N_a - N_b} = \left\{\exp\left[\frac{\hbar(\omega_1 - \omega_2)}{\kappa T}\right] - 1\right\}^{-1}, \tag{6.226}$$

T is the temperature of the matter, and we assume that the transition $a - b$ is not saturated.

For passing from N to F, one has to multiply the photon flux density in one mode, Nc/L^3 (L^3 is the quantization volume), by the effective number of modes for the amplifier or the detector,

$$\Delta F = \frac{Nc}{L^3}\Delta g = \frac{Nc\Delta^3 k}{(2\pi)^3} = N\frac{\Delta\omega\Delta\Omega}{2\pi\lambda^2}. \tag{6.227}$$

Here, we have considered a single polarization type, assumed the refractive index to be unity, and denoted the effective frequency and solid-angle bandwidths of the amplifier or the detector as $\Delta\omega$ and $\Delta\Omega$, respectively.

If the scattering volume has the shape of a thread, with $a \ll 1$ and $G \gg 1$, only longitudinal scattering ($\vartheta \approx 0$) or backscattering ($\vartheta \approx 180°$) into the solid angle $\Delta\Omega \approx A/l^2$ is considerable.[k] (Here, $A \equiv a^2$ is the cross-section area of the scattering volume and l is its length.) Then, from (6.227) we find the relation between the noise power per unit frequency band, $\mathcal{P}_\omega = \hbar\omega FA/\Delta\omega$, and the number of photons per mode,

$$\mathcal{P}_\omega = \hbar\omega n_F^2 N/2\pi, \tag{6.228}$$

where the dimensionless number $n_F \equiv A/\lambda l$ is called the wave parameter or the Fresnel number. Substitution of (6.225) into (6.228) yields ($\alpha \equiv \alpha_2$, $\omega \equiv \omega_2$)

$$\mathcal{P}_\omega = \hbar\omega n_F^2(\mathcal{N} + 1)(e^{\alpha l} - 1)/2\pi. \tag{6.229}$$

The total output power is found by integrating this expression in ω, which reduces to the multiplication by the effective noise band $\Delta\omega$ and the substitution of $\omega_1 - \Omega_0 \equiv \bar{\omega}$ for ω. In the case of weak pumping, $\Delta\omega$ is determined by the width

[k] In stimulated Mandelshtam-Brillouin scattering, it is only backscattering that is efficient, since $\Omega(0) = 0$ (see (6.93)).

of the Raman resonance, i.e., by the decay rate of molecular vibrations (usually, $\Delta\omega/2\pi c \sim 1 - 10\ \text{cm}^{-1}$). At high amplification, $\Delta\omega$ becomes $(\alpha l)^{1/2}$ times as narrow (Sec. 2.3)

Thus, according to (6.229), the scattered power depends on the pump intensity as $\exp(\beta l F_{10}) - 1$. At $\beta l F_{10} \ll 1$, the exponential dependence turns into a linear one, and then the scattering is called *spontaneous* (SpRS). Replacing A/l^2 in (6.229) by $\Delta\Omega$, we find the power of spontaneous Stokes scattering per unit frequency and solid angle,

$$\mathcal{P}_{\omega\Omega} = (\hbar c/\lambda^3)(\mathcal{N} + 1)\beta V F_{10}, \tag{6.230}$$

where $V = Al$ is the volume of the scattering area. It is not difficult to verify that this expression agrees with the above-calculated cross section (6.148) of a spontaneous-stimulated transition. In contrast to StRS, SpRS has a broad direction diagram, determined by the convolution of $\chi^{(3)}$ with the polarization unit vectors (6.220).

At $F_2 \gg F_1$, the role of the pump is played by the lower-frequency field, and the external (anti-Stokes) field $F_1(z)$, according to (6.217), (6.218), decays exponentially. This effect is used in the spectroscopic method of *inverse Raman scattering*. The noise field is then described by (6.225), (6.229) or (6.230) with the permutation of indices 1, 2 and the replacements $\mathcal{N} + 1 \to -\mathcal{N}, \beta \to -\beta$. Therefore, spontaneous anti-Stokes scattering with a given frequency is

$$\mathcal{N}/(\mathcal{N} + 1) = N_b/N_a = \exp(-\hbar\Omega_0/\kappa T)$$

times as weak as the Stokes scattering, provided that all other conditions are the same. With a sufficiently strong pumping, the power of the anti-Stokes component is saturated. The number of photons per mode tends then, according to (6.225), to $\mathcal{N}$, i.e., the anti-Stokes field acquires the brightness temperature

$$T_{ef} = T\omega_1/\Omega_0 \gg T. \tag{6.231}$$

6.4.5 *Self-focusing*

This single-frequency effect is connected with the 'self-action' of a quasi-plane quasi-monochromatic wave due to the real part of the cubic susceptibility $\chi^{(3)} \equiv \text{Re}\chi^{(3)}(\omega, -\omega, \omega)$. The main contribution in $\chi^{(3)}$ is from the electro-striction and orientational anharmonicity (then, $\chi^{(3)} > 0$). In an absorptive medium, temperature anharmonicity ($\chi^{(3)} < 0$) and saturation are added.

With an account for linear susceptibility, the polarization amplitude is (we assume that $\chi^{(1)\prime\prime} = \chi^{(3)\prime\prime} = 0$)

$$P = (\chi^{(1)} + \chi^{(3)}|E|^2)E \equiv \chi(|E|^2)E,$$

where $E \equiv E_1(\boldsymbol{r}, t)$ is the slowly varying amplitude, i.e., the envelope. As a result, the refractive index depends on the local intensity of the field,

$$n(\boldsymbol{r}, t) \equiv [1 + 4\pi\chi(|E|^2)]^{1/2} \approx n_0 + n_2|E|^2, \tag{6.232}$$

where $n_2 \equiv 2\pi\chi^{(3)}/n_0$.

Let $\chi^{(3)} > 0$, then a light beam entering a medium forms a dielectric waveguide with the profile $\Delta n(\boldsymbol{r}_\perp) = n_2|E(\boldsymbol{r}_\perp)|^2$ repeating the intensity distribution in the beam. The propagation velocity c/n at the beam axis will be less than at its edges, so that the side beams will bend towards the axis (*nonlinear refraction*), and the beam will shrink. This will lead to an additional increase in Δn near the axis, as well as in the rate of the shrinking, so that, as a result, the beam will 'collapse' to a certain minimal size, determined by competing processes (Fig. 6.11).

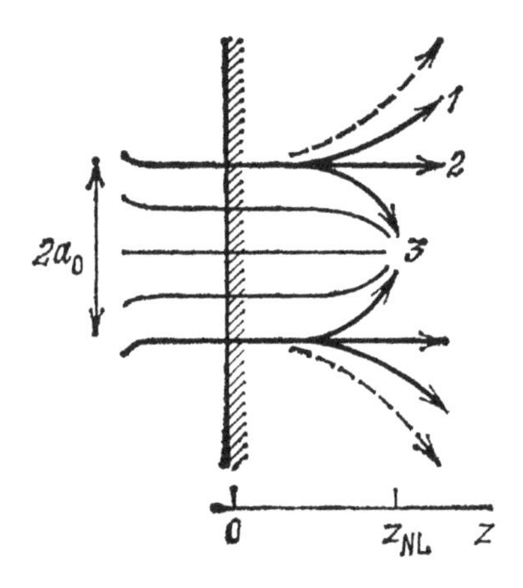

Fig. 6.11 Self-focusing and defocusing (dashed line): 1, a beam with low power, $\mathcal{P} \ll \mathcal{P}_0$, slowly diverges due to diffraction; 2, a beam with a critical power, $\mathcal{P}_0$, becomes 'channeled'; 3, a beam with $\mathcal{P} > \mathcal{P}_0$ 'collapses'.

One of the competing effects is diffraction. It is described by the divergence angle $\vartheta_d \sim \lambda_0/2n_0a$, where a is the initial beam radius and $\lambda_0 \equiv 2\pi c/\omega$, while the nonlinear refraction is characterized by the angle of total internal reflection due to the refractive index variation Δn,

$$\vartheta_{NL} \equiv \arccos\frac{n_0}{n_0 + \Delta n} \approx \sqrt{\frac{2\Delta n}{n_0}}, \tag{6.233}$$

where the beam profile is assumed to be rectangular. If refraction and diffraction compensate each other exactly, then waveguide propagation occurs, with a constant beam radius (*self-channeling*). Assuming $\vartheta_{NL} = \vartheta_d/2$, one can obtain a rough estimate for the self-channeling condition:

$$\Delta n_{min} \equiv n_2|E|^2_{min} \approx n_0\vartheta_d^2/8 \approx \lambda_0^2/32n_0a^2. \tag{6.234}$$

It follows that the waveguide cross-section πa^2 scales as the inverse intensity of light. In this case, the power of a self-channeling beam should be

$$\mathcal{P}_0 = cn_0|E|^2_{min}a^2/8 \approx c\lambda_0^2/256n_2. \tag{6.235}$$

This estimate differs $\pi^2/4$ times from a more accurate result, which will be obtained below from the nonlinear quasi-optical wave equation (6.239) in the *aberration-free approximation*, i.e., under the assumption that the wave front is spherical,

$$\mathcal{P}_0 = c\lambda_0^2/64\pi^2 n_2. \tag{6.236}$$

Let $\lambda_0 = 1\ \mu$, $\chi^{(3)} = 10^{-12}$ cm^2/erg (carbon sulfide), and $n_0 = 1.5$, then, according to (6.236), $\mathcal{P}_0 \sim 10^4$ W. This estimate shows that self-focusing can considerably change the features of other nonlinear effects, such as harmonics generation, StRS, and others, which are usually observed at strong pumping. Self-focusing is also related to the important problem of the optical strength of matter, which determines the maximal intensity of laser beams.

6.4.6 °*Self-focusing length*

Clearly, at $\mathcal{P} > \mathcal{P}_0$ the beam will compress (self-focus), and under the opposite condition it will diverge due to diffraction. The length of self-focusing will apparently depend on the value $(\mathcal{P} - \mathcal{P}_0)/\mathcal{P}_0$.

In order to find this dependence, consider the solution to the nonlinear wave equation (6.151) for a monochromatic field $\mathrm{Re}\tilde{\boldsymbol{E}}(\boldsymbol{r})e^{-i\omega t}$ in quasi-optical and aberration-free approximations (see Ref. [Vinogradova (1979)]). In a homogeneous isotropic medium without external sources, the field is transverse, so that (6.151) leads to the nonlinear Helmholtz equation,

$$\nabla^2\tilde{\boldsymbol{E}} + k^2(1 + (\epsilon_2/\epsilon_0)|\tilde{E}|^2)\tilde{\boldsymbol{E}} = 0, \tag{6.237}$$

where $k = n_0\omega/c$, $n_0 = \sqrt{\epsilon_0}$, $\epsilon_2 \equiv 4\pi\chi^{(3)} = 2n_0n_2$.

Let a polarized light beam with a limited cross-section but with a narrow direction diagram (narrow angular spectrum) be incident on a nonlinear transparent medium occupying a semispace $z > 0$. Then the field in the medium can be searched in the form

$$\tilde{\boldsymbol{E}}(\boldsymbol{r}) = \boldsymbol{e}E(\boldsymbol{r})\exp(ikz), \tag{6.238}$$

where $|\boldsymbol{e}|^2 = 1$, $e_z = 0$, and $E(\boldsymbol{r})$ is a slowly (compared to e^{ikz}) varying amplitude (SVA) describing the variation of the local amplitude of field oscillations along the beam and over its cross section.

In practice, the typical distances z_d, z_{NL} at which $E(z)$ changes noticeably due to diffraction and nonlinearity are much greater than λ; therefore, $\partial^2 E/\partial z^2$ is negligible compared to $2k\,\partial E/\partial z$ (the *SVA approximation*). In this case, from (6.237) and (6.238) we obtain a *nonlinear parabolic equation* for $E(\boldsymbol{r})$,

$$2ik\,\partial E/\partial z + \nabla_\perp^2 E + (k^2\epsilon_2/\epsilon_0)|E|^2 E = 0, \tag{6.239}$$

where $\nabla_\perp^2 \equiv \partial^2/\partial x^2 + \partial^2/\partial y^2$. At $\epsilon_2 = 0$, (6.239) is the basic equation of *quasi-optics*, which studies the propagation of directed wave beams.

If we neglect the variation of E over the cross-section, (6.239) leads to

$$E(z) = E(0)e^{i\Delta kz}, \ \Delta k/k = \Delta n/n_0 = 16\pi^2\chi^{(3)}I/cn_0^3, \tag{6.240}$$

i.e, the nonlinearity only changes the phase velocity $\omega/(k + \Delta k) = c/(n_0 + \Delta n)$ of the plane wave. For CS_2, at $I = 1$ GW/cm^2, $\Delta n = 2 \cdot 10^{-5}$.

Now, let the incident wave have a Gaussian profile and a plane wave front at the boundary: $E(x, y, 0) = E_0 \exp(-\rho^2/a_0^2)$, where $\rho^2 \equiv x^2 + y^2$ and a_0 is the initial radius of the beam. Let us try to find the solution to (6.239) in the form of a Gaussian beam with a variable radius, $a(z) \equiv a_0 f(z)$, and a parabolic wave front with a variable curvature $\beta(z)$ on the axis,

$$E(\boldsymbol{r}) = E_0 \exp[F(z)\rho^2 + i\varphi(z)]/f(z), \tag{6.241}$$

$$F(z) \equiv ik\beta(z)/2 - 1/a_0^2 f^2(z). \tag{6.242}$$

Here, $\varphi(z)$ is the additional to kz phase delay on the axis, and the factor $1/f$ provides the dependence of the wave power on z,

$$\mathcal{P} = (cn_0/8\pi)\int dx\,dy\,|E|^2 = (cn_0/16)E_0^2 a_0^2. \tag{6.243}$$

At small ρ, the wave (6.241) has approximately spherical wavefront with the radius $1/\beta(z)$. Below, it will be shown that the curvature of the front and the width of the beam at a given z are related through Eq. (6.247).

From (6.241), we find

$$\partial E/\partial z = (\rho^2 F' + i\varphi' - f'/f)E, \ \nabla_\perp^2 E = 4F(1 + \rho^2 F)E, \tag{6.244}$$

$$|E|^2 = \frac{E_0^2}{f^2}\exp\left(-\frac{2\rho^2}{a_0^2 f^2}\right) \approx \frac{E_0^2}{f^2}\left(1 - \frac{2\rho^2}{a_0^2 f^2}\right).$$

In the last equation, we have passed from a Gaussian profile to a parabolic one, which is possible for the paraxial part of the beam (the *aberration-free approximation*). By substituting (6.244) into (6.239), one can find the functions f, β, φ. Indeed, setting to zero the sum of the coefficients by ρ^2, we find

$$2ikF' + 4F^2 - k^2/z_{NL}^2 f^4 = 0, \tag{6.245}$$

where

$$z_{NL}^2 \equiv n_0 a_0^2 / 4\Delta n_0 \tag{6.246}$$

and Δn_0 is the refractive index variation at point $r = 0$. The imaginary part of (6.245) yields

$$\beta = d(\ln f)/dz, \tag{6.247}$$

and the real part,

$$f'' = 1/z_0^2 f^3. \tag{6.248}$$

Here, the following notation has been introduced:

$$\begin{aligned} z_0^2 &\equiv (z_d^{-2} - z_{NL}^{-2})^{-1} = z_d^2/(1 - \mathcal{P}/\mathcal{P}_0) = z_{NL}^2/(\mathcal{P}_0/\mathcal{P} - 1), \\ z_d &\equiv ka_0^2/2, \; z_d^2/z_{NL}^2 = \mathcal{P}/\mathcal{P}_0 = (\Delta n_0/n_0)(ka_0)^2. \end{aligned} \tag{6.249}$$

The z_0 parameter determines the length at which the beam 'collapses', and the z_d parameter has the meaning of the 'diffraction' length; here, $\mathcal{P}_0$ and $\mathcal{P}$ are given by Eqs. (6.236) and (6.243).

Hence, we finally obtain the z dependencies for the width and the curvature radius $R(z)$ of a parabolic beam due to its self-action,

$$a(z)/a_0 = f = (1 + \tilde{z}^2)^{1/2}, \tag{6.250}$$

$$R(z) = 1/\beta = z(1 + \tilde{z}^{-2}), \tag{6.251}$$

with $\tilde{z} \equiv z/z_0$. The function $\varphi(z)$ can be found from the coefficients of the zeroth order in ρ appearing after the substitution of (6.244) into (6.239). The field amplitude in the beam has the form

$$E(\boldsymbol{r}) = \frac{E_0}{(1 + \tilde{z}^2)^{1/2}} \exp\left[-\frac{\rho^2}{a_0^2(1 + \tilde{z}^2)} + \frac{ik\rho^2}{2z(1 + \tilde{z}^{-2})} + i\varphi(z)\right]. \tag{6.252}$$

This solution at $\mathcal{P} \ll \mathcal{P}_0$ (i.e., at $z_0 \approx z_d$) describes the so-called TEM_{00} wave, which is the simplest solution to the nonlinear parabolic equation,

$$E_{\mathrm{TEM}} = \frac{E_0}{1 + iz/z_d} \exp\left[-\frac{\rho^2}{a_0^2(1 + iz/z_d)}\right]. \tag{6.253}$$

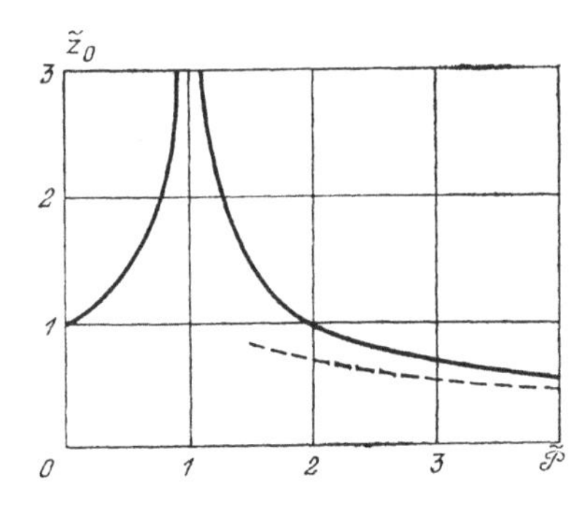

Fig. 6.12 Dependence of z_0 (in $ka_0^2/2$ units) on the beam power $\mathcal{P}$ (in the units of critical power $\mathcal{P}_0$), given by the expression $|1 - \tilde{\mathcal{P}}|^{-1/2}$. At $\tilde{\mathcal{P}} < 1$, z_0 has the meaning of the beam diffraction length, and at $\tilde{\mathcal{P}} > 1$, it is the 'collapse' length. The latter, at $\tilde{\mathcal{P}} \gg 1$, is approximately equal to the 'nonlinear' length z_{NL} (dashed line).

(Note that $1 + iz = (1 + z^2)^{1/2} \exp(i \arctan z)$, i.e., here, $\varphi(z) = -\arctan(z/z_d)$.) The diffraction length z_d determines the distance at which the cross-section of the beam is doubled and the front radius achieves its minimal value $2z_d = ka_0^2 \equiv b$ (the so-called *confocal parameter* of the beam). At $\mathcal{P} \to \mathcal{P}_0$, the role of z_d is played by z_0, and diffraction slows down. Finally, at $\mathcal{P} > \mathcal{P}_0$, the parameter z_0^2 and, hence, $\tilde{z}^2$, becomes negative, the beam is narrowed, and its width and the wavefront radius turn into zero at $z = |z_0|$; at $\mathcal{P} \gg \mathcal{P}_0$, $|z_0| \approx z_{NL}$ (Fig. 6.12). Certainly, quasi-optical aberration-free approximation describes correctly only the initial stage of self-focusing; however, the typical self-focusing length should be on the same order of magnitude as $|z_0|$.

In the vicinity of the focus, an important role can be played by higher-order nonlinearities $\chi^{(5)}, \ldots$ and multi-frequency nonlinear effects, for instance, stimulated scattering. Note that in the case of pulsed fields, the focus position $|z_0(t)|$ is time-dependent: as the intensity increases, the focus moves from infinity to some minimal distance, and then it goes to infinity again (the *moving focus*).

Next, let a wave with a plane wavefront and a complicated profile $|E(x, y, 0)|^2 \equiv I(x, y)$ (a wave with transverse amplitude modulation) be incident on a nonlinear medium. In the course of propagation, diffraction tends to smoothen the profile inhomogeneities, while the nonlinear refraction, in contrast, makes them more pronounced (*self-modulation*). Qualitatively, these effects are described by (6.250) if $a(z)$ is understood as the width of an inhomogeneity. As a result, a wave can split in numerous waveguided filaments, each of them carrying a power of $\mathcal{P}_0$. This splitting is an example of dynamic *transverse instability* for waves in a nonlinear medium and is similar to the formation of solitons in the case of *longitudinal instability* (see below).

Clearly, at $\chi^{(3)} < 0$, self-defocusing should take place (Fig. 6.11). It is described by the above-given theory with negative $\mathcal{P}_0$ and z_{NL}^2. This effect can be easily observed in continuous-wave laser beams by adding absorbing dyes to transparent solutions.

In two-frequency (and/or two-wave) experiments, *mutual focusing* or *defocusing* is observed due to $\mathrm{Re}\chi^{(3)}(\omega_2 = \omega_2 - \omega_1 + \omega_1)$: an intense light beam with a frequency ω_1 turns a plane-parallel plate into a collecting or dispersing lens for the wave ω_2. The same frequency component of the cubic susceptibility describes *induced dispersion* and the *optical Kerr effect*: a plane pump with the frequency ω_1 turns an isotropic medium into a birefringent one for the second wave, with the frequency ω_2. The optical Kerr effect and mode-locked lasers enable the manufacturing of optical shutters with switching times on the order of 1 ps.

Consider now a plane quasi-monochromatic wave with the time-modulated envelope $E(t)$. In a linear medium, within the first approximation of the dispersion theory, $E(t)$ propagates without changing its shape, with the group velocity $\boldsymbol{u} = \nabla\omega(\boldsymbol{k})$. In the framework of the second approximation, $d^2\omega/dk^2 \neq 0$, and the intensity spikes get spread; this effect is similar to the diffraction smoothing of transverse inhomogeneities. In a nonlinear medium, self-action can lead to an inverse effect: at $\chi^{(3)}d^2\omega/dk^2 < 0$, pulses with sufficient energy are compressed (*self-compression*, see Ref. [Vinogradova (1979)], Sec. 9.5). A quasi-monochromatic wave with a small initial modulation splits, as a result of self-modulation, in separate pulses with fixed energy and shape, the solitons,[1], which are analogues of waveguided filaments in the case of self-focusing. In practice, such effects are observed for picosecond pulses, so that inertial nonlinearity mechanisms (for instance, the temperature one) are not manifested.

6.5 Parametric interactions

In the case of a parametric interaction of m modes with positive frequencies ω_i, the stationarity condition has the form of the energy conservation law for an m-photon process,

$$\Delta\omega \equiv \sum_{i=1}^{m} s_i\omega_i = 0, \qquad (6.254)$$

where $s_i = \pm 1$, $\omega_i = \omega(\boldsymbol{k}_i) > 0$. According to the one-dimensional model, the waves interacting in an infinite layer should satisfy the transverse momentum con-

[1]Recall that light solitons (with the area 2π) are also formed in the range of resonance absorption due to the two-level anharmonicity (Sec. 5.1)

servation law,

$$\Delta k_x = 0,\ \Delta k_y = 0, \tag{6.255}$$

where $\Delta \boldsymbol{k} \equiv \sum s_i \boldsymbol{k}_i$. Formally, one obtains (6.254) and (6.255) by doing the three-dimensional Fourier transformation of Maxwell's equations (Sec. 6.3). In experiment, the interaction time τ and cross-section $A \equiv a^2$ are always limited; therefore, (6.254) and (6.255) are valid only up to an accuracy of $1/\tau$, $1/a$, respectively,

$$|\Delta\omega| \lesssim 1/\tau,\ |\Delta k_\perp| \lesssim 1/a. \tag{6.256}$$

In this section, as a rule, for simplicity we will neglect linear absorption ($\alpha_i = 0$). Then, interaction of m modes can be described within the linear approximation, according to (6.173), by a system of linear equations,

$$\frac{dE_1^*}{dz} = \frac{2\pi s_1\omega_1}{ic\overline{n}_1}P_1^*(z)e^{i\Delta z}, \tag{6.257}$$

where

$$P_1^*(z) = \chi^{(m-1)}(s_1\omega_1; s_2\omega_2; \ldots; s_m\omega_m)E_2(z)\cdots E_m(z), \tag{6.258}$$

$E_i = E_{k_i}$ at $s_i = 1$ and $E_i = E_{k_i}^*$ at $s_i = -1$, $\chi^{(m-1)}$ is the convolution of the nonlinear susceptibility tensor with the polarization unit vectors, $\overline{n} \equiv n\cos\theta\cos\rho$, n is the refractive index, θ and ρ are the angles formed by the Poynting vector with the z axis and the $\boldsymbol{k}$ vector, respectively, $\Delta \equiv \Delta k_z$. Equations for $E_2, \ldots, E_m$ have a similar form and can be obtained by subscript permutations.

It follows from (6.257) that parametric interaction is efficient only at sufficiently small $|\Delta|$, i.e., under the condition that the longitudinal momentum of the field is preserved,

$$|\Delta| \equiv |\sum s_i k_{iz}| < 1/l. \tag{6.259}$$

The phase matching condition $\Delta\boldsymbol{k} \sim 0$ reduces dramatically the number of essential 'interactions', i.e., combinations $\{\boldsymbol{k}_i, \nu_i, s_i\}$, especially in an anisotropic medium where the refractive index $n_\nu(\omega, \vartheta, \varphi) = ck/\omega_\nu(\boldsymbol{k})$ depends not only on the frequency, but also on the polarization index $\nu = 1, 2$ and the direction $\hat{\boldsymbol{k}} \equiv \boldsymbol{k}/k \equiv \{\vartheta, \varphi\}$.

6.5.1 *Undepleted-pump approximation — the near field*

Let $k_{1z} > 0$ and $E_1(0) = 0$, i.e., the mode labeled by 1 is not excited by an external source. If the nonlinearity, or the layer thickness, or the incident pump fields $E_i(0)$, $i = 2, \ldots, m$, are sufficiently small, one can neglect the effect of the nonlinearity on the pump, i.e, assume that the nonlinear polarization $P_1(z)$ at frequency ω_1 in the right-hand side of (6.257) is a given function of the coordinates, determined by the spatial distribution of free incident fields.

For a given polarization, taking into account definition (6.168), from (6.257) we find the expression for the amplitude of the mode $(\boldsymbol{k}, \nu)$ excited in a homogeneous transparent linear medium by a Fourier component of the polarization,

$$E_\nu(k) = \frac{8\pi^2 iku_k}{L^3 n_k^2 \cos\rho_k} \int_V d^3r\, e^{-i\boldsymbol{k}\cdot\boldsymbol{r}} \boldsymbol{e}_\nu \cdot \boldsymbol{P}(\omega_k, \boldsymbol{r}), \tag{6.260}$$

where u_k is the group velocity. Note that (6.260) determines the radiation field outside of the radiating area V regardless of its shape.

In the case of a plane-parallel layer with the thickness z, it follows from (6.257) that

$$E_1^*(z) = -\frac{2\pi s_1 \omega_1 \chi^{(m-1)}}{c\overline{n_1}} E_2(0) \ldots E_m(0) \frac{e^{i\Delta z} - 1}{\Delta}, \tag{6.261}$$

$$I_{1z}(z) = \left(\frac{8\pi}{c}\right)^m \frac{\omega_1^2 |\chi^{(m-1)}|^2 z^2}{16\overline{n_1} \ldots \overline{n_m}} \operatorname{sinc}^2\left(\frac{\Delta z}{2}\right) I_{2z}(0) \ldots I_{mz}(0), \tag{6.262}$$

where $I_{kz} = c\overline{n_k}|E_k|^2/8\pi = \hbar\omega_k F_{kz}$ is the longitudinal component of the energy flux density in mode k and $\operatorname{sinc}(x) \equiv (\sin x)/x$. Eqs. (6.261), (6.262) give the efficiency of parametric frequency conversion.

Note that the 'new' mode $(\boldsymbol{k}_1, \nu_1) \equiv k_1$ is determined through the pump modes $k_2, \ldots, k_m$ by conditions (6.254), (6.255) only up to the polarization type ν_1 and the sign σ_1 of the longitudinal component k_{1z} (see (6.165)). In other words, each combination of polarized pump waves $\sum s_i \boldsymbol{k}_i$, $i = 2, \ldots$ excites, in principle, four waves with the frequency ω_1, which differ by indices $\nu_1 = 1, 2$ and $\sigma_1 = \pm 1$. However, the factor $\operatorname{sinc}^2(\Delta z/2)$ in (6.262) usually strongly suppresses the wave with $k_{1z} < 0$ (about $(k_{1z}z)^2$ times, see Fig. 6.13).

Thus, *in the undepleted-pump approximation, excitation of new waves scales as the product of the incident intensities, the squared nonlinear susceptibility, and, provided that phase matching is satisfied* ($|\Delta|z \ll 1$)*, as the square of the interaction length.* In the case of the opposite inequality, z^2 is replaced by $(\Delta/2)^{-2}\sin^2(\Delta z/2) \sim 2/\Delta^2$, i.e., if the phase velocities are not matched, the amplitude of the new wave periodically turns to zero. The distance $\pi/|\Delta|$, at which the intensity of the wave varies monotonically, is called the *coherence length*.

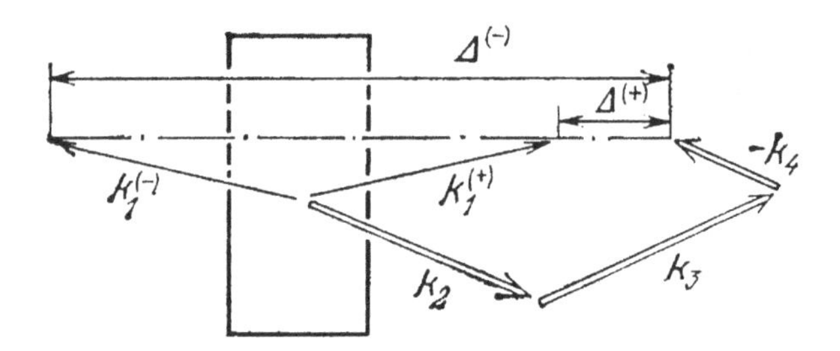

Fig. 6.13 In the case of parametric frequency conversion, $\omega_2 + \omega_3 - \omega_4 \to \omega_1$, there are two waves, $k_1^{(\pm)}$, satisfying the energy and transverse momentum conservation laws. However, the coherence length $\pi/\Delta^{(-)}$ for the $k_1^{(-)}$ wave is extremely small, on the order of the wavelength.

6.5.2 *The far field*

In practice, Eqs. (6.261), (6.262), as well as, generally, the one-dimensional approximation with a finite number of transverse modes, determine only the near field of the nonlinear sample, where the diffraction effects are not important due to the finite cross-section A of the interaction area. However, in the undepleted-pump approximation it is not necessary to use the mode expansion and the one-dimensional model. Indeed, the problem is reduced to solving a single wave equation (6.151) at a given distribution of monochromatic sources $\boldsymbol{P}(\omega, \boldsymbol{r})$, i.e., to finding the Green's function $\boldsymbol{G}(\omega, \boldsymbol{r})$ of the Helmholtz equation. This function has especially simple form in the case of a homogeneous isotropic medium and a large distance r from the sources to the observer.

We have already defined the Green's function of the Maxwell equations in the $\omega, \boldsymbol{k}$ representation (see (4.23)). One can show that its three-dimensional Fourier transformation yields, in the first order in λ/r, i.e., for the *wave field*, the expression

$$\boldsymbol{G}(\omega, \boldsymbol{r}) = \frac{\omega^2}{c^2 r} e^{ikr} \boldsymbol{\Pi}(\hat{\boldsymbol{r}}), \tag{6.263}$$

where $\Pi_{\alpha\beta} = \delta_{\alpha\beta} - \hat{r}_\alpha \hat{r}_\beta$ is the projector onto the plane orthogonal to the direction $\hat{\boldsymbol{r}} \equiv \boldsymbol{r}/r$ and $k \equiv n(\omega)\omega/c$. Let $r \gg kA$ (the *far field*), then (6.263) leads to the following relation between the field and the external polarization (compare with (6.260)):

$$\boldsymbol{E}(\omega, \boldsymbol{r}) = \frac{\omega^2}{c^2 r} e^{ikr} \int_V d^3r' \, e^{-i\boldsymbol{k}\cdot\boldsymbol{r}'} \boldsymbol{\Pi}(\hat{\boldsymbol{r}}) \cdot \boldsymbol{P}(\omega, \boldsymbol{r}'), \tag{6.264}$$

where $\boldsymbol{k} \equiv k\hat{\boldsymbol{r}}$ and $\boldsymbol{r}$ connects some point inside the V area containing the sources and the observation point. Thus, *at large V the far field scales as the Fourier transform of the external polarization.*

Consider generation of a field with the frequency ω_1 by an undepleted pump consisting of $m-1$ plane waves. At $V \to \infty$, the nonlinear polarization is also a plane wave, its frequency and wave vector being

$$\omega_1 = -\sum_{i=2}^{m} s_i\omega_i, \quad \boldsymbol{K} \equiv -\sum_{i=2}^{m} s_i\boldsymbol{k}_i. \tag{6.265}$$

Let the tensor $\chi^{(m-1)}$ be nonzero within a rectangular box with dimensions a, b, c, then, at $s_1 = 1$, (6.264) yields

$$\boldsymbol{E}^*(\omega_1, \boldsymbol{r}) = \frac{\omega_1^2}{c^2 r} e^{-ik_1 r}(\boldsymbol{\Pi} \cdot \chi^{(m-1)} \vdots \boldsymbol{E}_2 \ldots \boldsymbol{E}_m) V f(\hat{\boldsymbol{r}}), \tag{6.266}$$

$$f(\hat{\boldsymbol{r}}) \equiv \int d^3r'\, e^{i\Delta\boldsymbol{k}'\cdot\boldsymbol{r}'}/V = \mathrm{sinc}(\Delta k_x a/2)\mathrm{sinc}(\Delta k_y b/2)\mathrm{sinc}(\Delta k_z c/2), \tag{6.267}$$

$$\Delta\boldsymbol{k} \equiv n_1\omega_1\hat{\boldsymbol{r}}/c - \boldsymbol{K}. \tag{6.268}$$

The direction diagram of the field ω_1 at fixed $\boldsymbol{K}$ is mainly determined by the function $f(\hat{\boldsymbol{r}})$. For an efficient frequency conversion, the length of $\boldsymbol{K}$ should be close to $n_1\omega_1/c = k_1$; in this case, the radiation intensity is maximal in the direction $\boldsymbol{K}$ (for which $f \approx 1$) and has a considerable value within a solid angle of about λ_1^2/A, where A is the area of the V projection onto a plane orthogonal to $\boldsymbol{K}$ (Fig. 6.14).

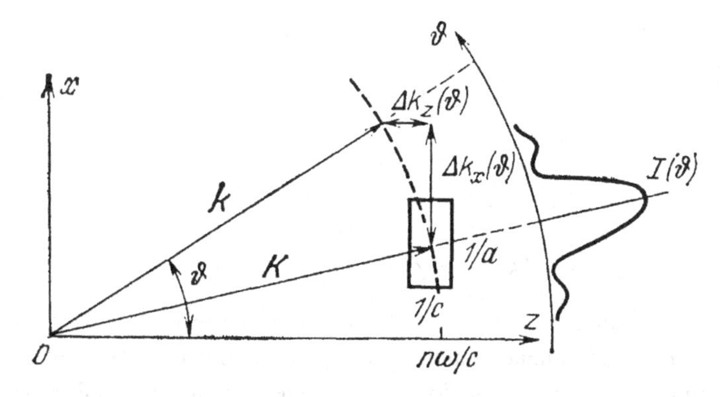

Fig. 6.14 To the definition of the direction diagram $I(\vartheta)$ for the field emitted by a polarization with the frequency ω and the wave vector $\boldsymbol{K}$ from a domain with the dimensions a, b, c: the radiation has a noticeable intensity only if the observed wave vector $\boldsymbol{k}$ belongs to a domain of dimensions $1/a, 1/b, 1/c$ near the point $\boldsymbol{K}$.

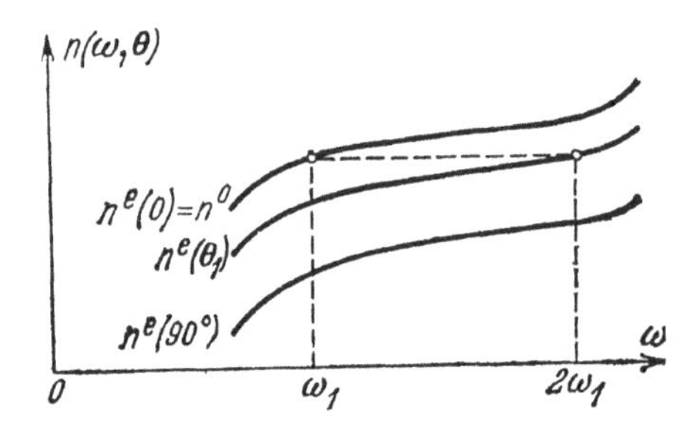

Fig. 6.15 Angular dispersion of the refractive index $n^e(\theta)$ for the extraordinary wave enables the frequency dispersion to be compensated and the phase matching condition $n^o(\omega_1) = n^e(2\omega_1, \theta_1)$ to be satisfied for second-harmonic generation, as well as for other three-wave interactions.

6.5.3 *Three-wave interaction*

Let us consider, in the framework of the one-dimensional model, the interaction of three waves or modes with the frequencies $\omega_1 + \omega_2 = \omega_3$, due to the quadratic susceptibility, which has a noticeable value only in piezoelectric crystals. For the phase matching condition, $\Delta \boldsymbol{k} \equiv \boldsymbol{k}_3 - \boldsymbol{k}_2 - \boldsymbol{k}_1 = 0$, to be satisfied in the transparency range, the crystal should be birefringent and have a certain orientation of optical axes with respect to the incident beams (Fig. 6.15).

Let us re-normalize the wave amplitudes so that their squares are equal to F_i, the longitudinal photon flux densities in $\hbar\omega_i$ units (see (6.178)). From (6.257) (or from (6.180)) it follows, at $s_1 = s_2 = -s_3 = -1$ and in the case of all waves co-propagating, $\overline{n_i} > 0$, that

$$da_1/dz = i\beta e^{i\Delta z} a_2^* a_3, \tag{6.269}$$

$$da_2/dz = i\beta e^{i\Delta z} a_1^* a_3, \tag{6.270}$$

$$da_3/dz = i\beta e^{-i\Delta z} a_1 a_2, \tag{6.271}$$

where

$$\beta \equiv (32\pi^3 \hbar\omega_1\omega_2\omega_3 / c^3 \overline{n}_1 \overline{n}_2 \overline{n}_3)^{1/2} \chi, \tag{6.272}$$

$$\Delta \equiv \Delta k_z, \ \chi \equiv \boldsymbol{\chi}^{(2)}(\omega_3 = \omega_2 + \omega_1) \vdots \boldsymbol{e}_3 \boldsymbol{e}_2 \boldsymbol{e}_1. \tag{6.273}$$

Here, the convolution χ can be considered as real and invariant to a simultaneous permutation of the frequencies and the polarization unit vectors $\boldsymbol{e}_i$ (Sec. 6.1).

After multiplying these equations by a_i^*, we see that the rates of flux variations, dF_i/dz, are equal for modes 1 and 2, while for modes 1 and 3 they have opposite signs. Thus, the energy flux $\sum \hbar\omega_i F_i$ is only redistributed, without absorption, between the three modes, and the share of each mode scales as its frequency (the

Manley-Rowe relations, see Sec. 6.3),

$$\Delta F_1 = \Delta F_2 = -\Delta F_3, \tag{6.274}$$

where

$$\Delta F_i \equiv F_i(z) - F_i(0). \tag{6.275}$$

There is also the third independent integral of motion (6.198) for Eqs. (6.269)–(6.271), determining the phase difference. These integrals allow one to solve Eqs. (6.269)–(6.271) in terms of a single quadrature. As a result, the $a_i(z)$ dependence is given by elliptic functions describing a periodic energy exchange between the three modes [Bloembergen (1965)]. Here, we will consider the limiting cases where one of the waves is much stronger than the other two (the undepleted-pump approximation, or the *parametric approximation*).

6.5.4 *Frequency up-conversion*

Let $F_{10} \gg F_2, F_3$ (undepleted low-frequency pump), $a_{10} = a_{10}^*$, and $\Delta = 0$, then (6.270), (6.271) yield

$$da_2/dz = i\gamma a_3, \tag{6.276}$$

$$da_3/dz = i\gamma a_2, \tag{6.277}$$

$$\gamma \equiv \beta a_{10} = (32\pi^3\omega_2\omega_3/c^3\bar{n}_1\bar{n}_2\bar{n}_3 I_{10})^{1/2}\chi. \tag{6.278}$$

These equations are easily solved with the help of the substitution $a_i = c_i e^{i\gamma z}$:

$$\begin{aligned} a_2 &= a_{20}\cos\gamma z + ia_{30}\sin\gamma z, \\ a_3 &= a_{30}\cos\gamma z + ia_{20}\sin\gamma z. \end{aligned} \tag{6.279}$$

Thus, modes 2 and 3, similarly to coupled pendulums, periodically (at a distance $\pi/2\gamma$) exchange energy (in $\hbar\omega_i$ units) according to (6.274) (Fig. 6.16).

Let us estimate a typical length of the parametric interaction, $z_{NL} \equiv 1/\gamma \approx \lambda_0 n/4\pi^2\chi E_{10}$. Let $\lambda_2 = \lambda_3 = 1\ \mu$, $n_i = 1, \chi = 10^{-8}\ \mathrm{G}^{-1}$, $I_{10} = 100\ \mathrm{MW/cm^2}$, then $z_{NL} = 0.3$ cm.

At $a_{30} = 0$, it follows from (6.279) that

$$F_3 = F_{20}\sin^2(z/z_{NL}), \quad F_2 = F_{20}\cos^2(z/z_{NL}). \tag{6.280}$$

This equation describes parametric *sum-frequency generation*, or *frequency up-conversion*, which is used, in particular, for visualizing IR radiation. At the interaction length $z = \pi z_{NL}/2$, the output intensity of the converter reaches its maximum, equal (in photon units) to the input intensity. The conversion efficiency is then equal, in ordinary units, to $\omega_3/\omega_2 > 1$.

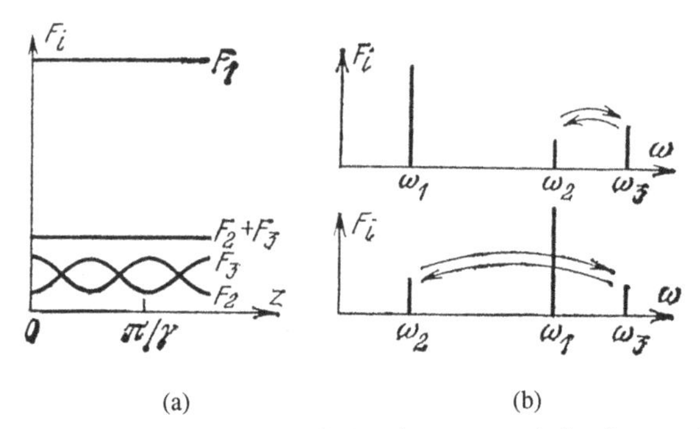

Fig. 6.16 Parametric interaction in the case of a low-frequency undepleted pump. (a) The intensity versus the distance z; (b) two versions of the radiation spectrum (arrows show the direction of the energy transfer).

Equation (6.280) can be easily generalized to the case $\Delta \neq 0$. The sum-frequency generation is then described by the equation

$$F_3 = F_{20}[\gamma \sin(\Gamma' z)/\Gamma']^2, \ \Gamma' \equiv (\gamma^2 + \Delta^2/4)^{1/2}. \tag{6.281}$$

Hence, at $\Delta^2 \gg 4\gamma^2$ we obtain (6.262) once again. Comparison of (6.280), (6.281) with (6.261), (6.262) shows that the approximation of two undepleted pumps, $a_{1,2}$ = const, is applicable only at $(2\gamma/\Delta)^2 \gg 1$. Then, the period of spatial modulation is determined by the wave mismatch Δ and not by the pump intensity γ.

6.5.5 *Parametric amplification and oscillation*

Let $F_{30} \gg F_1, F_2$ (high-frequency pump), then, instead of (6.276)–(6.278), we have

$$da_1/dz = i\gamma a_2^*, \ da_2/dz = i\gamma a_1^*,$$

$$\gamma \equiv \left(\frac{32\pi^3 \omega_1 \omega_2}{c^3 \bar{n}_1 \bar{n}_2 \bar{n}_3} I_{30}\right)^{1/2} \chi(-\omega_3; \omega_1, \omega_2) \vdots e_3 e_1 e_2. \tag{6.282}$$

The solution to this system is (compare with (6.279))

$$\begin{aligned} a_1 &= a_{10} \cosh \gamma z + i a_{20}^* \sinh \gamma z, \\ a_2 &= a_{20} \cosh \gamma z + i a_{10}^* \sinh \gamma z. \end{aligned} \tag{6.283}$$

If $a_{10} = 0$, then (compare with (6.280))

$$\begin{aligned} F_1 &= F_{20} \sinh^2 \gamma z \equiv (G-1) F_{20}, \\ F_2 &= F_{20} \cosh^2 \gamma z \equiv G F_{20}, \end{aligned} \tag{6.284}$$

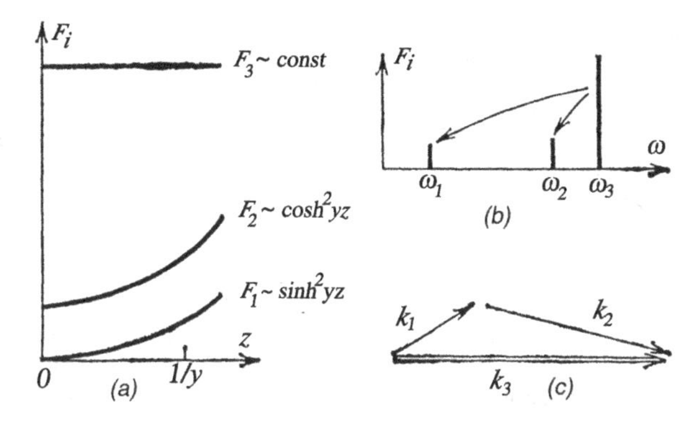

Fig. 6.17 Parametric amplification and difference-frequency generation (high-frequency pumping, forward interaction). (a) Dependence of the intensity on the distance; (b) typical frequency spectrum and the directions of the energy transfer; (c) relation between the wave vectors at phase matching.

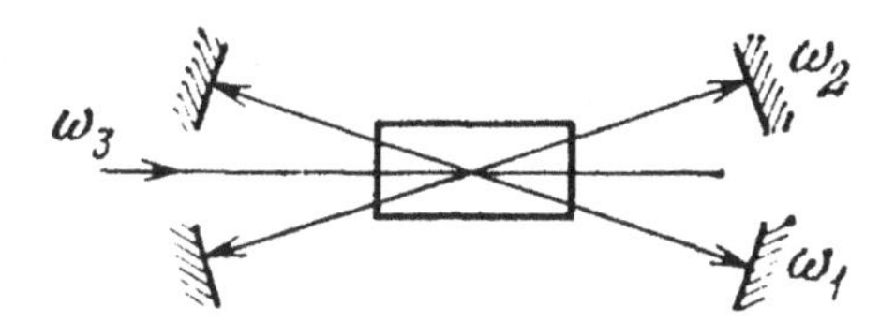

Fig. 6.18 Optical parametric oscillator.

where G is the parametric amplification coefficient. These equations describe the effects of *difference-frequency generation*, $\omega_3 - \omega_2 \to \omega_1$, and *parametric amplification*. In contrast to the case of a low-frequency pump, here the intensities grow exponentially (Fig. 6.17). Note that (6.284) satisfies the Manley-Rowe equations (6.274), $\Delta F_1 = \Delta F_2 = (G-1)F_{20}$.

It is not difficult to show that in the case of $\Delta \neq 0$, amplification and difference-frequency generation with high-frequency pumping are described by (compare with (6.281))

$$G = 1 + [\gamma \sinh(\Gamma z)/\Gamma]^2, \ \Gamma \equiv (\gamma^2 - \Delta^2/4)^{1/2}. \tag{6.285}$$

Note that at $\gamma^2 < \Delta^2/4$, the exponential growth turns into beats.

If there is a positive feedback at the frequency ω_1 or/and ω_2 (Fig. 6.18), the amplifier turns into an *optical parametric oscillator* (OPO). One of the frequencies (for instance, ω_1) is then called the *signal* one and the other one (ω_2), the *idler* one. Certainly, the oscillation requires that the absorption and other losses (which we ignore here) are compensated for by the parametric amplification. The OPO oscillation frequencies are mainly determined by the phase-matching condition, i.e., by the refractive indices n_i; therefore, at a fixed pump frequency ω_3

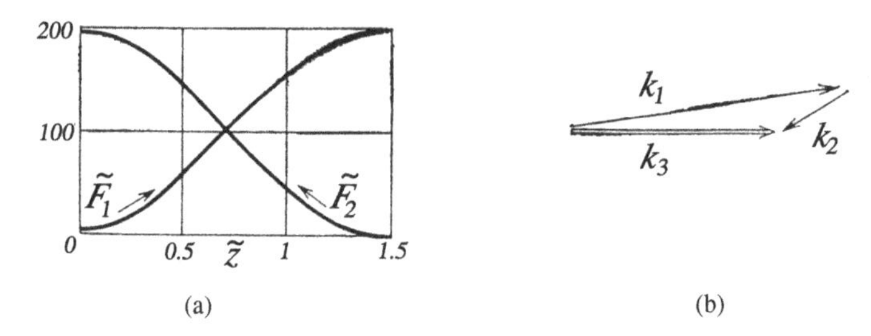

Fig. 6.19 Parametric amplification and difference-frequency generation in the case of backward interaction. (a) Intensity (in F_{10} units) as a function of the coordinate z (in $1/\gamma$ units) at $F_{2l} = 0$ and $\gamma l = 1.5$; the parametric amplification coefficient is $1/\cos^2(1.5) \approx 200$; (b) the phase-matching triangle for backward interaction.

the OPO can be smoothly tuned by varying the orientation or temperature of the crystal. The existing pulse-pumped OPOs cover the range $\lambda \approx 0.4$–$20\,\mu$, which conveniently complements the range of tunable dye lasers. By optimizing the parameters of the crystal, the focused pump beam, and the cavity, one can obtain oscillation even in the cw regime.

6.5.6 *Backward interaction*

Let the pump and signal waves propagate in the layer 'from left to right', $k_{1z}, k_{3z} > 0$, while the idler wave, 'from right to left', $k_{2z} < 0$. The phase matching condition can then be satisfied due to strong birefringence or anomalous dispersion. For the backward wave, $\vartheta_2 > \pi/2$ and $\overline{n_2} < 0$,[m] so that instead of (6.282) we have

$$da_1/dz = i\gamma a_2^*, \; da_2/dz = -i\gamma a_1^*. \tag{6.286}$$

Let the layer have a thickness l, then the boundary conditions have the form $a_1(0) = a_{10}$, $a_2(l) = a_{2l}$. It is easy to check that the solutions to (6.286) are (Fig. 6.19)

$$\begin{aligned} a_1 &= \{a_{10}\cos[\gamma(l-z)] + ia_{2l}^*\sin\gamma z\}/\cos\gamma l, \\ a_2 &= \{a_{2l}\cos\gamma z + ia_{10}^*\sin[\gamma(l-z)]\}/\cos\gamma l, \end{aligned} \tag{6.287}$$

where γ is assumed to be real. At $\gamma l \to \pi/2$, these solutions tend to infinity, i.e., oscillations are generated, despite the absence of the mirrors. One can say that backward interaction provides a distributed feedback (a similar effect takes place in a backward-wave tube).

[m]To be precise, the sign of $\overline{n} \equiv n\cos\theta\cos\rho$ is determined by the angle $\theta = \vartheta \pm \rho$ between the Poynting vector and the z axis; however, in the transparency range, the anisotropy angle does not exceed several degrees, and we neglect it here.

6.5.7 *Second harmonic generation*

Second harmonic generation (SHG) in the one-dimensional approximation and in the absence of linear absorption is described by equations (6.270), (6.270), with the subscript replacement $3 \to 2 \to 1$,

$$da_1/dz = i\beta e^{i\Delta z} a_1^* a_2, \tag{6.288}$$

$$da_2/dz = i\beta e^{-i\Delta z} a_1^2/2, \tag{6.289}$$

where

$$\beta \equiv (64\pi^3\hbar\omega_1^3/c^3\overline{n}_1^3\overline{n}_1^2\overline{n}_2)^{1/2}\chi, \;\; \Delta \equiv k_{2z} - 2k_{1z}, \tag{6.290}$$

$$\chi = \chi^* \equiv \chi^{(2)}(-\omega_1; 2\omega_1, -\omega_1) \vdots e_1 e_2 e_1 = 2\chi^{(2)}(-2\omega_1; \omega_1, \omega_1) \vdots e_2 e_1 e_1. \tag{6.291}$$

The factor 2 in the last expression was added according to relation (6.13).

One can easily see that these equations satisfy the condition of the field energy conservation, which in this case coincides with the Manley-Rowe equation,

$$|a_1|^2 + 2|a_2|^2 = C_1, \tag{6.292}$$

and the factor 2 is due to the fact that each second-harmonic photon has an energy twice as large as the pump photon energy. The second integral of equations (6.288), (6.289), according to (6.198), has the form

$$|a_1|^2[\Delta - 2\beta|a_2|\cos(\varphi_2 - 2\varphi_1 + \Delta z)] = C_2, \tag{6.293}$$

where φ_i are the phases of complex amplitudes a_i.

In the general case, the solutions to equations (6.288), (6.289) describe spatially periodic energy exchange between the modes of the pump and the second harmonic. (These solutions can be expressed in terms of the elliptic sine function [Bloembergen (1965); Dmitriev (1982)]).

Here, we will only consider the most important and simple case where $\Delta = 0$, $a_2(0) = 0$. Then, assuming in (6.293) $z = 0$, we find that $C_2 = 0$ and, hence, at any z we have $\cos\varphi = 0$, where $\varphi = \varphi_2 - 2\varphi_1$. Thus, the phase shift between the modes is constant. One can easily show that the phases are constant as well. Indeed, let us substitute into (6.289) $a_i \equiv b_i \exp(i\varphi_i)$, $b_i \geqslant 0$:

$$db_2/dz = (1/2)\beta b_1^2 \sin\varphi = (1/2)|\beta| b_1^2, \tag{6.294}$$

$$d\varphi_2/dz = (\beta b_1^2/2b_2)\cos\varphi = 0. \tag{6.295}$$

From the second equation, it follows that both φ_1 and φ_2 are independent of z, and from the first one, that the energy transfer direction (from the first harmonic into

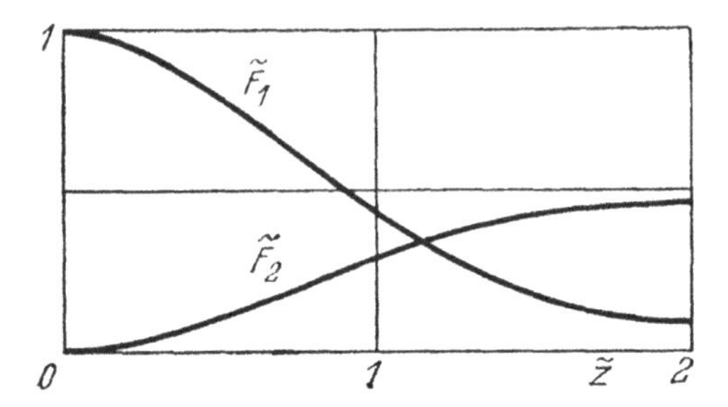

Fig. 6.20 Second harmonic generation. The horizontal axis shows the distance (in $1/\gamma$ units), the vertical axis corresponds to photon fluxes for the pump and the harmonic (in F_{10} units). At $\tilde{z} \gg 1$, every two pump photons turn into a single double-frequency photon.

the second one or vice versa) depends on the signs of β and φ. In the case considered, b_2 is increasing (from zero); therefore, $\beta \sin\varphi > 0$, i.e., $\varphi = (\pi/2)\,\mathrm{sign}\chi$.

Substituting (6.292) with $C_1 = b_{10}^2$ into (6.294), we easily find the solution

$$db_2/dz = |\beta|(b_{10}^2 - 2b_2^2)/2, \tag{6.296}$$

$$z = \frac{1}{|\beta|}\int \frac{db_2}{b_{10}^2/2 - b_2^2} = \frac{\sqrt{2}}{|\beta| b_{10}} \mathrm{arctanh} \frac{\sqrt{2} b_2}{b_{10}}. \tag{6.297}$$

Hence,

$$b_2 = (1/\sqrt{2}) b_{10} \tanh \gamma z, \;\; b_1 = b_{10}/\cosh \gamma z, \tag{6.298}$$

where

$$\gamma \equiv |\beta| b_{10}/\sqrt{2} = 2(2\pi/c\bar{n}_1)^{3/2} \omega_1 |\chi| I_{1z}^{1/2}. \tag{6.299}$$

Thus, *the intensity of phase-matched second harmonic $I_{2z}(z)$ increases monotonically as* $\tanh^2 \gamma z$, *and at $\gamma z \gg 1$ it achieves the initial intensity of the pump* $I_{1z}(0)$ (Fig. 6.20). The pump intensity goes in this case down to zero as $1/\cosh^2 \gamma z$, with its phase being constant. At $z = 1/\gamma$, the SHG efficiency η is 58% in the energy and 29% in the photon number. At $\Delta \neq 0$ or $a_2(0) \neq 0$, monotonic solutions are replaced by periodic ones, and η reduces.

In practice, powerful pulsed lasers provide η as high as several tens per cent. Note that phase-matched SHG in piezoelectric crystals has important applications, such as shifting laser frequencies up by as much as an octave.

Let us also mention much weaker effects that do not require phase matching: *SHG under the reflection from a medium without the center of symmetry* [Bloembergen (1965)] and incoherent *scattering of light at frequency* 2ω by non-centrosymmetric molecules [Kielich (1980)], as well as by any molecules, atoms, and free electrons, due to the 'magnetic' anharmonicity (Sec. 6.2).

Note that the condition $E_2(+0) = 0$ accepted above does not fully correspond to an ordinary experiment where it is the incident second-harmonic field that is zero, while the transmitted and reflected fields $E_2(+0)$, $E_2(-0)$ are non-zero due to the requirement that the tangent components of the fields E_2, H_2 should be continuous at the boundary [Bloembergen (1965)]. However, in practice $E_2(\pm 0)$ are indeed very small.

6.5.8 *The scattering matrix*

Parametric interaction between two modes in a nonlinear layer, considered in the undepleted-pump approximation, leads to a *linear relation between the input and output amplitudes*, which in the general case can be written as

$$a_1 = g_{11}a_{10} + g_{12}a_{20}^*, \quad a_2^* = g_{21}a_{10} + g_{22}a_{20}^*. \tag{6.300}$$

The coefficients g_{ij} form the two-dimensional *scattering matrix* of the sample, which depends on the layer length l, absorption α_i, nonlinearity χ, the pump amplitude a_{30}, the wave mismatch Δ and so on. If reflections from the layer boundaries are taken into account, the matrix becomes four-dimensional, while in the presence of diffraction (finite transverse dimensions of A) it becomes infinitely dimensional.

The scattering matrix should possess a certain symmetry, which follows from the general principles, in particular, from the Manley-Rowe relations (6.274). In the case where the input fields a_1 and a_2 are statistically independent, photon exchange between the modes is described by the *energy scattering matrix* $G_{ij} \equiv |g_{ij}|^2$,

$$F_1 = G_{11}F_{10} + G_{12}F_{20}, \quad F_2 = G_{21}F_{10} + G_{22}F_{20}. \tag{6.301}$$

Substitution of (6.301) into (6.274) yields

$$(G_{11} - 1)F_{10} + G_{12}F_{20} = (G_{22} - 1)F_{20} + G_{21}F_{10}.$$

Because F_{10} and F_{20} can be varied independently, the scattering matrix of a transparent layer should satisfy the relations

$$G_{11} - 1 = G_{21}, \quad G_{22} - 1 = G_{12}. \tag{6.302}$$

The system (6.282) is symmetric with respect to indices 1, 2; therefore, $G_{11} = G_{22} \equiv G$ and there is only a single independent element of the energy scattering matrix, the transfer coefficient G. As a result, (6.301) takes the form

$$\begin{aligned} F_1 &= GF_{10} + (G-1)F_{20}, \\ F_2 &= GF_{20} + (G-1)F_{10}. \end{aligned} \tag{6.303}$$

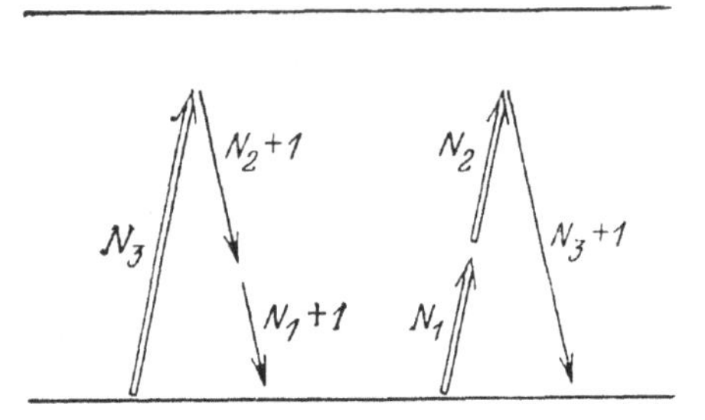

Fig. 6.21 Spontaneous multi-photon processes can be taken into account in the semiclassical theory, by adding a single extra photon into the output modes (the corresponding arrows are directed downwards). The figure shows the direct and inverse three-photon parametric processes.

6.5.9 °*Parametric down-conversion*

The quantum theory, in principle, allows a three-photon transition in which transparent matter (an atom or crystal) absorbs a photon from mode k_3 and emits a pair of photons into modes k_1 and k_2, returning into its initial state (Fig. 6.21). The probability of such a transition scales as $(N_1 + 1)(N_2 + 1)N_3$, where N_i are the initial photon numbers per mode and the unities appear due to the quantum fluctuations of the mode amplitudes in the ground state (Chapter 7). The probability of the inverse process, where photons in modes k_1, k_2 are absorbed and a photon is emitted into mode k_3, scales as $N_1N_2(N_3 + 1)$. As a result, the rate of pair generation scales as

$$\dot{N}_1 = \dot{N}_2 = -\dot{N}_3 \sim (N_1 + N_2 + 1)N_3 - N_1N_2. \tag{6.304}$$

Hence, in the first order in N_3, we have

$$N_1 - N_{10} = N_2 - N_{20} \sim N_{10} + N_{20} + 1. \tag{6.305}$$

Thus, due to the effect of the radiation at frequency ω_3, the matter generates pairs of photons with frequencies ω_1 and $\omega_2 = \omega_3 - \omega_1$ distributed within a broad range, from zero to the incident radiation frequency. In the case of macroscopic matter, the frequencies and directions of emission are related through the phase matching condition $\boldsymbol{k}_3 = \boldsymbol{k}_1 + \boldsymbol{k}_2$. This effect, observed in birefringent crystals, is called *parametric down-conversion* (PDC)[n] or *parametric fluorescence* [Zhabotinsky (1969); Klyshko (1980); Akhmanov (1971)]. PDC can be interpreted as the quantum noise of an optical parametric amplifier.

[n]Editors' note: in the original text, D. N. Klyshko used the term 'parametric scattering', suggested in his pioneering work (1967) on the theory of PDC [Klyshko (1980)]. Later, the term 'parametric down-conversion' became widely accepted.

The scaling factor in (6.305) should be equal to the conversion coefficient $G_{12} = G - 1$, which was found in (6.285) in the framework of classical nonlinear optics. Hence (compare with (6.303)),

$$N_1 = N_{10} + (G - 1)(N_{10} + N_{20} + 1) = GN_{10} + (G - 1)(N_{20} + 1). \qquad (6.306)$$

A similar expression can be obtained for N_2 through the permutation of indices $1, 2$. Here, $G - 1$ does not have to be linear in the pump intensity any more.

The obtained expression allows the following rule to be formulated. *In order to take into account spontaneous emission in the classical description of a parametric amplifier, one should add an extra single photon into each input idler mode.*[o] The same result follows from (6.306) in the case of a more general transformation,

$$N_{10} \to N_{10} + p, \; N_{20} \to N_{20} + q, \; N_1 \to N_1 + p,$$

where $p+q = 1$, $0 \leqslant p \leqslant 1$. In particular, one can add half a photon to all numbers of photons per mode.

According to (6.306), quantum noise yields $G-1$ photons in each output mode of a parametric amplifier, with G being the amplification coefficient for this mode. A similar result (Kirchhoff's law) is valid for quantum and Raman amplifiers (see (7.7) and (6.225)).

Note that at $N_2 = N_3 = 0$, it follows from (6.304) that three-frequency up-conversion with low-frequency pumping has no contribution from quantum fluctuations. In the case of four-frequency interaction, it is possible that two pump photons decay in a pair of photons with the frequencies ω and $2\omega_3 - \omega$, occupying the range $0 - 2\omega$ (*hyper-parametric scattering* or *light scattering by light*).[p]

Let us pass from photon numbers per mode to the spectral brightness. Using (6.306) and (6.227), we find that at $N_{i0} = 0$ (we assume that $\rho_1 = 0$)

$$I_{\omega\Omega}(\boldsymbol{k}_1) \equiv \hbar\omega_1 F_{\omega\Omega}(\boldsymbol{k}_1) = (\hbar\omega_1/2\pi\lambda_1^2)[G(\boldsymbol{k}_1) - 1], \qquad (6.307)$$

where $F_{\omega\Omega} \equiv dF/d\omega d\Omega$ is the flux of photons in the $\boldsymbol{k}_1$ direction with the frequency $\omega_\nu(\boldsymbol{k}_1)$ per unit solid angle, unit area (orthogonal to $\boldsymbol{k}_1$) and unit circular frequency. The value $I_{\omega\Omega}(\boldsymbol{k}, \boldsymbol{r})$ is called the *spectral brightness* (sometimes, simply the *intensity*) of incoherent radiation. Recall that the brightness, in the case of a transparent medium, does not vary in the direction of light propagation.

[o]An 'idler' mode, with respect to the considered 'signal' mode with the frequency ω_1, is the mode with the frequency $\omega_2 = \omega_3 - \omega_1$.

[p]Editors' note: now this process is commonly called *spontaneous four-wave mixing*. Starting from the beginning of the 2000 s, it is applied for the generation of nonclassical light, both faint one (photon pairs) and bright one (squeezed light).

Thus, the noise spectral brightness of an ideal amplifier, normalized to $G - 1$ (i.e., relative to the input), is

$$I^{vac}_{\omega\Omega} = \hbar\omega/2\pi\lambda^2 = \hbar\omega^3 n^2/8\pi^3 c^2. \tag{6.308}$$

This value can be naturally called the spectral brightness of zero-point fluctuations of macroscopic field (for a single polarization type) or 'the vacuum spectral brightness'. It corresponds to the presence of a single photon in each mode. At $\lambda = 1\,\mu$ and $n = 1$, the brightness of fluctuations per unit wavelength interval is

$$I^{vac}_{\lambda\Omega} = |d\omega/d\lambda| I^{vac}_{\omega\Omega} = \hbar c^2/\lambda^5 \approx 0.6\text{W}/(\text{Å}\cdot\text{cm}^2\cdot\text{sr}). \tag{6.309}$$

Thus, the relation between the spectral brightness and the photon number per mode is

$$I_{\omega\Omega}(\boldsymbol{k}) = I^{vac}_{\omega\Omega} N(\boldsymbol{k}). \tag{6.310}$$

Let us also estimate the brightness (effective) temperature of superfluorescence at the output of an ideal amplifier. From the Planck formula (2.102) and the Kirchhoff law $N = G - 1$, it follows that

$$\begin{aligned} T_{eff}(\boldsymbol{k}) &\equiv \hbar\omega/\kappa \ln(1 + N_k^{-1}) = -\hbar\omega/\kappa \ln(1 - G_k^{-1}) \\ &\approx (\hbar\omega/\kappa)G \sim 10^6 K. \end{aligned} \tag{6.311}$$

The last estimate was made for $\lambda = 1\,\mu$ and $G = 100$.

Substituting (6.285) and (6.308) into (6.307), we obtain the intensity of the parametric amplifier quantum noise,

$$I_{\omega\Omega}(\boldsymbol{k}) = I^{vac}_{\omega\Omega} \sinh^2[(\gamma^2 - \Delta^2(\boldsymbol{k}/4))^{1/2} l]/(1 - \Delta^2(\boldsymbol{k})/4\gamma^2), \tag{6.312}$$

where l is the thickness of the nonlinear layer, $\Delta(\boldsymbol{k})$ is the deviation from phase matching for mode $\boldsymbol{k}$, and $\gamma \sim \chi^{(2)} E_3$ (see (6.282)). According to (6.312), the PDC intensity has a sharp maximum at frequencies and directions $\{\omega, \vartheta, \varphi\}$ satisfying the phase matching condition $\Delta\boldsymbol{k} = 0$.

This condition determines the *frequency-angular spectrum* of PDC $\omega(\vartheta)$, where ϑ is the *angle of scattering*, i.e., the angle between the observed wave vector $\boldsymbol{k}$ and the pump wave vector $\boldsymbol{k}_3$. The dependence of the frequency on the angle φ, as a rule, can be neglected, i.e, the PDC spectrum is axially symmetric w.r.t. the direction $\boldsymbol{k}_3$. Field with a given frequency ω is emitted along a cone with a certain angle $\vartheta(\omega)$. As one can see from Fig. 6.22, the frequency spectrum of PDC from a blue pump covers a broad range of IR and visible wavelengths. The visible radiation is directed mostly forward, at angles not exceeding several degrees, which is due to the fact that crystals have small birefringence in the transparency range.

When the idler frequency approaches the eigenfrequencies of the crystal lattice, which are usually in the region of hundreds of inverse centimeters, PDC

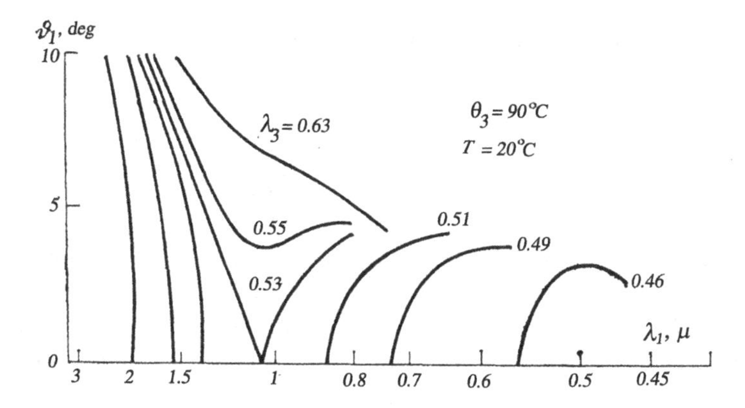

Fig. 6.22 Frequency-angular spectrum of parametric down-conversion in lithium niobate for different pump wavelengths λ_3: ϑ_1 is the angle of the cone along which the wave λ_1 is emitted. The angle θ_3 between the pump beam and the crystal axis is 90°. When θ_3 is reduced, the gap in the spectrum (for $\lambda_3 < 0.53\,\mu$) disappears.

continuously turns into the Raman scattering by polaritons and optical phonons. Then, χ has a resonance growth due to the contribution from electron-nuclear anharmonicity (Sec. 6.2), but this growth is accompanied by an increase in the linear absorption α_2 at the idler frequency, so that the brightness integrated over the frequency, I_Ω, does not change much. Certainly, when the idler frequency is small and $\alpha_2 l \gg 1$, (6.312) has to be multiplied by $\mathcal{N}(\omega_2/T) + 1$, where T is the lattice temperature and $\mathcal{N}$ is the Planck function.

By analogy with the scattering by polaritons, PDC can be defined as the scattering of the pump radiation by the field fluctuations in the idler modes, i.e., as *light scattering by light*, similarly to the way Mandelshtam-Brillouin scattering can be called light scattering by sound.

The important features of PDC, distinguishing it from other types of light scattering in matter, are, first, a broad continuous spectrum, not related directly to the eigenfrequencies of the matter, and, second, the two-photon structure of the emitted radiation: at small pumping ($\gamma l \ll 1$), signal and idler photons are emitted only in pairs, practically simultaneously.

Note that, in addition to coherent (forward) emission, there is incoherent PDC by separate non-centrosymmetric molecules (to be precise, by the density and orientation fluctuations of such molecules). Then, the phase matching condition plays no important role, since the momentum deficit of the field is provided by the molecule.

Let $\gamma l \ll 1$ and $n \approx 1$, then from (6.312) we find the intensity of spontaneous

PDC (SPDC) in the direction $\boldsymbol{k}$ with the frequency $\omega(\boldsymbol{k})$,

$$I_{\omega\Omega}(\boldsymbol{k}) = I_{\omega\Omega}^{vac}\gamma^2 l^2 \mathrm{sinc}^2[\Delta(\boldsymbol{k})l/2]$$
$$= 4\hbar c^{-5}\omega^4\tilde{\omega}\chi^2 l^2 I_{30}\mathrm{sinc}^2[\Delta(\boldsymbol{k})l/2], \tag{6.313}$$

where $\tilde{\omega} \equiv \omega_3 - \omega$ and the polarization indices are assumed to be chosen so that Δ is minimal. Note that the PDC intensity in the directions of exact phase matching depends on the nonlinear layer thickness quadratically, which is typical for SPDC. Let $I_{30} = 1$ W/cm^2 and $l = 1$ cm, then the amplification in the phase-matched direction is $G = 1 + \gamma^2 l^2 \sim 1 + 10^{-7}$ (see the estimate after (6.279)), and it follows from (6.311) that at $\lambda = 0.5\,\mu$, $T_{eff} \sim 1800$ K. Such radiation can be easily seen by eye and looks like colored rings. Note that the actual transfer coefficient of the sample G' with such pumping is always less than the unity because of the reflection, absorption, and scattering losses.

The effective frequency band of the SPDC spectrum $\Delta\omega$ at a fixed observation direction is determined, similarly to the band of parametric amplification, by the phase matching width, i.e., by the condition $\Delta l = \pm\pi$. Considering only the linear expansion of $\Delta(\omega)$, we can write

$$\Delta\omega = (2\pi/l)|\partial\Delta/\partial\omega|^{-1} \approx 2\pi/|\tau_1 - \tau_2|, \tag{6.314}$$

where $\tau_i \equiv l/u_i$. According to Eq. (6.314), which is valid in the case of collinear phase matching with $\omega_1 \neq \omega_2$,[q] *the width of the SPDC spectrum (in Hz) is equal to the inverse time delay (in seconds) between the signal photon and the idler one during their passage through the interaction region.* Usually, $\Delta\omega/2\pi c \approx 10$ cm^{-1} at $l = 1$ cm.

At $\gamma l \gg 1$, (6.312) describes *parametric super-fluorescence*, or *stimulated parametric down-conversion*. According to the estimates made above, stimulated PDC can be observed from pulsed pump with the intensity about 100 MW/cm^2 or higher. Certainly, stimulated PDC, similarly to parametric oscillation, is observable in practice only in the direction $\vartheta \sim 0$, at the frequencies of collinear phase matching, $\omega_i(0)$. This is due to the fact that, as it follows from the geometry of the experiment, the effective interaction length l_{ef} in the case of a narrow pump beam ($a/l \ll 1$) is dramatically reduced (as $a/\sin\vartheta$) at $\vartheta \gg a/l$. The total power of stimulated PDC can be estimated as

$$\mathcal{P} = I_{\omega\Omega}\Delta\omega\Delta\Omega A = I_{\omega\Omega}^{vac}e^{2\gamma l}\Delta\omega\Delta\Omega A/4, \tag{6.315}$$

where we have introduced the effective frequency and solid-angle bands and the effective cross-section (see the estimate of $\mathcal{P}$ for the StRS (6.229)). Stimulated

[q]Editors' note: It is also valid in the case of type-II phase matching, in which the signal and idler photons are orthogonally polarized and therefore have different group velocities.

PDC is used, similarly to parametric oscillation, for a controllable shift of laser radiation frequencies.

6.5.10 °*Light scattering by polaritons*

This process is described by the system (6.269)–(6.271) with linear absorption introduced into one of the equations. In the approximation of undepleted high-frequency pump, the interaction between the Stokes (signal) and the polariton (idler) waves is determined by the equations

$$da_1/dz = i\gamma a_2^* e^{i\Delta z}, \tag{6.316}$$

$$da_2/dz \pm \delta a_2 = \pm i\gamma a_1^* e^{i\Delta z}, \tag{6.317}$$

where $\delta \equiv \alpha_2/2$, $\gamma = \beta a_{30}$ (see (6.282)), and the lower signs correspond to the case of the backward idler wave. In the polariton range of the PDC spectrum, i.e., in the vicinity of lattice eigenfrequencies, the idler wave is strongly absorbed, so that usually $\delta \gg |\gamma|$, $\delta l \gg 1$, and the amplitude of the idler wave is determined by the local amplitude of the signal wave.

For this reason, let us search the solution to (6.317) in the form $a_2(z) = b_2 \exp(i\Delta z)$,

$$db_2/dz \pm \delta b_2 + i\Delta b_2 = \pm i\gamma a_1^*. \tag{6.318}$$

Here, the derivative db_2/dz does not much exceed $|\gamma b_2|$; therefore, if $|\gamma| \ll |\delta + i\Delta|$, it can be neglected. Then,

$$b_2 = i\gamma a_1^*/(\delta \pm i\Delta). \tag{6.319}$$

In this approximation, the incident idler radiation does not influence the output field, due to the high absorption, which is much stronger than the parametric amplification. Substituting (6.319) into (6.316), we obtain

$$da_1/dz = (g/2)a_1, \tag{6.320}$$

where

$$g \equiv 2|\gamma|^2/(\delta \mp i\Delta) = 64\pi^3\omega_1\omega_2|\chi^{(2)}|^2 I_{30} c^3 \overline{n}_1\overline{n}_2\overline{n}_3(\delta \mp i\Delta).$$

Thus, strong absorption of one of the two interacting waves leads to the exponential increase, similarly to the case of usual Raman interaction due to $\chi^{(3)}$ (Sec. 6.4), with the only difference that here, phase matching condition $|\Delta| \ll \delta$ is important.

Comparing g at $\Delta = 0$ with the Raman amplification coefficient (6.224), with the definition of α_2 (6.167) taken into account, we find the equivalent cubic susceptibility (assuming $\bar{n}_i = 1$),

$$\chi^{(3)''}_{equiv} = -|\chi^{(2)}|^2/\chi^{(1)''}. \tag{6.321}$$

The same relation between the resonance susceptibilities of the first, second, and third order follows from the microscopic theory [Klyshko (1980)].

The spectral brightness of the Stokes field $I_{\omega\Omega}(\boldsymbol{k}_1)$ in spontaneous or stimulated light scattering by polaritons can be found through the Kirchhoff law,

$$I_{\omega\Omega}(\boldsymbol{k}_1) = I^{vac}_{\omega\Omega}[\mathcal{N}(\omega_2/T) + 1]\{\exp[g'(\boldsymbol{k}_1)z] - 1\}, \tag{6.322}$$

or by using the results of Sec. 6.4. Note that if a_2 is understood as the amplitude of the sound wave, the above analysis will describe the Mandelshtam-Brillouin scattering (spontaneous and stimulated).

6.5.11 *Four-wave interactions*

In centrally symmetric media, macroscopic quadratic susceptibility $\chi^{(2)}$ is equal to zero, and therefore the simplest parametric process involves four field modes interacting due to the cubic susceptibility $\chi^{(3)}(\omega_4 = \omega_3 + \omega_2 + \omega_1)$. In the corresponding quantum transition, four photons are absorbed or emitted and the state of the matter remains the same.

The most important four-wave parametric effects are (Fig. 6.23) *generation of combination frequencies* (for instance, the third harmonic) for converting laser radiation to the UV and IR ranges, *coherent anti-Stokes Raman scattering* (CARS), and *optical phase conjugation* (OPC). Let us also mention three-wave effects due to $\chi^{(3)}(\omega_4 = \omega_3 + \omega_2 + 0)$ in a constant field E_0, which breaks the central symmetry of the medium.

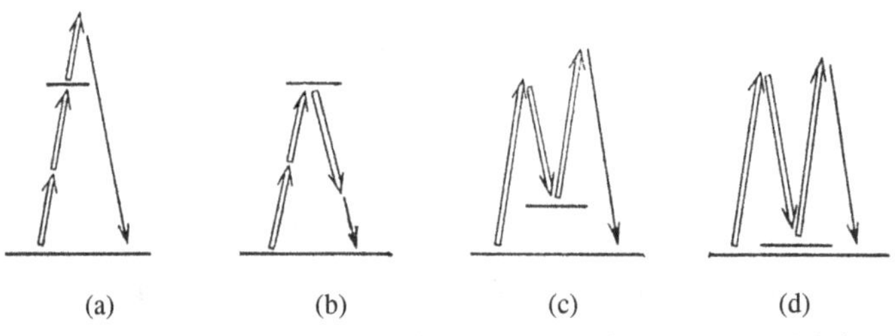

Fig. 6.23 Basic types of four-wave single-resonance parametric processes and their applications: (a,b) generation of coherent UV and IR radiation; (c) coherent anti-Stokes Raman scattering (active, or CARS, spectroscopy); (d) phase conjugation.

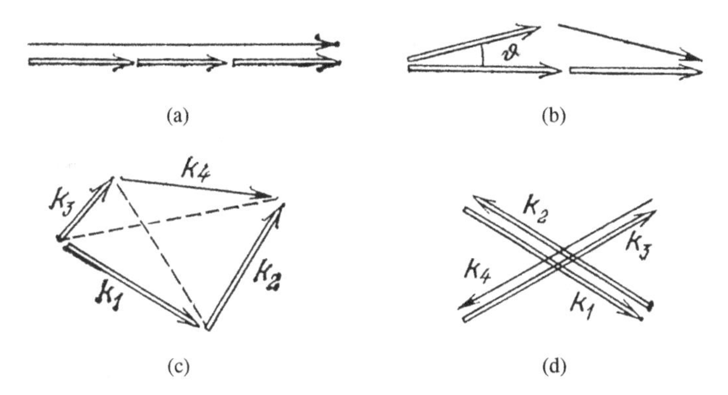

Fig. 6.24 Phase matching conditions in four-wave interactions: (a) generation of UV radiation is usually performed under collinear phase matching; (b) in CARS, all frequencies are close and phase matching is satisfied at small angles of scattering ϑ; (c) in the general case of frequency conversion of the form $\omega_1 + \omega_2 - \omega_3 \to \omega_4$, the phase matching has the form $\boldsymbol{k}_1 + \boldsymbol{k}_2 = \boldsymbol{k}_3 + \boldsymbol{k}_4$. The wave vectors in this case can be non-planar: the phase matching quadrangle can be folded along the dashed lines; (d) in OPC, all four frequencies are equal and the phase matching condition is satisfied for any two standing waves.

In the frequency conversion, it is usually the electron anharmonicity that 'works', CARS is due to the mixed electron-nuclear (Placzek's) anharmonicity, and OPC, similarly to self-focusing, is due to the inertial types of anharmonicity: the orientation and sriction ones (Sec. 6.2).

Note that in OPC, which will be described in more detail below, all frequencies are degenerate, as in non-parametric interactions. Nevertheless, we classify it as a parametric process, since it preserves the field energy and leads to the coherent excitation of new modes with the phase matching condition $\boldsymbol{k}_1 + \boldsymbol{k}_2 = \boldsymbol{k}_3 + \boldsymbol{k}_4 = 0$.

In the case of frequency conversion in gases, collinear phase matching is satisfied due to anomalous dispersion (it is convenient to use mixtures of gases for this purpose). CARS in many condensed materials requires non-collinear phase matching with the angles of scattering about 1° (in the case of a two-beam pump, see Fig. 6.24).

In all effects mentioned so far, the incident (input) field has, in the general case, three modes excited ($N_{i0} \neq 0$, $i = 1, 2, 3$), and the output radiation has 'new' photons in the fourth mode, $\boldsymbol{k}_4(\omega_1 + \omega_2 + \omega_3 \to \omega_4)$. Such effects can be called *stimulated*, they are well described by classical electrodynamics. In addition, there are *spontaneous* effects, in which the pump contains, in the general case, two modes: $\omega_1 + \omega_2 \to \omega_3 + \omega_4$. Then, photons in the output modes 3, 4 appear simultaneously, in pairs, due to spontaneous-stimulated transitions (compare with the parametric down-conversion). Such processes, which include *hyperpara-*

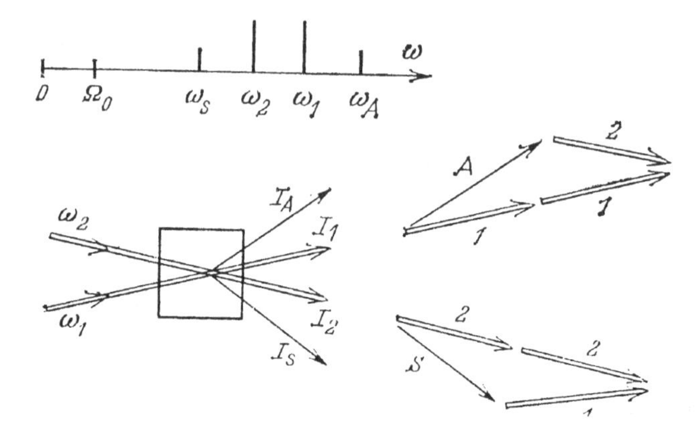

Fig. 6.25 Active spectroscopy. The matter is excited by two lasers with frequencies ω_1 and ω_2 such that $\omega_1 - \omega_2$ equals the frequency Ω_0 of molecular vibrations. The resonance can be detected by an increase in the intensity I_i, polarization, or phase of any one of the four frequencies observed while scanning ω_1 or ω_2.

metric scattering,[r] or *light scattering by light*, and 'spontaneous CARS', are only described in terms of nonlinear quantum optics and, in some cases, equations like the Kirchhoff law [Klyshko (1980)].

6.5.12 *Nonlinear spectroscopy*

The absolute value of $\chi^{(3)}$ and, correspondingly, the efficiency of four-wave parametric interactions increase dramatically near intermediate resonances, where one of the virtual levels coincides with a real one (Fig. 6.23). Most interesting are two-photon resonances like $\omega_1 \pm \omega_2 = \omega_3 \pm \omega_4 \approx \Omega_0$, in which neither the pump not the output field undergo single-photon resonance absorption.

Observation of $\chi^{(3)}$ resonances is the main instrument for several methods of *nonlinear spectroscopy*, the most important of which is, probably, *active spectroscopy*, based on CARS [Akhmanov2 (1981); Bloembergen (1977)]. The scheme of a nonlinear spectroscope for the study of $\chi^{(3)}$ dispersion is shown in Fig. 6.25. It uses two lasers with the frequencies ω_1, ω_2 (let $\omega_1 > \omega_2$), one of them (say, ω_2) tunable within the range $\omega_1 - \Omega_0$, where Ω_0 is the frequency of the molecular vibration under study. At $\omega_2 = \omega_1 - \Omega_0$, the frequency components of the cubic susceptibilities $\chi(\omega_1 = \omega_1 - \omega_2 + \omega_2)$, $\chi(\omega_2 = \omega_2 - \omega_1 + \omega_1)$, $\chi(\omega_A = 2\omega_1 - \omega_2)$, $\chi(\omega_S = 2\omega_2 - \omega_1)$ have resonances (see (6.88)) caused by the excitation of atom oscillations in a molecule by bi-harmonic light due to the mutual influence of electronic and nuclear degrees of freedom (Sec. 6.2).

[r]Editors' note: called now *spontaneous four-wave mixing*.

In the case of active spectroscopy, the resonance manifests itself in the intensity variation at the 'new' frequencies ω_A or ω_S. (It is convenient to use the anti-Stokes range, ω_A, where there is less stray light caused by the fluorescence of the sample and the optical elements.) According to (6.261), (6.262), in the undepleted-pump and plane-wave approximations, the intensity of CARS is

$$I_A(\omega_2) = \left(\frac{4\pi}{c}\right)^4 \frac{\omega_A^2 l^2 I_1^2 I_2}{n_1^2 n_2 n_A} |\chi_{NR}^{(3)} + \chi_R^{(3)}(\omega_A = 2\omega_1 - \omega_2)|^2, \tag{6.323}$$

where the non-resonance real part of $\chi^{(3)}$, $\chi_{NR}^{(3)}$, has a weak dependence on the frequency and causes the asymmetry of the observed spectrum. *Active spectroscopy has advantages over usual Raman spectroscopy in sensitivity, resolution, and the amount of available information.* Note that using tunable lasers allows one to do without dispersive elements.

In the case of *Raman amplification spectroscopy*, one registers the intensity increase, ΔI_2, at the output, equal, according to (6.224), to

$$\Delta I_2(\omega_2) = -\frac{32\pi^2 \omega_2 l I_1}{c^2 n_1 n_2} \mathrm{Im}\chi^{(3)}(\omega_2 = \omega_2 - \omega_1 + \omega_1). \tag{6.324}$$

One can also observe the decrease of the intensity I_1 of the field with the higher frequency ω_1 near the resonance. This phenomenon is called the *inverse Raman effect*.

The resonance can be also registered in other ways: by measuring the phase delay of the incident field ω_1 or ω_2, which scales as $\mathrm{Re}\chi^{(3)}$ (the Raman Kerr-effect method), or by measuring the polarization parameters of the fields ω_i, $i = 1, 2, 3, 4$ (*nonlinear ellipsometry*) [Akhmanov2 (1981)].

All these nonlinear spectroscopy methods, together with the two-photon spectroscopy and saturation spectroscopy, considered in Sec. 6.4, became a considerable extension of the traditional pre-laser spectroscopy, which mainly used linear effects and spontaneous Raman scattering. *A significant broadening of the scope of spectroscopy due to the use of lasers together with the abilities of nonlinear spectroscopy allowed one to speak about the 'laser revolution' in spectroscopy.*

6.5.13 *Dynamical holography and phase conjugation*

The idea of the optical phase conjugation (OPC) method based on the four-wave interaction is clear from Figs. 6.24(d) and 6.26. Let a standing monochromatic pump wave (a 'reference wave', according to holography termonology), i.e., $\boldsymbol{k}_2 = -\boldsymbol{k}_1$, be excited in a medium with the cubic nonlinearity $\chi^{(3)}(\omega = \omega + \omega - \omega)$. Then, if a third plane wave $\boldsymbol{k}_3$ with the same frequency ω and an arbitrary direction is

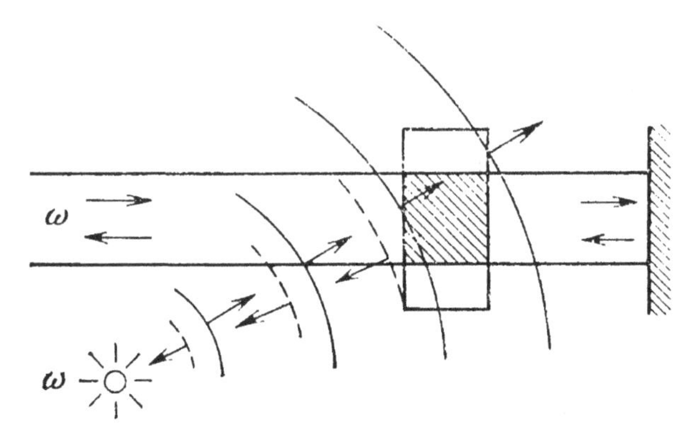

Fig. 6.26 Phase conjugation using degenerate four-wave interaction. A nonlinear medium excited by a standing pump wave 'reflects back' all waves of the same frequency along the way of their incidence. As a result, a divergent spherical wave becomes a wave convergent to the source (dashed line).

incident on the medium, the fourth wave will appear, with the frequency ω and the wavevector $\boldsymbol{k}_4 = \boldsymbol{k}_1 + \boldsymbol{k}_2 - \boldsymbol{k}_3 = -\boldsymbol{k}_3$. Thus, the medium excited by the pump acts as a mirror reflecting all plane waves back along the path of their arrival, in contrast to a usual mirror, which performs the transformation $k_z \to -k_z$.

In the case of an arbitrary spatial distribution of the signal (*object*) field $E_3(\boldsymbol{r})$ it will contain many Fourier components $\{\boldsymbol{k}_3\}$, each of them creating its own conjugate component. As a result, the initial object wave will be restored around the medium, $RE_3^*(\boldsymbol{r})$, with the same wave front shape, but propagating in the opposite direction (from the medium) and, of course, having a different energy ($R \neq 1$). It is important that due to the backward parametric amplification (see (6.287)), $|R|$ can considerably exceed the unity. (Usually, this is only achieved in the pulsed regime.)

The possibility of the OPC effect for an arbitrary optical field, which, in a sense, provides the time inversion,[s] looks striking from the viewpoint of nonlinear optics. In fact, this effect has been discovered in the optical range as early as in 1949, long before the birth of lasers and nonlinear optics, by Gabor, the author of holography. In holography, OPC manifests itself in the appearance of twin images, which were considered by Gabor only as sources of noise. The perspectives of OPC applications were understood only much later, mainly in the seventies, when the practical methods of *dynamical holography* have been invented, which enabled OPC to work 'in real time', without the delay for photographic film developing. These methods, in addition to four-wave parametric interaction,

[s]OPC is equivalent to the $t \to -t$ transformation only for a strictly monochromatic field. In the case of a quasi-monochromatic field, OPC does not change the shape of the envelope $E_0(t)$.

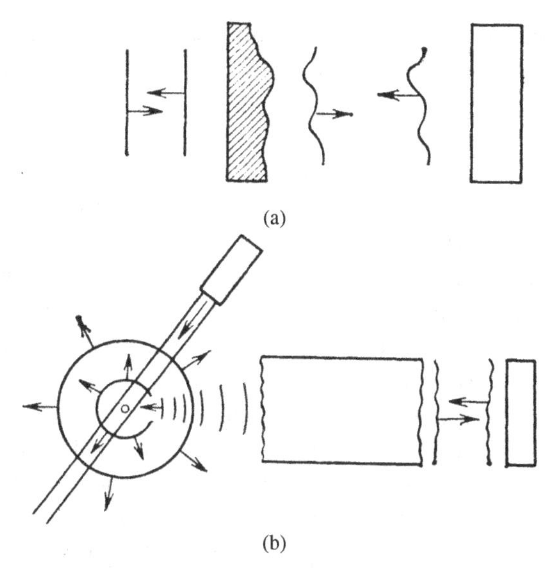

Fig. 6.27 Applications of the OPC effect: (a) correcting the wave front distortions. The plane front of the wave incident from the left becomes distorted, but after being reflected by a 'phase conjugating mirror' and passing the same medium again it restores its initial shape. (Reflection from a usual mirror would double the distortions.) (b) Focusing of strong laser radiation on small targets: light from a weak laser (top) is scattered by a target, part of the scattered field is amplified by a strong amplifier, gets reflected by the 'mirror', is amplified once again and returns to the target.

use three-wave degenerate interaction effects, 180° light scattering, and superfluorescence [Zeldovich (1985); Bespalov (1979)]. OPC methods have been also developed for acoustic waves.

OPC is an example of *adaptive optics*, aiming for the automatic correction of optical systems. OPC allows one to correct the distortions of the wave front (phase distortions) appearing due to the signal wave passing through an optically inhomogeneous medium, like an opaque glass or a quantum amplifier. To correct for this, it is sufficient to reflect the wave back at the output of the inhomogeneous medium by means of a phase-conjugating mirror and this way to make it travel along the same path in the opposite direction. Then, all distortions of the wave front formed on the way will be 'straightened' and the front will restore its original shape (Fig. 6.27(a)). (Of course, amplitude distortions caused by irreversible absorption or amplification will not be compensated but accumulated.) This effect allows powerful but inhomogeneous amplifiers to be used for increasing the energy of weak lasers providing single-mode beams with the minimal (diffraction) divergence and 'natural' bandwidth. This technique provides record values of spectral brightness.

Another application of OPC, important for solving the problem of laser thermonuclear synthesis, is automatic focusing of strong laser radiation onto small targets (Fig. 6.27(b)).

Using a resonance susceptibility $\chi^{(3)}$ (in particular, enhanced by two-photon resonances in the vapors of alkali metals) allows one to reduce the pump power in four-wave interactions (down to 1 W/cm^2, with the reflection coefficient $|R| \approx 1$).

In *OPC via stimulated scattering*, a sufficiently strong monochromatic wave $\mathrm{Re}E(\boldsymbol{r})\exp(-i\omega t)$ serves as a pump causing 180° stimulated Raman scattering or stimulated Mandelshtam-Brillouin scattering in a nonlinear medium. If the wave front of the pump is rather non-uniform, the back-reflected Stokes radiation $\mathrm{Re}E_S(\boldsymbol{r})\exp(-i\omega_S t)$ (with $\omega_S = \omega - \Omega_0$) has approximately the same wave front shape, $E_S(\boldsymbol{r}) \approx RE^*(\boldsymbol{r})$, where $|R|$ is close to unity.

At first stages of backscattering ($z \lesssim l$) the Stokes field is chaotic, it has all modes with different wavevectors $\boldsymbol{k}$ excited independently and homogeneously. (Since the wavevector length $k = n(\omega_S)\omega_S/c$ is fixed, the modes are determined by the transverse component $\boldsymbol{k}_\perp \equiv \boldsymbol{q}$). In the case of a multi-mode pump, various Stokes modes $(\boldsymbol{q}, \omega_S)$ have different Raman amplification coefficients $\alpha(\boldsymbol{q})$. Moreover, one can show [Zeldovich (1985)] that if some mode $(\boldsymbol{q}, \omega)$ is present in the pump spectrum, then the Stokes mode $(-\boldsymbol{q}, \omega_S)$ has, on the average, an amplification coefficient twice as large: $\alpha(-\boldsymbol{q}) \approx 2\bar{\alpha}$. Because at $\bar{\alpha}l \gg 1$ this difference is amplified exponentially, it is sufficient that the part of the Stokes field repeating the pump spectrum is considerably more intense than the noise part.

The advantages of 'Brillouin' or 'Raman' mirrors are the absence of the pump (they are analogues of reference-free holography) and almost 100% efficiency. The disadvantages are the existence of a threshold, the necessity to have multiple modes in the signal field, and the frequency shift of the reflected radiation. The latter restricts the accuracy of the reconstruction. Of much interest are lasers in which one of the mirrors is phase conjugating (a 'Brillouin' one) and the other one, a usual mirror, plane or concave. This scheme simultaneously provides Q-switching (due to the threshold behavior of the stimulated scattering) and correction for the optical inhomogeneity of the active medium.

Let us consider OPC via four-wave interaction in a little more detail. An arbitrary signal field (for simplicity, considered as scalar) can be represented as

$$E(\boldsymbol{r}, t) = \mathrm{Re}E_0(\boldsymbol{r})e^{-i\omega t} = |E_0(\boldsymbol{r})|\cos[\omega t + \varphi(\boldsymbol{r})]. \tag{6.325}$$

The phase-conjugated field describing monochromatic waves with the inverse directions of wavefront propagation, by definition, differs by the sign of the time,

$$\tilde{E}(\boldsymbol{r}, t) \equiv E(\boldsymbol{r}, -t). \tag{6.326}$$

The amplitude and phase of the phase-conjugated field will be defined by analogy with (6.325),

$$\tilde{E}(\boldsymbol{r},t) = \mathrm{Re}\tilde{E}_0(\boldsymbol{r})e^{-i\omega t} = |\tilde{E}_0(\boldsymbol{r})|\cos[\omega t + \tilde{\varphi}(\boldsymbol{r})]. \tag{6.327}$$

Certainly, the transformation $E \to \tilde{E}$ is possible in practice only in the absence of irreversible processes. From (6.325)–(6.327), we find the relations between the spectral amplitudes of the incident field and the phase-conjugate one,

$$\tilde{E}_0(\boldsymbol{r},t) = E_0^*(\boldsymbol{r}),\ \tilde{\varphi}(\boldsymbol{r}) = -\varphi(\boldsymbol{r}). \tag{6.328}$$

Hence, the wave surfaces of the monochromatic fields determined by the equations $\varphi(\boldsymbol{r}) = \text{const}$ and $\tilde{\varphi}(\boldsymbol{r}) = \text{const}$ coincide. Thus, the OPC effect reverses not the wave fronts but the propagation directions.[t] Note that reflection by a usual mirror, plane, spherical, or of a more complicated shape, also leads to the transformation $E \to \tilde{E}$, but only in trivial cases where the mirror surface coincides with the wave surface.

The reversed field $\tilde{E}$ copies the signal one in all space outside of the nonlinear medium, including any transparent or scattering bodies of arbitrary shapes (provided of course that the scattered field fits the 'mirror' aperture, see Fig. 6.26). However, as we have already mentioned, if real, irreversible absorption (or amplification) is present, these are only phase surfaces that are restored; the amplitude of the reversed field will be further reduced (or increased) on the way back through the absorber (amplifier).

Let a signal field, $\mathrm{Re}E_3\exp(-i\omega_3 t)$, and a pump field, $\mathrm{Re}E_1\exp(-i\omega_1 t)$, be excited in a nonlinear medium, which for the sake of simplicity will be considered isotropic. An electromagnetic field present in a medium is accompanied by other fields, for instance, pressure $p(\boldsymbol{r},t)$, temperature, vibrations of molecules, excited electrons, and so on (Sec. 6.2). In photographic materials, there are 'fields' of metallic silver or other products of photochemical reactions. In the simplest cases, the amplitudes of these fields scale as the constant or slowly varying part of the squared local field $\overline{E^2(\boldsymbol{r},t)}$. For instance, due to the optical electrostriction,

$$p(\boldsymbol{r},t) \sim \mathrm{Re}E_1(\boldsymbol{r})E_3^*(\boldsymbol{r})\exp[i(\omega_3 - \omega_1)t]. \tag{6.329}$$

The pressure field (6.329) is a bulk hologram, it contains full information about the signal provided that the pump field is known. Certainly, a record made by pressure will be erased soon (within a relaxation time, λ/v, where v is the sound velocity) after the signal is turned off, in contrast to the fields of photochemical reactions. However, if the signal is changed, the record will stabilize fast as well. In

[t]Editors' note: the author used the term 'wave front reversal', which was more popular earlier.

the case of monochromatic fields with the same frequencies, $p(\boldsymbol{r})$ is a static field materializing the spatial distribution of the interference field $E_1(\boldsymbol{r})E_3^*(\boldsymbol{r})$. This simple model illustrates the idea of dynamical holography, which is used for the study of fast processes.

The record (6.329) can be read out with the help of a second reference wave $\mathrm{Re}E_2 \exp(-i\omega_2 t)$, which will be scattered by the field $\Delta n(\boldsymbol{r}, t)$ of the refractive index induced by the pressure field. In other words, the 'readout' field E_2 induces polarization scaling as pE_2, and the field it emits is

$$E_4(\boldsymbol{r}) \sim P_4(\boldsymbol{r}) = \chi^{(3)} E_1(\boldsymbol{r}) E_2(\boldsymbol{r}) E_3^*(\boldsymbol{r}), \tag{6.330}$$

where $\omega_4 = \omega_1 + \omega_2 - \omega_3$. The relation between E_4 and P_4 in the Born approximation is given by Eqs. (6.260), (6.264). One can see from (6.330) that if the product $E_1 E_2$ has a weak dependence on $\boldsymbol{r}$, the OPC effect takes place, $E_4 \sim E_3^* = \tilde{E}_3$. In particular, this is the case for a plane monochromatic standing-wave pump, for which $\omega_2 = \omega_1$ and $\boldsymbol{k}_2 = -\boldsymbol{k}_1$. Note that in usual, static holography, since the nonlinearity is inertial, it is necessary that $\omega_1 = \omega_3$ and $\omega_2 = \omega_4$.[u]

Note that OPC can be also obtained via three-wave interaction,

$$P_4(\boldsymbol{r}) = \chi^{(2)} E_1(\boldsymbol{r}) E_3^*(\boldsymbol{r}). \tag{6.331}$$

However, in this case, only a part of the signal field E_3 is reversed, whose frequency and angular spectrum is within the phase matching band $|\Delta \boldsymbol{k}| l < 1$, $\Delta \boldsymbol{k} \equiv \boldsymbol{k}_1 - \boldsymbol{k}_3 - \boldsymbol{k}_4$, which leads to a loss of fine spatial and temporal details of the signal. In addition, the phase-conjugated field E_4 propagates 'from left to right', similarly to the signal field, and an additional mirror is required for reflecting it back. A huge advantage of four-wave interaction is the possibility to obtain phase matching automatically using standing plane waves.

Let us estimate the efficiency $|R|$ of OPC via degenerate ($\omega_i \equiv \omega$) interaction using one-dimensional (diffraction-free) approximation in the undepleted-pump regime. To this end, let us use Eqs. (6.286) for backward interaction, with an obvious replacement of frequency subscripts. From (6.287) at $z = 0$, we find the ratio of the incident and reflected amplitudes in modes $-\boldsymbol{k}$ and $\boldsymbol{k}$,

$$|R| \equiv |E_4(-\boldsymbol{k})/E_3(\boldsymbol{k})| = \tan \gamma l, \tag{6.332}$$

where, according to (6.278), with the replacement $\chi^{(2)} \to \chi^{(3)} E_2$ and $E_1 = E_2$,

$$\gamma = \frac{2\pi\omega}{cn} |\chi^{(3)} E_1 E_2| = \frac{32\pi^3}{cn^2 \lambda_0} |\chi^{(3)}| I_1. \tag{6.333}$$

Let $\chi^{(3)} = 10^{-12}$ cm^3/erg, $l = 1$ cm, $\lambda_0 = 1\,\mu$, $n = 1.5$, then $\gamma l = 1$ at $I_1 \approx 1$ GW/cm^2.

[u]In the case of a bulk hologram ($kl \gg 1$), both a static or a dynamic one, the phase matching condition (Fig. 6.24(c)) leads to the equality of all four frequencies.

It is important that at $\gamma l = \pi/2$, the conversion coefficient turns to infinity, and then the fields E_3, E_4 are emitted spontaneously. Thus, a standing wave in a cubic medium has instability with respect to the parametric generation of backward waves, — in addition to the instability of traveling waves with respect to self-focusing, self-modulation, and stimulated scattering. Note that at $\gamma l = 1$ self-focusing, according to (6.236), will be not significant at

$$a^2 \ll \mathcal{P}_0/I = \lambda l/4, \tag{6.334}$$

i.e., in the case where the pump has considerable diffraction divergence in the layer. (In this case, the validity of the one-dimensional approximation is also violated.)

Chapter 7

Statistical Optics

In classical electrodynamics, the electric field $E_\alpha(\boldsymbol{r}, t) \equiv E(x)$, $x \equiv \{\boldsymbol{r}, t, \alpha\}$ is assumed to be deterministic and, in principle, measurable with an accuracy as good as required. (We speak only about the electric field as it is the field that determines the observable effects.)

In classical statistical optics, which includes the coherence theory as its important part, $E(x)$ for every x is considered as a random variable, with x being a parameter. It is convenient to split space and time in numbered cells and to consider x as a discrete parameter spanning a countable number of values x_i. Thus, a fluctuating light field is described by a set of random variables $E_i \equiv E(x_i)$. (Another way of 'discretizing' the field, the mode decomposition, is described in Sec. 7.3.) All properties of the random set ensemble $\{E_i\}$ are given by a multi-dimensional *distribution function*, or a set of *moments* (correlation matrices) $\langle E_1 E_2 \cdots E_m \rangle$ of various orders m (the angular brackets denote averaging over the distribution function.) In experiment, the averaging is of course not over an ensemble of fields but over a certain spatial and temporal interval V_{det}. In addition, the field is filtered in frequency and in the propagation direction.

From the viewpoint of statistical optics, the macroscopic Maxwell's equations are kinetic equations for the first moments $\langle E_i \rangle$, $\langle H_i \rangle$. The intensity and spectrum of light are determined by the second moments $\langle E_i E_j \rangle$, while n–quantum processes are given by moments of order $2n$.

However, classical statistics is, strictly speaking, not applicable to the optical range, since the degeneracy factor $\langle N \rangle = [\exp(\hbar\omega/\kappa T_{ef}) - 1]^{-1}$, which has the meaning of the average number of photons in one mode, or the spectral brightness $I_{\omega\Omega}$ in $\hbar c/\lambda^3 \equiv I_{\omega\Omega}^{vac}$ units, is usually much less than unity. For instance, for the green part of the sunlight spectrum ($T_{ef} \approx 6000\,\text{K}$, $\lambda \approx 0.5\,\mu$), $\langle N \rangle \approx 0.01$, and it reaches the unity only in the IR range, at $\lambda = 3.5\,\mu$. Among the few exceptions there are laser fields, for which effective (brightness) temperature exceeds the one of the sunlight by many orders of magnitude. In connection with this, let us mention one of the paradoxes in the history of physics: quantum optics started its rapid development only during the laser age, when light fields with $\langle N \rangle \gg 1$ appeared (although the general principles of quantum electrodynamics were developed much earlier).

In quantum optics and electrodynamics, an ensemble of fields is given by a wavefunction Ψ or a density operator ρ. The angular brackets in the definition of the correlation function denote now quantum averaging over Ψ or ρ; then, E_i are operators acting on Ψ according to certain rules. It is important that, in the general case, the fields at the neighbouring points of space-time do not commute, which leads to the quantum fluctuations of the field, to the moments $\langle E_i E_j \rangle$ being nonzero even for the vacuum, to the spontaneous emission of excited atoms, and to the noise of quantum amplifiers and generators.

The statistical theory of laser radiation studies the most important parameters of lasers, such as the maximal possible monochromaticity of oscillators and the sensitivity of amplifiers. Similarly to a consistent theory of thermal light emission by heated bodies, this theory should be based on quantum electrodynamics and non-equilibrium thermodynamics [Zubarev (1971); Klimontovich (1982)]. A special difficulty in the quantum-statistical analysis of a laser is that the nonlinearity plays a principle role, as it determines, due to the saturation effect, the stationary amplitude of oscillations (the limiting cycle of a classical auto-oscillator).

The most important achievement in the nonlinear quantum theory of lasers [Loudon (2000); Klimontovich (1980); Lax (1968); Arecchi (1974); Klauder (1968)] is the conclusion that in the cavity of a laser operating well above threshold, the field is in *a coherent state*, a concept introduced into quantum optics by Glauber [Glauber (1965)]. There is a close analogy between a field in a coherent state with a large amplitude and a classical harmonic oscillation and, since the saturation effect manifests itself only in the case of high amplitudes, the nonlinear regime of a laser is rather accurately described by the semi-classical theory. Nonlinear theories predict all statistical characteristics of laser radiation: the intensity, the spectral width, the coherence radius, and the higher moments.

Even more crude, but still useful is the approximation considered in the linear theory of noise in quantum amplifiers and oscillators (Sec. 7.1), which ignores the saturation effect and is therefore valid only below the oscillation threshold. The most important results of this theory are *the Kirchhoff law, giving the noise intensity of an amplifier in terms of its amplification coefficient, and the Townes formula, relating the 'natural' bandwidth of an oscillator to its power.* According to the linear theory, the radiation has Gaussian statistics, and therefore these two parameters, the intensity and the spectral width, fully describe the statistics of the field.

This chapter is devoted to the foundations of quantum optics. The consideration starts with the linear theory of noise in quantum amplifiers, which does not require the quantization of the field (Sec. 7.1). In Sec. 7.2, basic notions of classical statistical optics are considered. The next section (Sec. 7.3) describes the initial stage of the field quantization, i.e., writing the Maxwell equations in the canonical form, which allows the quantization to be performed in an easy way (Sec. 7.4). Section 7.5 considers the basic classes of quantum states of the field, and Sec. 7.6 describes the statistics of photons and photoelectrons for these states. Finally, in Sec. 7.7 we once again return to the question about the probability of a transition due to a noise field (Sec. 2.4), but this time we consider it in the framework of the quantum theory.

7.1 The Kirchhoff law for quantum amplifiers

The intensity of the noise radiation of a laser or a maser operating in a linear stationary regime can be calculated without using quantum electrodynamics, from the general considerations based on certain rules of non-equilibrium thermodynamics like the Nyquist formula (or the fluctuation-dissipation theorem, FDT) and the Kirchhoff law for the thermal radiation.

7.1.1 *The Kirchhoff law for a single mode*

Let an ideal waveguide be filled by homogeneous matter of temperature T, thermodynamical or effective. Consider the electromagnetic energy carried by a single waveguided field type (for instance, H_{01} in a rectangular waveguide) in the stationary case. Let the waveguide be connected with a matched source and a matched load, so that the backward wave is independent of the forward one (Fig. 7.1).

We will be calculating the power spectral density $\mathcal{P}_f(\omega, z) \equiv \Delta\mathcal{P}/\Delta f$, i.e., the energy of incoherent radiation with the frequency $\omega = 2\pi f > 0$ transferred through a cross-section z in 1 s within the frequency bandwidth 1 Hz. It is easy to show that in $\hbar\omega = hf$ units, $\mathcal{P}_f(\omega, z)$ corresponds to the mean number of photons $N(\omega, z)$ per one longitudinal mode of the waveguide,

$$\mathcal{P}_f = \frac{u}{L}\mathcal{E}_f = \frac{hfu}{L}N_f = \frac{hfu}{L}g_f N = hfN. \tag{7.1}$$

Here, $u = d\omega/dk$ is the group velocity; k is the propagation constant; L is the length of some part of the waveguide, which is much greater than the wavelength

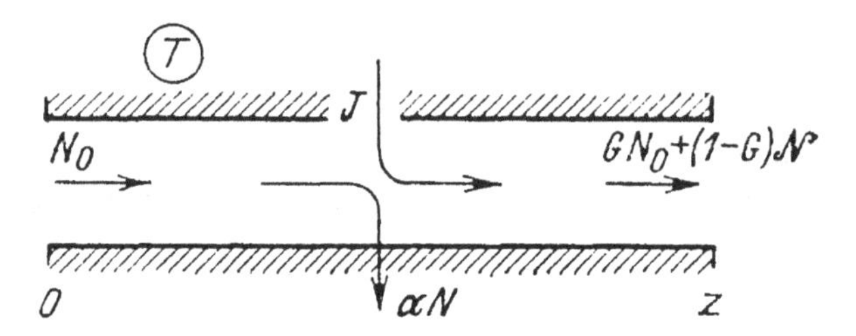

Fig. 7.1 Waveguide Kirchhoff law. The noise of the waveguide is determined by the competition between the absorption α and the spontaneous emission $j = \alpha_0 \mathcal{N}$, where $\mathcal{N} = N_2/(N_1 - N_2)$ is the Planck function and N_i are the level populations. As a result, the noise can be expressed in terms of the transfer coefficient $G = e^{-\alpha z}$ and the temperature T.

$\lambda = 2\pi/k$ but much less than the distance at which $\mathcal{P}_f(z)$ changes noticeably; $\mathcal{E}_f$ is the spectral energy density of the macroscopic field within L; $N_f \equiv \mathcal{E}_f/hf$ is the number of photons within L per 1 Hz, which is equal to N times the spectral density of the longitudinal modes $g_f = Ld\lambda^{-1}/d\omega = L/u$. The inverse value, u/L, is equal to the interval between the eigenfrequencies of the neighboring modes (for the notion of a mode in more detail, see Sec. 7.3).

Let us start from the linear kinetic equation for $N(\omega, z)$ of the form

$$dN/dz = -\alpha N + j. \tag{7.2}$$

Here, $\alpha(\omega)$ is the absorption or amplification coefficient due to stimulated transitions and $j(\omega) \geqslant 0$ is a distributed source of noise due to spontaneous emission. This equation provides a phenomenological description of the stationary interaction of matter with a single transverse mode of the field, not necessarily in a waveguide.[a] It is useful to compare this equation with one-dimensional equations (6.171) for the 'slow amplitudes' and with Einstein's relations (2.99) for non-stationary interactions. Apparently, $\alpha \sim B(N_1 - N_2)$ and $j \sim AN_2$, where A, B are the Einstein coefficients and N_i are the populations of a pair of levels separated by an interval $\hbar\omega$.

The solution to (7.2) has the form

$$N = N_0 G + (j/\alpha)(1 - G), \quad G(\omega, z) \equiv \exp[-\alpha(\omega)z], \tag{7.3}$$

where N_0 is the signal at the input of the amplifier and G is the *transfer coefficient* of the waveguide. Here, the second term, which is independent of the input signal, is the noise of the waveguide.

[a]Such equations are called *transfer equations* (for photons, neutrons, etc.), they imply that a space coordinate z, defined up to an accuracy of $\pm L/2$, can be attributed to a photon. In the general case, a function $N(\boldsymbol{k}, \boldsymbol{r}, t)$ is used, which has the meaning of *local spectral brightness* in $\hbar c/\lambda^3$ units.

Let us express the ratio j/α in terms of the temperature of the matter. Let $\alpha z \gg 1$, then, according to (7.3), $N = j\alpha$. On the other hand, in a sufficiently long waveguide with damping, equilibrium radiation should be formed, with the mean number of photons per mode given by the Planck function $\mathcal{N}$. Hence,

$$j/\alpha = \mathcal{N} \equiv [\exp(\hbar\omega/\kappa T) - 1]^{-1}$$
$$= (1/2)[\coth(\hbar\omega/2\kappa T) - 1] = N_2^{(0)}/(N_1^{(0)} - N_2^{(0)}). \quad (7.4)$$

This conclusion is rigorous only for a completely equilibrium system where the populations N_i and the photon number N obey the Boltzmann and Planck distributions, respectively.

However, it is reasonable to assume that (7.4) is approximately valid even in the absence of equilibrium radiation, for instance, at $\alpha z \lesssim 1$ and $N_0 = 0$. Then, the parameter T relates only to the matter, whose temperature is maintained constant and uniform, despite the radiation cooling ('quasi-equilibrium'). For this, it is necessary that the degrees of freedom that emit and absorb radiation at frequency ω interact with the thermostat much stronger than with the field.[b] In this approximation, Eq. (7.3) at $N_0 = 0$ describes the spatial spreading of equilibrium Planck's field (with a single propagation direction) in a waveguide of a layer with the length z, temperature T and absorption coefficient $\alpha(\omega)$,

$$N = \mathcal{A}\mathcal{N}, \ \mathcal{A} \equiv 1 - G. \quad (7.5)$$

Here, $\mathcal{A}$ is the absorptivity of the waveguide or the layer (in the general case, with an account for reflections at the input and output). It can be measured, according to (7.3), using an external signal: $\mathcal{A} = 1 - dN/dN_0$.

Equation (7.5) is the Kirchhoff law relating the thermal radiation of a heated body to its thermodynamic parameter T and kinetic parameter $\mathcal{A}$. It can be easily derived in the form (7.5) without using the kinetic equation [Landau (1964)]. Note that this law is used in the construction of photometry reference sources for light intensity, although, in principle, it is only applicable within the approximations of strong coupling with the thermostat, linear optics, and geometric optics. For its generalizations, see Ref. [Klyshko (1980)].

7.1.2 *The Kirchhoff law for a negative temperature*

The phenomenological relation of spontaneous and stimulated effects with the level populations, $j/\alpha = N_2/(N_1 - N_2)$, is confirmed by calculations in the frame-

[b] Let the time T_1 characterize the interaction of the particles with the thermostat; then the condition for the Kirchhoff law (and, generally, FDT-type equations) to be applicable to non-equilibrium problems is, apparently of the form $1/T_1 \gg A(2N + 1)$, compare with (2.115).

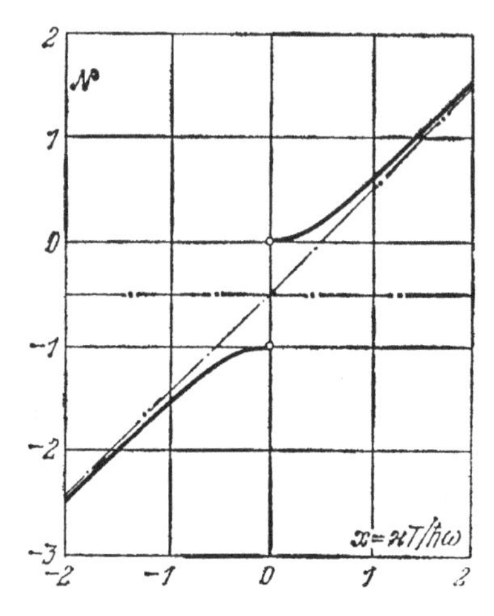

Fig. 7.2 The Planck function determining the mean number of photons with the frequency $\omega > 0$ in a single mode of an equilibrium field, as well as the number of excited atoms N_2 divided by the population difference $N_1 - N_2$. Here, T is the effective temperature; in the case of a field, $T > 0$.

work of the two-level model. It is then generalized to the case of a non-equilibrium matter with non-Boltzmann N_i. In this approximation, it is also convenient to keep relation (7.4) and the Kirchhoff law in the form (7.5), but to interpret T as the effective (*spin*) temperature T_{ef}, which is determined by the actual population ratio, $\exp(\hbar\omega/\kappa T_{ef}) \equiv N_1/N_2$ (Sec. 3.2).

From the definition of the function $\mathcal{N}(T)$, it follows that in the case of population inversion, where α and T are negative, $\mathcal{N}$ is also negative (Fig. 7.2), $\mathcal{N}(-T) = -[\mathcal{N}(T) + 1]$; therefore, the Kirchhoff law can be written as

$$N = [\mathcal{N}(-T) + 1](G - 1),$$

or (see Fig. 7.3)

$$N = \mathcal{N}\{1 - \exp[-\alpha_0(\omega)z/(2\mathcal{N} + 1)]\}, \tag{7.6}$$

since $\alpha = \sigma(N_1 - N_2) = \alpha_0 \tanh(\hbar\omega/2\kappa T)$, where α_0 is the absorption coefficient at $T = +0$.

In particular, in the case of full inversion, $N_1 = 0$, $T = -0$, $\mathcal{N} = -1$, so that the spectral noise density of a single-mode ideal quantum amplifier in hf units is simply equal to the transfer coefficient with the unity subtracted,

$$N = G - 1. \tag{7.7}$$

The -1 appears here because the noise source is a distributed one. Hence, the noise relative to the input, N/G, is $1 - G^{-1}$, which at $G \gg 1$ yields one photon per mode, i.e., one photon in a unity frequency band per unit time.

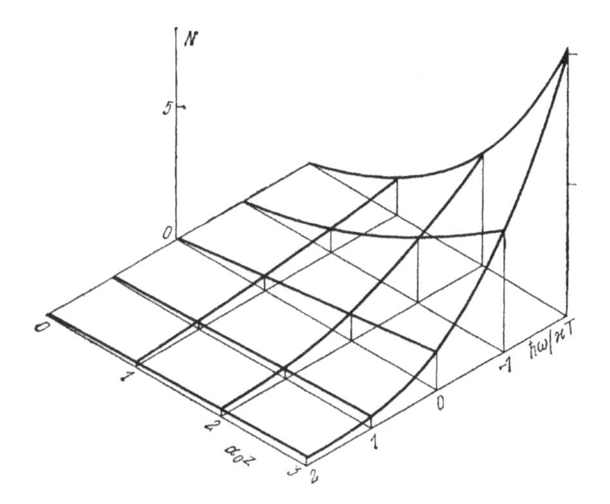

Fig. 7.3 The spectral density of the thermal radiation emitted by a single-mode attenuator ($T > 0$) or amplifier ($T < 0$) as a function of its length z and temperature T, according to the Kirchhoff law. α_0 is the absorption coefficient at $T = +0$.

If an incoherent signal with the power spectral density hfN_0 is present at the input, one should add GN_0 to (7.7). The result can be represented in the form

$$N + 1 = G(N_0 + 1). \tag{7.8}$$

This equation provides the following algorithm for taking into account the spontaneous emission (or, in other words, the quantum fluctuations) of an ideal amplifier: a unity is added to the number of photons at the input, the result is multiplied by the classical amplification coefficient, and then the unity is subtracted, which yields the output number of photons. Thus, *the signal-to-noise ratio at the output of an ideal quantum amplifier with large amplification is equal to N_0, the number of signal photons per mode.*

The unities in (7.8) can be interpreted as a result of zero-point field fluctuations; however, this interpretation should be treated with caution [Ginzburg (1983)]. Quantum fluctuations determine the limiting sensitivity and accuracy of quantum amplifiers, as well as any measurement devices. At $N_1 \neq 0$, 'thermal' fluctuations, scaling as $\mathcal{N}(-T)$, are added to the quantum ones (see (7.6)).

In the microwave range, one often uses the concept of the *noise temperature of an amplifier*, T_n. By definition, T_n is the brightness temperature of the noise radiation at the output relative to the input,

$$\mathcal{N}(T_n) \equiv N/G = \mathcal{N}(T_{ef})(e^{\alpha z} - 1), \tag{7.9}$$

where $\mathcal{N}$ is the Planck function for the chosen frequency ω. This equation relates

T_n with the effective (spin) temperature of matter, T_{ef}. The notion of T_n is convenient only in the classical temperature range, where $\mathcal{N} \approx \kappa T/\hbar\omega$ and (7.9) takes the form

$$T_n = (e^{\alpha z} - 1)T_{ef},$$

which in the two limiting cases of small absorption and large amplification yields

$$T_n = \alpha z T_{ef}, \quad |\alpha z| \ll 1, \tag{7.10}$$

$$T_n = -T_{ef}, \quad \alpha z \ll -1. \tag{7.11}$$

Thus, the spin temperature determines not only the ratio of the populations but also the limiting noise temperature of a quantum amplifier.

In paramagnetic amplifiers, $|T_{ef}|$ is on the order of the lattice temperature of the active crystal, which is cooled down to helium temperatures for the sake of lowering T_n, reducing spin-lattice relaxation, and increasing the equilibrium population difference. As a result, the noise temperature of such amplifiers is as low as several Kelvins. At the same time, in ordinary electronic amplifiers, $T_n \sim 10^3$ K, due to the high temperature of the cathode; in parametric microwave amplifiers based on semiconductor diodes, $T_n \sim 100$ K.

In practice, T_n in paramagnetic amplifiers is determined by the thermal emission not from the active medium, according to (7.11), but from the elements of the input channel (the aerial, waveguides, and ferrite devices), according to (7.10). Indeed, an input waveguide with the walls at room temperature and the losses of only 1% makes a noticeable contribution, $T_n \approx 3$ K.

It is not difficult to generalize the Kirchhoff law (7.5) to the case where the waveguide has several sources of homogeneous linear loss, α_i, and noise, j_i, with different effective temperatures T_i $(i = 1, 2, \ldots)$. Assuming again $j_i = \alpha_i \mathcal{N}_i$, we obtain the kinetic equation

$$dN/dz = \sum_i \alpha_i(\mathcal{N} - N), \tag{7.12}$$

which only differs from (7.2) by the replacements $\alpha \to \sum \alpha_i$ and $\alpha\mathcal{N} \to \sum \alpha_i \mathcal{N}_i$. As a result, (7.5) takes the form

$$N = (1 - e^{-\alpha z}) \sum_i (\alpha_i/\alpha)\mathcal{N}_i, \tag{7.13}$$

with $\alpha \equiv \sum \alpha_i$. Thus, *the contribution of each element to the total thermal emission scales as its contribution α_i/α to the total absorption coefficient.*

Consider now the shape of the thermal emission spectrum in the vicinity of a single narrow resonance at various optical densities αz (Fig. 2.4). When $|\alpha|z$ is small, the emission spectrum repeats the absorption spectrum, $N(\omega) = \alpha(\omega)z\mathcal{N}$ (the dependence $\mathcal{N}(\omega)$ is too slow to be taken into account here.) At large positive αz, the line gets flattened and broadened, tending eventually to the equilibrium spectrum $\mathcal{N}(\omega)$, and at large negative αz the linewidth is reduced (similarly to the amplification band, see (2.68)) and tends to zero as $\sqrt{\ln G}$.

7.1.3 *Noise of a multimode amplifier*

We have so far discussed a single transverse mode of a waveguide. In the general case, where the radiation is delivered to the detector as several independent modes, (7.5) or (7.12) should be summed over all essential modes. If the waveguide cross-section A is much greater than λ^2, summation can be replaced by integration.

In free space, the number of transverse modes (i.e., modes with the same frequency $\omega_k = ck$) observed by a pointlike detector in the near-field zone scales as the spatial aperture of the emitter, $A \equiv ab$, and the angular aperture of the detector, $\Delta\Omega = \Delta\vartheta_x \Delta\vartheta_y$,

$$\Delta g_\perp = (a\Delta\lambda_x^{-1})(b\Delta\lambda_y^{-1}) = A\Delta\Omega/\lambda^2, \tag{7.14}$$

where $\lambda = 2\pi k$, we consider only a single polarization type and assume that the observation direction is orthogonal to A. According to (7.14), *the angular interval between the neighboring transverse modes is on the order of the diffraction angle* λ/a.

Multiplying (7.5) by the photon energy $\hbar\omega$, the detector band $\Delta f = \Delta\omega/2\pi$, and the number of transverse modes $\Delta g_\perp$, we find the power of multimode thermal radiation with a single polarization type,

$$\Delta\mathcal{P} = (\hbar\omega\Delta\omega\Delta\Omega A/2\pi\lambda^2)\mathcal{N}(1 - G), \tag{7.15}$$

where we assume that the transfer coefficient $G = e^{-\alpha z}$ and the effective temperature T are the same for the modes having close directions.

The ratio $\Delta\mathcal{P}/\Delta\omega\Delta\Omega A$ is called the *spectral brightness*; this is the main energy characteristic of incoherent multi-mode radiation. In the general case, $I_{\omega\Omega} \equiv I(\boldsymbol{k}, \boldsymbol{r})$ depends on the frequency, direction, and the observation point; however, in the absence of scattering, emission, and absorption between the points $\boldsymbol{r}$ and $\boldsymbol{r} + \boldsymbol{a}$, the spectral brightness at these points is the same provided that the argument $\boldsymbol{k}$ is parallel to $\boldsymbol{a}$,

$$I(\boldsymbol{k}, \boldsymbol{r} + \boldsymbol{a}) = I(\boldsymbol{k}, \boldsymbol{r}), \quad \boldsymbol{k} \parallel \boldsymbol{a}. \tag{7.16}$$

Certainly, it makes sense to speak about the displacement $\boldsymbol{a}$ of the observation point only at $a \gg \lambda$.

In photometry, there are special terms for the integrals of I w.r.t. various variables: *the brightness, the radiance, and the luminous intensity*,[c]

$$B(\hat{\boldsymbol{k}},\boldsymbol{r}) \equiv \int_0^\infty d\omega_k I(\boldsymbol{k},\boldsymbol{r}), \quad E(\boldsymbol{r}) \equiv \int d\Omega B(\hat{\boldsymbol{k}},\boldsymbol{r})\cos\vartheta_k,$$
$$\mathcal{P}(z) \equiv \int dxdyE(\boldsymbol{r}). \tag{7.17}$$

Here, $\hat{\boldsymbol{k}} \equiv \boldsymbol{k}/k$, ϑ_k is the angle between $\boldsymbol{k}$ and the z axis. The spectral and volume energy density is

$$\rho_\omega(\omega,\boldsymbol{r}) = \frac{1}{u(\omega)}\int_{4\pi} d\Omega I(\boldsymbol{k},\boldsymbol{r}), \tag{7.18}$$

where u is the group velocity, $\omega \equiv \omega_k$. In the case of the thermal radiation from a heated body, $I(\boldsymbol{k},\boldsymbol{r})\cos\vartheta_k$ is called the *emittance* of a body (at point $\boldsymbol{r}$ of its surface).

According to (7.15), the Kirchhoff law for the multimode emission from a matter layer with the thickness z and the effective temperature T has the form (compare with (6.225), (6.307))

$$I_{\omega\Omega}(\boldsymbol{k}) = I^{vac}_{\omega\Omega}N(\boldsymbol{k}), \tag{7.19}$$

where

$$I^{vac}_{\omega\Omega} \equiv \hbar\omega/2\pi\lambda^2, \quad N(\boldsymbol{k}) = \mathcal{N}(1 - e^{-\alpha(k)z}), \tag{7.20}$$

where we take into account only a single polarization type. The value $I^{vac}_{\omega\Omega}$ corresponds to the spectral brightness of radiation having a single photon in each mode (i.e., $I^{vac}_{\omega\Omega}$ is the doubled 'brightness of zero-point vacuum fluctuations'), it is a natural unit for the measurement of $I_{\omega\Omega}$: the mean number of photons per mode N is equal to the spectral brightness in $I^{vac}_{\omega\Omega}$ units.

7.1.4 *Equilibrium and spontaneous radiation; superfluorescence*

Consider three typical cases:

1. At $\alpha z \gg 1$, (7.19) leads to $I_{\omega\Omega} = I^{vac}_{\omega\Omega}\mathcal{N}$, which, in the case of isotropic radiation, corresponds to the blackbody radiation and, according to (7.18), at $u = c/n$, to the Planck formula,

$$2\rho_\omega = (8\pi n/c)I_{\omega\Omega} = (8\pi\hbar/\lambda^3)\mathcal{N}; \tag{7.21}$$

the factor 2 accounts here for two polarization types.

[c] Sometimes the attribute 'energy' is added, indicating that they are measured in physical units and not in light-engineering ones. The latter take into account the spectral characteristic of the human eye and are based on the *candela* unit.

2. If $|\alpha|z \ll 1$, then (7.19) leads to

$$I_{\omega\Omega} = I^{vac}_{\omega\Omega}N\alpha z = I^{vac}_{\omega\Omega}N_2\sigma l/\cos\vartheta_k, \tag{7.22}$$

where the optical density of the layer αz for the mode $\boldsymbol{k}$ was assumed to be $\sigma(N_1 - N_2)l/\cos\vartheta_k$. Let us write the transition cross-section σ in terms of the transition dipole element $\boldsymbol{d}_0$ according to (2.52) and find, using (7.17), the total power of spontaneous emission. Let $n = 1$, then

$$\mathcal{P} = \int_{4\pi} d\Omega \int_A dxdy\cos\vartheta_k \int_0^\infty d\omega I_{\omega\Omega} = (4\pi^2/\hbar c)I^{vac}_{\omega\Omega}N_2V\omega_0 \int_{4\pi} d\Omega(\boldsymbol{d}_0\cdot\boldsymbol{e}_k)^2, \tag{7.23}$$

where $V = Al$ is the matter volume. The integral here is $8\pi d_0^2/3$, so that $\mathcal{P}/VN_2$ coincides with the expression found above for the power of the spontaneous emission per one molecule, $4\omega_0^4 d_0^2/3c^3$ (see (5.32)).

3. Now, let $\alpha z \ll -1$, then $N = G\mathcal{N}$, and

$$\mathcal{P} = N_2(N_2 - N_1)^{-1}I^{vac}_{\omega\Omega}G_0A_{ef}\Delta\Omega_{ef}\Delta\omega_{ef}, \tag{7.24}$$

where we have introduced the effective apertures A_{ef}, $\Delta\Omega_{ef}$ and the frequency bandwidth $\Delta\omega_{ef}$, which can be rigorously defined in terms of the corresponding integrals, and the amplification coefficient at the center of the line, G_0. This equation determines the power of *superfluorescence*, the amplified spontaneous emission. Assume, for a crude estimate, that A_{ef} coincides with the cross-section A of the amplifier, $\Delta\omega_{ef}$ coincides with the amplification bandwidth $\Delta\omega$, and $\Delta\Omega_{ef} \approx A/l^2$, where l is the length of the amplifier, then at $N_1 = 0$

$$\mathcal{P} = I^{vac}_{\omega\Omega}G_0A^2\Delta\lambda/l^2. \tag{7.25}$$

At $\lambda = 0.5\,\mu$, $\Delta\lambda = 10$ nm, $A = 1$ cm^2, $l = 10$ cm, and $G = 10^3$, we obtain $\mathcal{P} = 2\cdot 10^4$W.

For many transitions in gases, inversion can be obtained only for a short time, $\tau \lesssim 10^{-9}$s, even with the help of very powerful lasers. This is due to the extremely long lifetime of the lower level. Then, if the active area length is larger than $c\tau \sim 30$ cm, a mirror-based feedback is too slow to have an effect. The radiation of lasers using such 'self-restricted transitions' is, in fact, superfluorescence. An important example is the nitrogen laser with the wavelength 330 nm.

Certainly, at sufficiently high G_0 the superfluorescence power (7.25) can be large enough for saturation; then the problem becomes nonlinear and the Kirchhoff law is not valid any more.

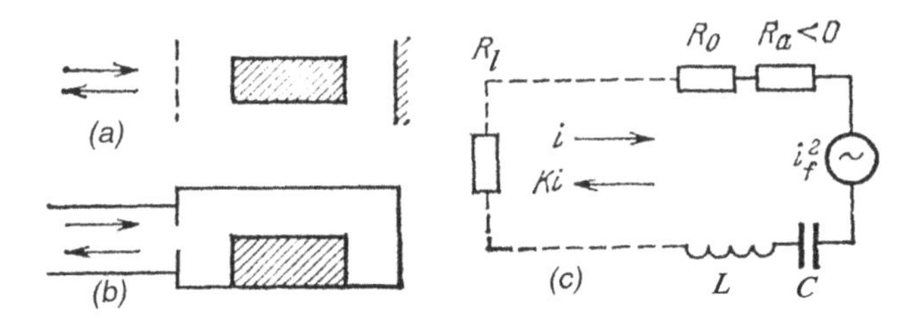

Fig. 7.4 Cavity quantum amplifier: (a) in the optical range; (b) in the microwave range; (c) an equivalent circuit.

7.1.5 *Gain and bandwidth of a cavity amplifier*

For increasing the gain at a given length of the active medium, one can use a positive feedback provided by a bulk cavity. With the help of an equivalent circuit, let us find the gain coefficient $G(\omega)$ and the transmission band $\Delta\omega$ of a *reflective cavity amplifier*, i.e., an amplifier with a single coupling waveguide or a single semi-transparent mirror (Fig. 7.4). In the microwave range, the amplified reflected wave is separated from the incident weak signal by means of non-reciprocal ferrite devices (circulators).

Consider the spectral range in the vicinity of some eigenfrequency ω_0 of the cavity. If the corresponding oscillation is non-degenerate and ω_0 is well separated from other frequencies, then the field at an arbitrary point of the cavity depends on the time similarly to the parameters of any other oscillator system with a single degree of freedom, for instance, like the current in an LC contour. Then the electric field has the form $\boldsymbol{E} = \boldsymbol{u}(\boldsymbol{r})q(t)$, where $\boldsymbol{u}(\boldsymbol{r})$ is a known function determined by the shape of the cavity and $q(t)$ obeys the equation of motion for a harmonic oscillator. This allows the cavity response to be calculated using the equivalent circuit shown in Fig. 7.4(c). Damping (and, according to the Nyquist theorem, noise as well) are introduced into the circuit by three resistances: the external load R_l (which, by definition, coincides with the impedance of a transmission line corresponding to the coupling waveguide), the resistance R_0 scaling as the losses in the cavity walls, and the negative resistance R_a scaling as the emission of the active medium. The width of the chosen spectral line is assumed to be much greater than $\Delta\omega$, and therefore R_a is a real parameter.

In the case of a transmission line, the reflection coefficient, i.e., the ratio of the complex amplitudes for the backward and forward waves, is determined by the ratio of the load (cavity) resistance Z_c to the impedance of the line R_l. For instance, the reflection coefficient for the current is

$$K(\omega) = \frac{R_l - Z_c}{R_l + Z_c} = \frac{R_l - R_c - iX}{R_l + R_c + iX}, \tag{7.26}$$

where

$$R_c \equiv R_0 + R_a,\ X \equiv \frac{1}{\omega C} - \omega L = \omega L\left(\frac{\omega_0^2}{\omega^2} - 1\right) \approx 2L(\omega_0 - \omega). \qquad (7.27)$$

The relation between R_l and the other parameters depends on the coupling between the cavity and the waveguide.

Let us pass from the equivalent parameters R_i to dimensionless variables that have a direct physical meaning: the Q-factors $Q_i \equiv \omega_0 L/R_i$ or the damping constants $d_i \equiv 1/Q_i$. The value Q_a is called the (magnetic) *quality factor of the active medium*. After dividing both the numerator and the denominator of (7.26) by $\omega_0 L$, we find

$$K(\omega) = \frac{d_l - d_c - 2ix}{d_l + d_c + 2ix}, \qquad (7.28)$$

where $x \equiv (\omega_0 - \omega)/\omega_0$. The gain (or attenuation) of the reflected wave power is

$$G(\omega) = |K(\omega)|^2 = \frac{(d_l - d_c)^2 + 4x^2}{(d_l + d_c)^2 + 4x^2}. \qquad (7.29)$$

It is important that the coupling between the waveguide and the cavity can be varied, for instance, by changing the orientation of the feedback loop or the transmission of the mirror. Then, the normalized losses $d_l = R_l/\omega_0 L$ due to the external circuit will also vary from zero (no feedback, a totally reflecting mirror) to infinity (maximal feedback, no mirror). In the case of a resonance ($\omega = \omega_0$) and positive losses in the cavity ($d_c > 0$), variation of the feedback strength from -1 to $+1$ leads to the variation of the amplitude reflection coefficient for the current K_0. The dependence passes through zero at the point of load matching, $d_l = d_c$ (Fig. 7.5).

If $d_c = d_0 + d_a < 0$, then by decreasing the feedback strengths one makes K_0 and G_0 vary from 1 to ∞. Thus, *a cavity quantum amplifier with a sufficiently high-quality cavity ($Q_0 > -Q_a$) enables a weak signal to be amplified as much as possible due to the regeneration.*

However, an increase in K_0 is accompanied by a decrease in another parameter, important for many applications, namely, the amplification bandwidth $\Delta\omega$. As we will show below, at $K_0 \gg 1$ the product of the amplitude gain and the bandwidth does not depend on the feedback,

$$K_0 \Delta\omega = 2\omega_0/|Q_a|. \qquad (7.30)$$

The product $K_0\Delta\omega$ is called the regeneration parameter. It characterizes the quality of the active medium for paramagnetic amplifiers.

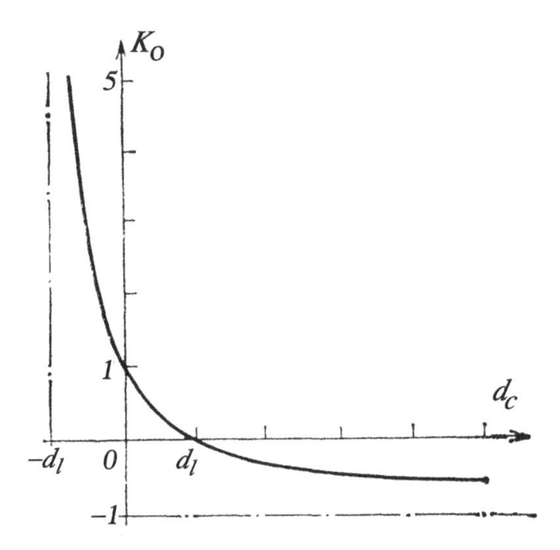

Fig. 7.5 The gain coefficient K_0 of a cavity amplifier as a function of the cavity quality factor $1/d_c$. By reducing the feedback d_l, at a fixed $d_c < 0$ one can make K_0 as high as desired. Here, d_c is the cavity damping constant with an account for the negative contribution from the active medium; d_l is the damping of the load, i.e., the relative losses due to the external circuit.

According to (7.29),

$$G - 1 = \frac{-d_c d_l}{d^2/4 + x^2}, \tag{7.31}$$

where $d \equiv d_l + d_c \equiv 1/Q$ is the total damping constant and Q is the quality factor of the loaded cavity. It follows that at $G \gg 1$ the frequency characteristic of the amplifier has a Lorentzian shape and the *amplification bandwidth is determined by the quality factor of the loaded cavity with an account for all losses*,

$$\frac{\Delta\omega}{\omega_0} = d = \frac{1}{Q_l} + \frac{1}{Q_0} + \frac{1}{Q_a}. \tag{7.32}$$

For achieving $G \gg 1$, a nearly full compensation of all losses is necessary, $-d_a \approx d_0 + d_l$, in which case $K_0 \approx 2|d_p|/d$, so that the product of the gain and the bandwidth is a constant, $K_0\Delta\omega = 2\omega_0|d_c|$. Usually, the quality of the cavity, $Q_0 \gtrsim 10^4$, is much greater than the quality of the active medium, $|Q_a| \lesssim 10^3$, hence, in practice, $Q_c = Q_a$, which yields (7.30).

Let us write Q_a in terms of the parameters of the matter by using the general definition of the quality factor as the ratio of the energy stored in an oscillation contour,

$$\mathcal{E} = \int_{V_c} d^3r(\epsilon'\overline{E^2} + \overline{H^2})/8\pi = \int_{V_c} d^3r\epsilon' u^2(\mathbf{r})/8\pi, \tag{7.33}$$

to the energy of losses during the time ω_0^{-1},

$$\mathcal{P}/\omega_0 = \int_{V_c} d^3r\epsilon''(\omega_0)u^2(\mathbf{r})/8\pi. \qquad (7.34)$$

Here, we have used the equalities $\mathcal{E}_{el} = \mathcal{E}_{mag}$, $\overline{q^2(t)} = 1/2$, and Eq. (4.15) for the losses per unit volume. Hence,

$$Q_a^{-1} = \eta\epsilon''/\epsilon' = \eta\alpha/k, \qquad (7.35)$$

where $\eta \approx V/V_c \leqslant 1$ is the factor showing the proportion of the cavity filled by the active medium and determined by the ratio of the integrals in (7.33) and (7.34); $\epsilon = 1 + 4\pi\chi$, α is the linear amplification factor for plane waves with the wavevector k; in the case of a magnetic transition, ϵ'' should be understood as the magnetic permeability. At electron paramagnetic resonance, $\alpha \approx -5 \cdot 10^{-2}$ cm^{-1} (Sec. 2.3), which yields, with $\lambda = 3$ cm and $\epsilon' = \eta = 1$, $Q_a = -40$. Let $K_0 = 100$, then it follows from (7.30) that $\Delta f = 5$ MHz. This bandwidth is not sufficient for many applications (communications, radar, radio astronomy); for this reason, one uses several coupled cavities and slowing-down systems.

7.1.6 *The Kirchhoff law for a cavity amplifier. The Townes equation*

In order to take into account the noise of the active medium, let us add to the equivalent circuit (Fig. 7.4) a source of current with the spectral density given by the Nyquist equation (see Ref. [Akhmanov (1981)]),

$$i_f^2 = 4\kappa T R_a/|Z|^2, \qquad (7.36)$$

where $Z = R_l + R_c + iX$ and T is the effective (spin) temperature of the active medium. (We neglect the noise related to Q_0.) For taking into account quantum noise, κT should be replaced by $\hbar\omega\mathcal{N}$. The load resistance will produce the spectral power density $\mathcal{P}_f = i_f^2 R_l$, which is, in $\hbar\omega$ units,

$$N = 4\mathcal{N}d_a d_l/(d^2 + 4x^2) = \mathcal{N}(1 - G). \qquad (7.37)$$

Here, we used (7.31) in the approximation $Q_0 \gg |Q_a|$. Thus, we have again, according to the Kirchhoff law, expressed the noise in terms of the effective temperature and the gain. The width of the noise spectrum $\Delta\omega_n$ at half maximum, according to (7.37), coincides with the amplification bandwidth $\Delta\omega$ (see (7.32)), and hence, it tends to zero at $d_c \to -d_l$. Similarly to the case of a traveling-wave amplifier, at $G \gg 1$ and $\hbar\omega \ll \kappa T$ the noise temperature has the same absolute value as the spin one.

Let us find the full noise power at $d_c \approx d_a \approx -d_l$ and $\Delta\omega \ll \omega_0$,

$$\mathcal{P} = \int_0^\infty df \hbar\omega N = -\frac{\hbar\omega_0^3 N}{2\pi Q_a^2}\int_0^\infty \frac{d\omega}{\Delta\omega^2/4 + (\omega_0 - \omega)^2} = \frac{\hbar\omega_0^3|N|}{\Delta\omega Q_a^2}. \tag{7.38}$$

This leads to the *Townes equation* giving the spectral width of the emission of a cavity amplifier in terms of the total radiation power,

$$\Delta\omega = \frac{\hbar\omega_0^3 N_2}{\mathcal{P} Q_a^2 (N_2 - N_1)}. \tag{7.39}$$

Let $\lambda = 1\,\mu$, $\mathcal{P} = 1$ mW= 10^4 erg/s, $N_1 = 0$, $\alpha \approx 1/\lambda\!\!\!^- Q_a = -0.01$ cm^{-1}, then

$$\Delta f = \hbar c^3 \alpha^2 / \mathcal{P}\lambda \approx 1\text{Hz}. \tag{7.40}$$

Certainly, this calculation of $\Delta\omega$ is valid only for the linear regime of the amplifier below its oscillation threshold ($Q_l < -Q_a$). However, one can expect that even above the threshold, (7.39) provides a correct order of magnitude for the bandwidth of a quantum oscillator (in the limiting case where the 'technical' noise, like mirror vibration, is absent). More accurate calculations, taking into account the nonlinearity due to the saturation [Loudon (2000); Klimontovich (1980); Lax (1968); Arecchi (1974); Klauder (1968)], which restricts the noise amplitude, lead to a factor 1/2 appearing in (7.39) for the regime high above the threshold. This can be explained by the suppression of amplitude fluctuations, which give the same contribution to the spectral width as the phase fluctuations. One should keep in mind that the random slow phase variation (phase drift) of a harmonic signal leads to the fluctuations of its frequency and to a finite width of its spectrum. The difference between the oscillation shapes of a quantum amplifier (oscillator) below threshold and above it is explained in what follows (Fig. 7.7).

It is only the absence of amplitude fluctuations that qualitatively distinguishes the radiation of a single-mode laser from a narrow-band noise. In the case of a multi-mode laser with independent modes, this difference vanishes as well.

7.2 Basic concepts of the statistical optics

Above, we have found, in the form of the Kirchhoff law, the spectral brightness of the radiation $I(\boldsymbol{k})$ as a function of the frequency and the observation direction in the far-field zone. Although the intensity is an important characteristic of the radiation, clearly, it does not provide the complete statistical information about the electromagnetic field.

In the present section, we will consider, using several typical experimental schemes, the observed quantities and the corresponding convenient theoretical tools, the correlation functions $G^{(n)}_{1\ldots 2n}$.

Because the most part of observable optical effects do not require the quantization of the field for their interpretation, we will restrict our consideration to the more simple classical theory, and only in some cases the results of the quantum theory (Secs. 7.4–7.7) will be given without the derivation.

7.2.1 *Analytical signal*

In statistical optics, it is convenient to use, instead of the real field $E_\alpha(\boldsymbol{r}, t)$, a complex function $E^{(+)}_\alpha(\boldsymbol{r}, t)$, which is called *the analytical signal* or the *positive-frequency field*. It is unambiguously defined as

$$E^{(+)}(t) \equiv \int_0^\infty d\omega e^{-i\omega t} E(\omega), \tag{7.41}$$

where $E(\omega)$ is the Fourier transform of the real field,

$$E(\omega) \equiv \int_{-\infty}^{\infty} dt e^{i\omega t} E(t)/2\pi = E^*(-\omega). \tag{7.42}$$

By definition, *the spectrum of the analytical signal $E^{(+)}(t)$ contains only positive frequencies ($\omega > 0$) while the spectrum of the complex conjugate (negative-frequency) field $E^{(-)} = E^{(+)*}$, only negative frequencies.* From the definition, the relation follows (Fig. 7.6)

$$E(t) = E^{(+)}(t) + E^{(-)}(t) = 2\mathrm{Re}E^{(+)}(t). \tag{7.43}$$

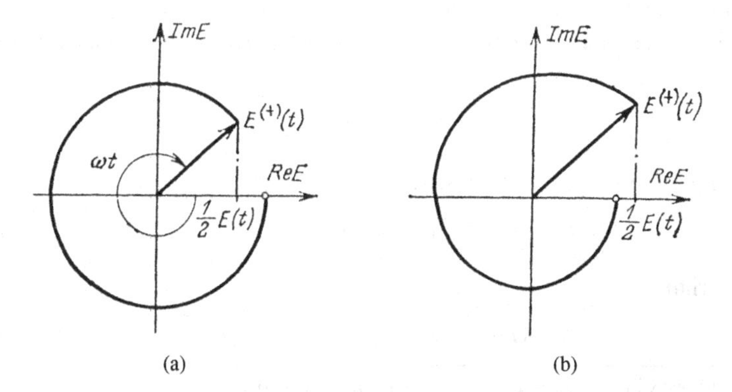

Fig. 7.6 The analytical signal $E^{(+)}(t)$ and the real field $E(t)$ in the case of a monochromatic spectrum (a) and in the general case (b).

In the case of a quasi-monochromatic field with a narrow spectrum, it is convenient to introduce a 'slow' complex amplitude $E_0(t)$, defined as

$$E(t) \equiv \mathrm{Re} E_0(t) e^{-i\bar{\omega}t}, \qquad (7.44)$$

where $\bar{\omega}$ is some mean frequency. The absolute value $|E_0(t)|$ is called *the envelope* and the argument is called the slowly varying phase.The spectrum of $E_0(t)$ is located within the interval $\pm\Delta\omega/2$ around the zero frequency. Apparently, one can write that

$$E^{(+)}(t) = (1/2)E_0(t)e^{-i\bar{\omega}t}. \qquad (7.45)$$

The readings of optical detectors, such as PMTs, bolometers, photographic films, etc. depend on the square (or higher powers) of the field averaged over a certain time T due to the non-instantaneous response of the detector. (So far, we do not take into account the finite spatial dimensions of the detector.) Assume that $\Delta\omega \ll 1/T \ll \bar{\omega}$, i.e., the envelope is much broader than the detector response time, which, in its turn, includes a large number of light periods. Then the reading of the detector will scale as the 'instantaneous intensity'[d]

$$I(t) \equiv \overline{E^2(t)}/2 \approx |E^{(+)}(t)|^2 = |E_0(t)|^2/4. \qquad (7.46)$$

In Sec. 7.7, this statement will be justified more rigorously. By using here the analytical signal, we automatically eliminate the terms oscillating with the double optical frequency.

7.2.2 *Random intensity*

In a random field, the instantaneous intensity $I(t)$ at point $\boldsymbol{r}$ varies: it fluctuates both in time and in space.[e] Often, one measures only the simplest characteristics of the field: the mean (ensemble-averaged or time-averaged) intensity at a single point ($E^{(+)} \equiv E_\alpha^{(+)}(\boldsymbol{r}, t)$),

$$\langle I \rangle = \langle E^{(-)} E^{(+)} \rangle, \qquad (7.47)$$

and the mean square of the intensity,

$$\langle I^2 \rangle = \langle E^{(-)2} E^{(+)2} \rangle, \qquad (7.48)$$

or the variance,

$$\langle \Delta I^2 \rangle \equiv \langle (I - \langle I \rangle)^2 \rangle = \langle I^2 \rangle - \langle I \rangle^2. \qquad (7.49)$$

[d] Sometimes we will omit the scaling factor, which in the case of a plane wave is $c/2\pi$.

[e] Strictly speaking, one should consider the random tensor $I_{\alpha\beta} \equiv E_\alpha^{(-)} E_\beta^{(+)}$, where $\alpha, \beta = x, y, z$ or, in the case of directed radiation, $\alpha, \beta = x, y$.

In principle, one can measure the parameters (7.47)–(7.49) by observing the mean value and the variance of the readings of a broadband single-quantum detector (see below). The readings of an n-quantum detector directly yield the n-th moment, $\langle I^n \rangle \equiv G^{(n)}$.

We will assume the field to be stationary, so that the mean values in (7.47)–(7.49) do not depend on the time, and ergodic. The angular brackets may then denote the averaging either over time or over an ensemble (with some distribution function $P(I)$).

After passing to the quantum theory (Sec. 7.4), $E^{(+)}$ is replaced by the operator $\hat{E}^{(+)}$, which is written in terms of the photon annihilation operators $\hat{a}_k$, while $E^{(-)}$ is replaced by the operator $\hat{E}^{(-)}$, which is written in terms of the photon creation operators $\hat{a}_k^\dagger$. Complex conjugation is then replaced by Hermitian conjugation, and the angular brackets denote quantum averaging with the help of a wave function or a density matrix. Most quantum states of the field also allow averaging using the *quasiprobability distributions* $P(z)$ (Sec. 7.5), which is similar to the classical averaging. In quantum averaging, the order of the operators matters; the ordering presented in (7.47), (7.48) is called the *normal* one. Of course, in the classical theory the normal moment $G^{(n)} \equiv \langle E^{(-)n} E^{(+)n} \rangle$ can be replaced by $\langle |E^{(+)}|^{2n} \rangle$.

In the vast majority of cases, the optical field is emitted by many independent sources with random amplitudes and phases, such as, for instance, in the case of the thermal radiation from heated matter, a quantum amplifier (Sec. 7.1), or a multimode laser with independent modes. Then, the distribution of the complex amplitude $E_0 \equiv E_0' + iE_0''$ is a normal (Gaussian) one, with independent E_0' and E_0'', while the intensity distribution is exponential (Fig. 7.7),

$$P_T(I) = \langle I \rangle^{-1} \exp(-I/\langle I \rangle). \tag{7.50}$$

Thus, *the mean intensity $\langle I \rangle$ fully determines the statistics of a stationary chaotic field (at a single point and for a single polarization type).* Note that the most probable intensity value is zero, $P(0) \gg P(I)$. Using (7.50), one can easily find the moments and the variance of the intensity,

$$G_T^{(n)} = n! \langle I \rangle^n, \; \langle \Delta I^2 \rangle_T = \langle I \rangle^2, \tag{7.51}$$

where the subscript T indicates that the field is chaotic (thermal).

Another typical case is the radiation of a single-mode laser with a stabilized amplitude (Fig. 7.7). Then,

$$P(I) = \delta(I - I_0), \; G^{(n)} = I_0^n, \; \Delta I = 0. \tag{7.52}$$

Equations (7.50)–(7.52) do not take into account the discreteness of possible energy values, i.e., they ignore the photon structure of the field; therefore, they

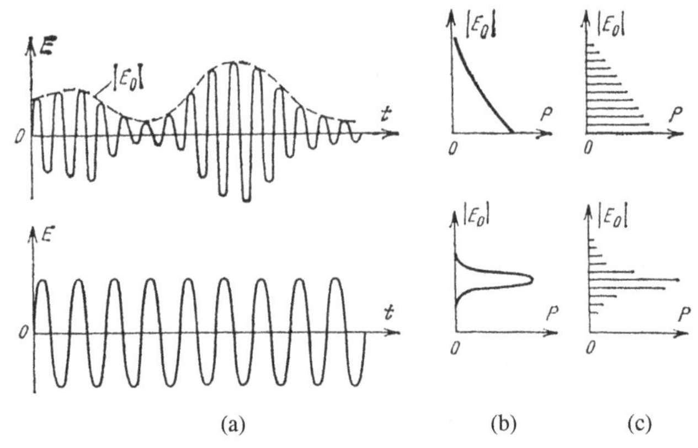

Fig. 7.7 Two basic types of the field states. The figure shows the approximate shape of the field variation in time from the classical viewpoint (a) and the corresponding distribution functions $P(|E_0|)$: the classical ones (b) and the quantum ones (c). Top: the quasi-monochromatic field of thermal radiation, of a quantum amplifier or oscillator below the threshold (both the amplitude and the phase fluctuate); bottom: the field of an above-threshold quantum oscillator (the amplitude fluctuations are suppressed due to the saturation effect).

are only valid for classical fields for which the degeneracy factor $\langle N\rangle$ (see below) is much greater than the unity. The general case will be considered in Sec. 7.6. So far, let us only note that in the quantum theory, the continuous distribution (7.50) is replaced by the 'discrete exponential' one, while (7.52) is replaced by the Poissonian distribution (Fig. 7.7(c)).

7.2.3 *Correlation functions*

The distributions or intensity moments considered above provide no information about the correlations between the fields at neighboring points in space and time, or between the different Cartesian components of the field. Complete information is given by the set of multi-dimensional distributions or tensor *correlation functions* (CF). The latter, according to Glauber [Glauber (1965)], are defined as

$$G^{(n)}_{1\dots 2n} \equiv \langle E^{(-)}_1 \dots E^{(-)}_n E^{(+)}_{n+1} \dots E^{(+)}_{2n}\rangle. \tag{7.53}$$

(Here we use the notation that is also valid in the quantum description.) Each subscript denotes a set of arguments, for instance, $E_1 \equiv E_{\alpha_1}(\boldsymbol{r}_1, t_1)$. For simplicity, we only consider CFs with even numbers of fields, since in optics the moments of the form $\langle E_1E_2E_3\rangle$ are usually equal to zero[f] and, in addition, are difficult to measure.

[f]Exceptions are fields at the output of nonlinear media excited by external radiation [Perina (1972); Akhmanov (1971)] or heating [Klyshko (1980)].

The CFs of a stationary field are invariant to the choice of the initial time moment, i.e., to the replacement of time arguments

$$t_1, \ldots, t_{2n} \rightarrow t_1 + \Delta t, \ldots, t_{2n} + \Delta t, \tag{7.54}$$

where Δt is arbitrary. It is convenient to choose $\Delta t \equiv -t_1$.

As a result, $G^{(n)}$ depends on $2n-1$ time arguments, while its Fourier transform, the spectral CF $\tilde{G}^{(n)}$, depends on $2n-1$ frequencies. For instance, the first-order CF has the form $G(\boldsymbol{r}_1, \boldsymbol{r}_2, \tau)$; it is called the *mutual coherence function* of the field at points $\boldsymbol{r}_1, \boldsymbol{r}_2$, while its Fourier transform, $\tilde{G}(\boldsymbol{r}_1, \boldsymbol{r}_2, \omega)$, is called the *mutual spectral density*. At $\boldsymbol{r}_1 = \boldsymbol{r}_2$, the first-order CF, $G(\boldsymbol{r}, t)$, is called the *autocorrelation function*, while $\tilde{G}(\boldsymbol{r}, \omega)$ is the *spectral density*. Traditional polarization characteristics of directed radiation, such as degree of polarization or the Stokes parameters, are also determined by the first-order CF with an account for tensor indices [Perina (1972)]. With all arguments coinciding, CFs become intensity moments (we omit the tensor indices),

$$G^{(n)}_{1\ldots1} = \langle |E^{(+)}(\boldsymbol{r}_1, t_1)|^{2n} \rangle = \langle I^n(\boldsymbol{r}_1) \rangle. \tag{7.55}$$

Among the various statistical models of the field, a special role belongs to the Gaussian model, in which all CFs are expressed in terms of the first-order CF [Glauber (1965)],

$$G^{(n)}_{1\ldots n1'\ldots n'} = \sum{}' G^{(1)}_{11'} \cdots G^{(1)}_{nn'}, \tag{7.56}$$

where $\sum'$ denotes the sum of all $n!$ permutations of the primed indices. For instance,

$$G^{(2)}_{1234} = G^{(1)}_{13} G^{(1)}_{24} + G^{(1)}_{14} G^{(1)}_{23}. \tag{7.57}$$

All information about a Gaussian (chaotic, thermal) field is contained in the first-order correlation function $G_{\alpha\beta}(\boldsymbol{r}_1, \boldsymbol{r}_2, \tau)$ or the mutual spectral density $\tilde{G}_{\alpha\beta}(\boldsymbol{r}_1, \boldsymbol{r}_2, \omega)$.

For a rough characterization of a quasi-monochromatic directed non-polarized Gaussian field, it is sufficient to fix at each point the intensity $G_\nu(\boldsymbol{r})$ and the coherence parameters $\tau_{coh} \sim 1/\Delta\omega$, ρ_{coh} for both polarization types $\nu = 1, 2$. The meaning of these parameters will be explained in the next sections.

7.2.4 *Temporal coherence*

Consider the field $E'(t)$ at the output of a Michelson interferometer (Fig. 7.8). It consists of two terms differing by a certain time delay, $\tau = t' - t$,

$$E'(t) = [E(t) + E(t')]/2,$$

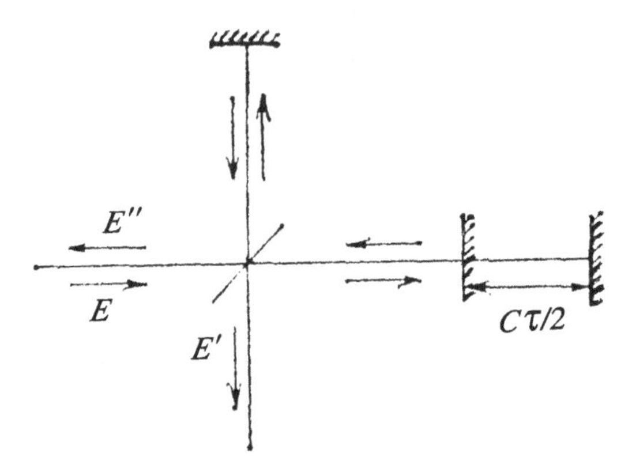

Fig. 7.8 The Michelson interferometer and the measurement of the longitudinal coherence length.

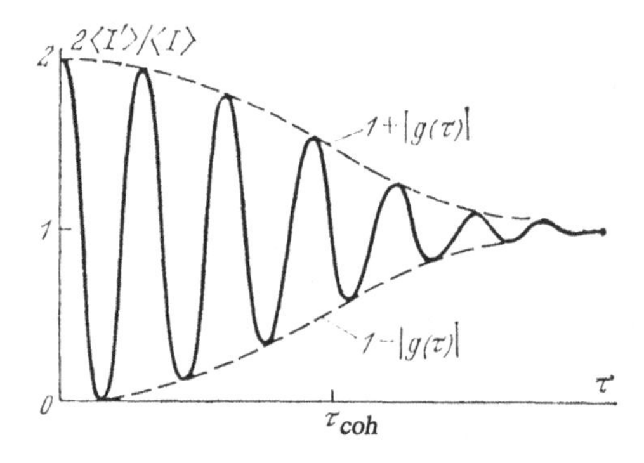

Fig. 7.9 Typical dependence of the mean intensity at the output of a Michelson interferometer, $\langle I' \rangle$, on the position of the moving mirror: $g(\tau)$ is the normalized autocorrelation function of the field at the input.

where $E(t)$ is the field of the plane wave at the input of the interferometer. According to (7.46), the intensity at the output is

$$I'(t) = [I(t) + I(t') + 2\text{Re}E^{(-)}(t)E^{(+)}(t')]/4.$$

Hence, in the case of stationary radiation,

$$\langle I' \rangle = [\langle I \rangle + \text{Re}G(\tau)]/2 = \langle I \rangle[1 + \text{Re}g(\tau)]/2, \tag{7.58}$$

where

$$G(\tau) \equiv \langle E^{(-)}(0)E^{(+)}(\tau)\rangle = \langle E_0^*(0)E_0(\tau)\rangle e^{-i\omega t}/4 \tag{7.59}$$

is the auto-correlation function of the field at the output, which is related, according to the Wiener-Khinchine theorem, to the spectral density $\tilde{G}(\omega)$ via the Fourier-transformation, $g(\tau) \equiv G(\tau)/\langle I \rangle$ is the normalized CF, and $\langle I \rangle = G(0)$. Note that in optics, CFs are usually normalized to the mean values and not to the standard deviations ΔI.

According to (7.58), *the dependence of the intensity at the output of a Michelson interferometer on the delay time determines the real part of the first-order CF*. One can show that the function $G(\tau)$, similarly to $E^{(+)}(\tau)$, is analytic in the lower semi-plane. Therefore, its real and imaginary parts are related via the Hilbert transformation, so that, in principle, from the interference pattern one can restore the radiation spectrum. This method forms the basis for the *Fourier spectroscopy*. Besides, the interference of the field under study with a reference coherent wave is the source of information in *holographic interferometry*.

A typical interference pattern in the case of a single spectral line is shown in Fig. 7.9. The relative amplitude of the oscillations in the output intensity is called the *interference visibility* or the *degree of coherence*. According to (7.58), (7.59), the visibility coincides with $|g(\tau)|$.

If we define the *coherence time* τ_{coh} by the condition $|g(\tau_{coh})| = 1/2$, then it follows from the properties of the Fourier transformation that $\tau_{coh} \sim 2\pi/\Delta\omega$. Hence, the *longitudinal coherence length* $c\tau_{coh}$ *is on the order of the inverse spectral width in inverse centimeters*,

$$l_{coh} \sim 2\pi c/\Delta\omega \equiv 1/\Delta\nu. \tag{7.60}$$

For instance, for lines with natural broadening in the visible range, $\Delta\omega \sim 2 \cdot 10^6 s^{-1}$, $l_{coh} \sim 10^5$ cm. This can be viewed as the length of the wave train emitted by an atom during spontaneous emission. In the case of a single-mode laser, estimation with the Townes equation (7.40) yields a coherence length on the order of a light second.

7.2.5 *Spatial coherence*

The correlation of fields at two points $\boldsymbol{r}_1$ and $\boldsymbol{r}_2$ can be measured with the help of the Young interferometer, which is a screen with two pinholes placed orthogonally to the direction towards the source (Fig. 7.10).

The interference pattern at an arbitrary point behind the screen, similarly to (7.58), is determined by the first-order CF of the general form:

$$G_{12}(\tau) \equiv G(\boldsymbol{r}_1, \boldsymbol{r}_2, \tau) \equiv \langle E^{(-)}(\boldsymbol{r}_1, t) E^{(+)}(\boldsymbol{r}_2, t+\tau) \rangle, \tag{7.61}$$

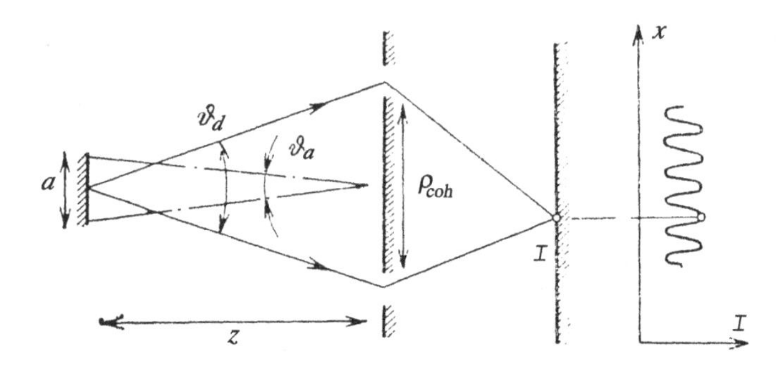

Fig. 7.10 Young's interferometer and the measurement of the transverse coherence radius.

while the pattern on the symmetry axis is determined by $G_{12}(0)$. The *coherence radius* ρ_{coh} is determined by the distance between the pinholes at which the visibility is 50% reduced.

An ideal laser with a single transverse mode emits plane or spherical waves, for which $\rho_{coh} = \infty$. In the case of a multi-mode laser, or a chaotic (thermal) light source, ρ_{coh} in the far-field zone ($z \gg a^2/\bar{\lambda}$) is determined by the transverse size a of the source and the distance z from the source (Fig. 7.10),

$$\rho_{coh} \sim \bar{\lambda} z/a \equiv z\vartheta_d \equiv \bar{\lambda}/\vartheta_a, \tag{7.62}$$

where ϑ_d is the diffraction angle and ϑ_a is the angular size of the source. *Due to diffraction, the transverse coherence radius increases in the course of light propagation.* The shape of the phase front in this case increases and tends to the spherical one. The van Cittert-Zernicke theorem [Akhmanov (1981); Klauder (1968); Perina (1972)], which describes this effect quantitatively, states that *the dependence of the first-order CF on $(\boldsymbol{r}_1 - \boldsymbol{r}_2)_\perp$ is given by the Fourier transform of the brightness distribution over the source cross-section.*

Relation (7.62) enabled Michelson to measure, with the help of his stellar interferometer, the angular diameters of several stars, for which $\vartheta_a \gtrsim 10^{-7}$ rad and $\rho_{coh} \lesssim 10$ m. At smaller ϑ_a, the coherence radius is determined by the atmospheric distortions of the wave front, which hinder the operation of the interferometer. The Hanbury Brown–Twiss interferometer (see below) is free of this drawback.

7.2.6 *Coherence volume and the degeneracy factor*

The coherence volume is defined as the product of the coherence area, ρ^2_{coh}, and the coherence length, l_{coh}. For the far-field zone of a chaotic source, it follows from (7.60) and (7.62) that

$$V_{coh} = \bar{\lambda}^4/\Delta\lambda\Delta\Omega_a, \tag{7.63}$$

where $\Delta\lambda = \bar{\lambda}^2/l_{coh}$ and $\Delta\Omega_a \equiv (a/z)^2$ is the solid angle at which the observer sees the source.

An important dimensionless statistical parameter of the radiation is the *degeneracy factor*, which is the mean energy of the field (for a single polarization mode), in $\hbar\bar{\omega}$ units, contained in the coherence volume. In other words, the degeneracy factor is the number of photons crossing the coherence area during the coherence time,

$$\delta \equiv \langle \mathcal{E} \rangle_{coh}/\hbar\bar{\omega}. \tag{7.64}$$

Let us write $\langle \mathcal{E} \rangle_{coh}$ in terms of the radiation spectral brightness,

$$\langle \mathcal{E} \rangle_{coh} = I_{\omega\Omega}\Delta\omega\Delta\Omega_a\tau_{coh}\rho_{coh}^2 = 2\pi\bar{\lambda}^2 I_{\omega\Omega}. \tag{7.65}$$

Hence,

$$\delta = \bar{\lambda}^3 I_{\omega\Omega}/\hbar c \equiv \langle N \rangle. \tag{7.66}$$

Thus, *the degeneracy factor is equal to the spectral brightness in $\hbar c/\bar{\lambda}^3$ units. In other words, the mean photon number per coherence volume coincides with the mean number of photons per single mode, $\langle N \rangle$.* In an equilibrium field, $\delta = \mathcal{N}(\bar{\omega}) \equiv [\exp(\hbar\bar{\omega}/\kappa T) - 1]^{-1}$. In a non-equilibrium field, the temperature should be interpreted as the brightness temperature, T_{ef}.

The degeneracy factor $\delta = \langle N \rangle$ is not only a convenient measure of the spectral brightness; it also defines the applicability of classical statistics: at $\delta \lesssim 1$ (i.e., $\hbar\bar{\omega} \lesssim \kappa T$) the photon structure of the field becomes important.[g]

Let us divide the far-field space of a quasi-monochromatic source into cells with volumes V_{coh}. By definition, the radiation at any two points of the same cell is mutually coherent. Hence, the field within one cell can be approximately considered as single-mode, i.e., assumed to be a spherical monochromatic wave with a definite amplitude $|E_0|$ and phase φ. Passing from one cell to another, we will observe random fluctuations of $|E_0|$ and φ. Thus, the spatial distribution of a stationary field forms the ensemble of harmonic-oscillator states.

Fields belonging to different coherence volumes are, by definition, uncorrelated and, generally, independent. Hence, by virtue of the central limit theorem of

[g]Editors' note: It may seem that the degeneracy factor δ can be considered as a measure of nonclassicality. However, this is not true, and this is not stated in the book. At $\delta \ll 1$, the *shot noise*, which is related to the photon structure of light, prevails over excess intensity fluctuations and therefore becomes important. At the same time, there are nonclassical states of light with $\delta \gg 1$ (squeezed states, squeezed vacuum).

Table 7.1 Basic distribution types for the photon number and the energy.

State	Quantum theory		Classical theory	
	Number of modes			
	1	$\gg 1$	1	$\gg 1$
Coherent (laser)	Poissonian	Poissonian	$\delta(\mathcal{E} - \mathcal{E}_0)$	Gaussian
Chaotic (thermal)	Geometric	Poissonian	Boltzmann's	Gaussian
K-photon (mixed)	$P_K = \langle N \rangle / K$, $P_0 = 1 - P_K$	Poissonian		

the probability theory, the energy distribution in a volume V that is much greater than V_{coh} will be Gaussian, with the variance inversely proportional to the number of cells V/V_{coh}. In the quantum case, where $\delta \lesssim 1$, the Gaussian distribution for a multi-mode field is replaced by the Poissonian one (Sec. 7.6)

Table 7.1 shows some types of distributions for the energy or the number of photons. The states that are called 'K-photon' ones have no classical analogues and manifest photon anti-bunching and bunching (Sec 7.6).

7.2.7 *Statistics of photocounts and the Mandel formula*

The intensity distribution for the field at one 'point' can be measured by means of a PMT operating in the *photon-counting mode*.[h] Then the mean intensity of light $\langle I \rangle$ should be sufficiently low, so that the photocurrent pulses at the PMT output do not overlap (Fig. 7.11). By repeatedly counting the number of pulses m coming within a certain fixed time interval T, one can find the distribution $P(m)$ of the number of primary photoelectrons released from the photocathode by the incident light. (Of course, the total duration of the measurement should considerably exceed the coherence time τ_{coh}.)

Let us find the relation between the statistics of photocounts and the field statistics. Suppose that $T \ll \tau_{coh}$ and $A \ll A_{coh}$, then one can neglect the intensity variation during the sampling time T and over the photocathode area A. Under this condition, the $P(m)$ distribution will be determined by the $P(I)$ distribution regardless of the detection volume $V_{det} \equiv cTA$. Here, τ_{coh} and $A_{coh} \equiv \rho_{coh}^2$ are the typical scales of the field fluctuations, and the above-given inequalities enable us to consider the detector as 'pointlike' and 'single-mode', i.e., measuring a single degree of freedom of the field. (This also implies that the detector measures a single polarization type.)

[h]Editors' note: nowadays, much more convenient for single-photon counting are avalanche photodiodes (APDs) (see also Sec. 1.3).

Assume first that the intensity I is constant, i.e., it does not vary from sample to sample. It is important that even in this case, the number of photoelectrons in a single sample is, according to the quantum mechanics, random, unpredictable. The very process of energy measurement necessarily introduces additional Poissonian stochasticity into the detector readings. (We do not consider the non-realistic case of a detector with a 100% quantum efficiency and a pure energy state of the field with a fixed number of photons.) Using the semi-classical (Sec. 2.1) or purely quantum (Sec. 7.7) perturbation theory, we can only find the ionization probability $W_1 \Delta t \sim I$ for a single atom of the photocathode during a small time interval Δt.

In the case of a sufficiently small Δt, the probability for any on the N independent atoms of the photocathode to be ionized is N times as large, and also scales as I,

$$W = NW_1 \equiv (\alpha/T)I. \tag{7.67}$$

Here α is the scaling factor, which can be represented as

$$\alpha = \eta V_{det}/2\pi\hbar\bar{\omega}, \tag{7.68}$$

where $V_{det} \equiv cTA$ is the effective detection volume, $\eta \equiv \sigma l N_0 = \sigma N/A$ is the quantum yield of a thin photocathode with the thickness l, σ is the ionization cross-section, and $N_0 \equiv N/Al$ is the concentration of atoms. We assume that σ is constant within the spectral width of the field.

By definition, all time moments within T are equivalent, since the wave incident on the photocathode is a 'pure' sine one, and an electron can appear within any time interval Δt with the same probability $\alpha I \Delta t/T$. This statistical model, as one can easily show (see, for instance, Ref. [Rytov (1976)]), leads to the Poissonian distribution with the parameter αI,

$$P(m|I) = C(\alpha I)^m/m!, \quad C \equiv e^{-\alpha I}. \tag{7.69}$$

Intensity fluctuations from sample to sample can be taken into account by averaging (7.69) with the $P(I)$ distribution,

$$P(m) = \int_0^\infty dI P(m|I)P(I) \equiv \langle P(m|I)\rangle. \tag{7.70}$$

As a result, we obtain the *semi-classical Mandel formula* for the photocount distribution, i.e., for the probability of discovering m pulses at the PMT output,

$$P(m) = \langle(\alpha I)^m e^{-\alpha I}\rangle/m!. \tag{7.71}$$

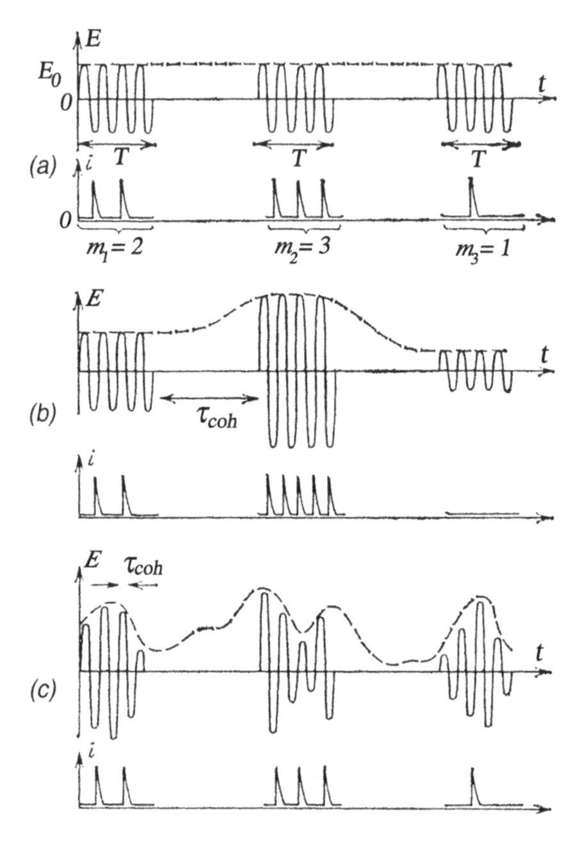

Fig. 7.11 Relation between the statistics of the field and the photocount statistics from the semi-classical viewpoint: E is the field, i is the PMT output current; (a) in the case of a field with a constant amplitude E_0, the number of photoelectrons m_i emerging during a certain time interval T has a Poissonian distribution; (b) the amplitude fluctuations of the field at $T \ll \tau_{coh}$ cause additional fluctuations in the number of the photoelectrons (the bunching effect); (c) a multi-mode ($T \gg \tau_{coh}$) detector averages over the field fluctuations, and the bunching effect is not observed.

The quantum-theoretical approach, mainly developed by Glauber (Sec. 7.6), yields an expression of the similar form, with the only difference that the probability $P(I)$ is replaced by the *quasiprobability*, a function that takes negative values or has singularities (like delta-function derivatives) for some states of the field.

By expanding (7.71) in powers of the detector quantum efficiency α, one can express $P(m)$ in terms of higher-order intensity moments, $G^{(k)} \equiv \langle I^k \rangle$, with $k \geqslant m$,

$$P(m) = \sum_{k=m}^{\infty} (-1)^{k-m} \alpha^k G^{(k)} / m!(k-m)! \qquad (7.72)$$

For a sufficiently small detection volume ($\alpha \sim TA \to 0$), as a rule, one can take into account only the first term in this expansion, i.e., neglect the exponent in (7.71). Then $P(m)$ can be represented as

$$P(m) = \frac{T^m}{m!}W^{(m)} = \frac{(NT)^m}{m!}\langle W_1^m \rangle, \tag{7.73}$$

where $W^{(m)}$ is the mth derivative of $P(m)$ in T, W_1 is the ionization probability per unit time for a single atom, N is the total number of atoms in the detector.

The Mandel formula (7.71) describes the ensemble of random samples differing by a *shift in time* (due to the assumed stationarity and ergodicity of the field). One can easily see that in the case of radiation that is homogeneous and 'spatially ergodic' along x, y and propagates approximately along z, the same formula describes an ensemble of samples differing by a *spatial shift in the* (x, y) *plane*. This, in principle, allows one to study the statistics of non-stationary fields in real time, with the help of a large number of photon counters placed at different points of the beam cross-section over an area much greater than A_{coh}.

Further, it is clear from the derivation of formula (7.71) that it is also valid in the case of an arbitrary coherence volume $V_{det} = cTA$ provided that I is understood as the intensity averaged over V_{coh},

$$\bar{I}(\boldsymbol{r}, t) \equiv \frac{1}{AT}\int dx'dy'dt' I(x', y', z'), \tag{7.74}$$

where the integration limits, $x \pm a/2$, $y \pm b/2$, $t \pm T/2$, are determined by the sizes and the time constant of the detector. In this case, the distribution $P(I)$ should be replaced by $P(\bar{I})$. In the limiting case of $A \gg A_{coh}$ or/and $T \gg T_{coh}$ (multi-mode detection), the fluctuations of I are completely eliminated due to averaging: $P(I) = \delta(\bar{I} - \langle I \rangle)$, so that we once again obtain the Poissonian distribution $P(m|\langle I \rangle)$ from (7.71), but now it is independent of the field statistics. By observing the dependence of $P(m)$ on A, T in the intermediate case, one can, in principle, obtain information about the coherence time and area of the field.

7.2.8 *Photon bunching*

Relation (7.71) between the distributions also determines the relations between the moments of the photocount numbers,

$$\langle m^k \rangle \equiv \sum_{m=0}^{\infty} m^k P(m), \tag{7.75}$$

and the intensity moments,

$$\langle I^k \rangle \equiv \int_0^{\infty} dI I^k P(I). \tag{7.76}$$

Using the generating functions method (Sec. 7.6), one can easily find the general rule,

$$\langle m(m-1)\ldots(m-k+1)\rangle = \alpha^k \langle I^k \rangle. \tag{7.77}$$

The linear combination of moments in the left-hand side is called the *kth-order factorial moment*. From this, in particular, it follows that $\langle m \rangle = \alpha \langle I \rangle$ and

$$\langle \Delta m^2 \rangle = \langle m \rangle + \alpha^2 \langle \Delta I^2 \rangle, \tag{7.78}$$

$$\langle \Delta I^2 \rangle \equiv \langle I^2 \rangle - \langle I \rangle^2, \tag{7.79}$$

and similarly for $\langle \Delta m^2 \rangle$. Thus, *fluctuations of the photoelectron number contain, in addition to the usual Poissonian (shot-noise) part, a contribution from the fluctuations of the light intensity.* Only in the case of a single-mode stabilized laser ($\Delta I = 0$) this contribution is absent. In other cases, fluctuations of the photoelectron number, according to (7.78), should at first sight exceed the shot noise, since from the definition (7.79) and from the condition $P(I) \geqslant 0$ it follows that $\langle \Delta I^2 \rangle \geqslant 0$.

The existence of these 'excess' fluctuations has been called the *photocount bunching effect*, since in a Poissonian sequence of pulses, by definition, the appearance of a single pulse has no effect on the appearance of the next one, and the inequality $\langle \Delta m^2 \rangle > \langle m \rangle$ means that the pulses have a tendency to bunch. A similar effect for the photon numbers, $\langle \Delta N^2 \rangle > \langle N \rangle$, is called *photon bunching* or *photon correlation*. A close effect has been discovered by Hanbury Brown and Twiss in 1956 in the chaotic light of a mercury lamp.

In chaotic light, it follows from (7.51) and (7.78) that

$$\langle \Delta m^2 \rangle_T = \langle m \rangle (1 + \langle m \rangle), \tag{7.80}$$

i.e., the excess part of the variance is $\langle m \rangle$ times as large as the Poissonian one, so that *the photon bunching effect is more pronounced in classical fields.* From the classical viewpoint, strong fluctuations of the amplitude $|E_0|$ of a wave formed by many independent sources with random phases are quite obvious. More surprising for the classical theory is the *anti-bunching of photons* and, correspondingly, photocounts, so that $\Delta m^2 < \langle m \rangle$, in contradiction with (7.78) and the initial Mandel formula (7.71) if $P(I \geqslant 0)$ is assumed. (As we have already mentioned, the last condition is violated in quantum theory.)

7.2.9 *Intensity correlation*

The photocount number distribution (7.71) does not provide direct information about the temporal or spatial radiation spectrum, as it only contains 'single-point'

CFs with all arguments coinciding (see (7.72)). A complete characterization of an nth-order CF requires the measurement of the field at $2n$ points of space-time. In the case of the first-order CF, this can be done with the help of interferometers, as it was schematically described above.

Consider now the measurement of the intensity correlation function, namely, the second-order CF of the following particular form:

$$G^{(2)}(x_1, x_2, x_1, x_2) \equiv G_{12}^{(2)}(\tau) = \langle : I(x_1)I(x_2) : \rangle, \tag{7.81}$$

where $x_1 \equiv \{\boldsymbol{r}_i, t_i\}$, $\tau \equiv t_2 - t_1$, and the $\boldsymbol{r}_2 - \boldsymbol{r}_1 \equiv \boldsymbol{\rho}$ vector is orthogonal to the propagation direction. The colons remind that in the quantum calculation, all $E^{(+)}$ operators should be put on the right of $E^{(-)}$ operators before the averaging.

In the case of a Gaussian field, with the help of (7.57) we find

$$G_{12}^{(2)}(\tau) = \langle I_1 \rangle \langle I_2 \rangle [1 + |g_{12}^{(1)}(\tau)|^2], \tag{7.82}$$

where $g_{12}^{(1)}(\tau)$ is the normalized first-order CF. Thus, *in a chaotic field, by measuring the intensity correlation one also gains information about the amplitude correlations.* This relation between $G^{(2)}$ and $G^{(1)}$ forms the basis for the *optical mixing spectroscopy* [Cummins (1974)], also called *spectroscopy of intensity fluctuations* or the *method of photon correlations.*

In the general case, $G^{(2)}$ is not related to $G^{(1)}$, and the second-order coherence parameters may differ from $\rho_{coh}^{(1)}, \tau_{coh}^{(1)}$. For instance, in two-photon light, $\rho_{coh}^{(2)} \gg \rho_{coh}^{(1)}$.[i]

The time dependence of $G^{(2)}(\tau)$ at $\rho = 0$ is measured by means of a single detector with a delay line and an electronic correlation circuit.[j] One can also use the spectral analysis of photocurrent fluctuations. The first experiment of this type has been performed by Forrester, Gudmundsen and Johnson as early as in 1955, before the advent of lasers.[k] Modern technique enables one to achieve the spectral resolution much less than 1 Hz.

In order to study the spatial second-order coherence $G^{(2)}(\rho)$, one has to use two detectors with a variable distance between them.[l] According to (7.82), when ρ is varied from ∞ to 0, $G^{(2)}$ is increased, in an ideal case, by a factor of two (Fig. 7.12). This effect has been discovered by Hanbury Brown and Twiss in 1956 and used for the measurement of the angular diameters of stars [Hanbury Brown

[i]Editors' note: This is the case in one of the possible definitions of second-order coherence parameters, see the last subsection.

[j]Editors' note: see Fig. 7.14(a).

[k]The possibility of similar experiments on the 'heterodyning' of light had been discussed even earlier by Gorelik [Gorelik (1948)].

[l]Editors' note: see Fig. 7.14(b).

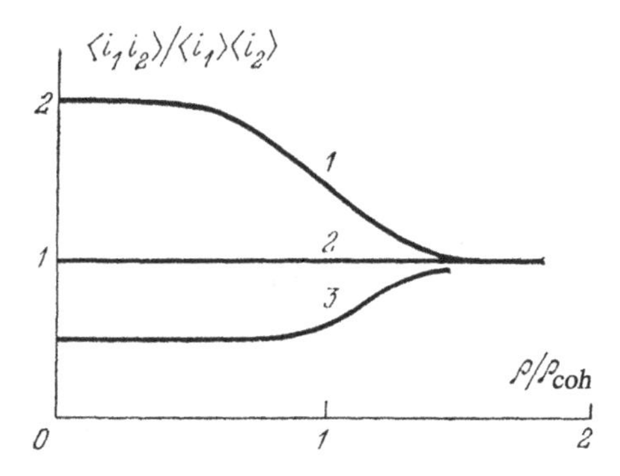

Fig. 7.12 Correlation and anti-correlation of light intensities: i_1, i_2 are the photocurrents of the detectors; ρ is the displacement of one of the detectors from the symmetric position; 1, thermal radiation; 2, laser radiation; 3, two-photon radiation.

(1971)] whose intensities are correlated within distances on the order of hundreds of meters.

The Hanbury Brown–Twiss experiment on the measurement of $G^{(2)}(\rho)$ for the light from a mercury lamp is shown in Fig. 7.13. Splitting of the beam with the help of a semi-transparent mirror allows one to measure the correlation at points that are arbitrarily close to each other. Let the PMTs operate in the photon counting regime, then the correlation between the photocounts in the two channels, $\langle m_1 m_2 \rangle$, is $\alpha_1 \alpha_2 \langle I_1 I_2 \rangle$. Hence, with the help of (7.82), we find

$$\langle m_1 m_2 \rangle = \langle m_1 \rangle \langle m_2 \rangle (1 + |g_{12}^{(1)}(0)|^2). \tag{7.83}$$

This result is only valid in the case of single-mode detectors, where the time constant of the detector T is much less than τ_{coh} and the detector aperture A is much less than A_{coh}. If, for instance, $T \gg \tau_{coh}$, then a small factor on the order of τ_{coh}/T appears by the second term of (7.83), which reduces the observed effect.

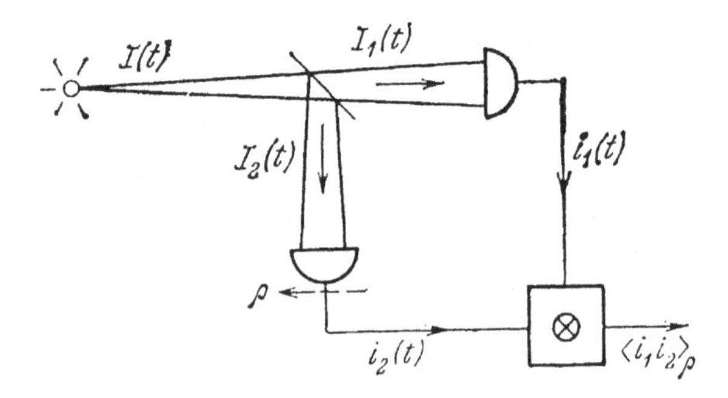

Fig. 7.13 Hanbury Brown–Twiss experiment on the observation of light intensity correlation.

The intensity correlation effect is closely connected with the intensity fluctuations of light at the input of Hanbury Brown–Twiss interferometer. Indeed, let I_1 and I_2 be random intensities in the arms of the interferometer and ΔI_1 and ΔI_2 be their fluctuations ($\Delta I_n \equiv I_n - \langle I_n \rangle$). The output signal $\langle i_1 i_2 \rangle$ scales as

$$\langle I_1 I_2 \rangle = \langle I_1 \rangle \langle I_2 \rangle + \langle \Delta I_1 \Delta I_2 \rangle. \tag{7.84}$$

The second term here characterizes the mutual correlation of intensities. From the condition $I_1 + I_2 = I$, we find the relation

$$\langle \Delta I^2 \rangle = \langle (\Delta I_1 + \Delta I_2)^2 \rangle = \langle \Delta I_1^2 \rangle + \langle \Delta I_2^2 \rangle + 2\langle \Delta I_1 \Delta I_2 \rangle. \tag{7.85}$$

Thus, the *correlation is determined by the variances*,

$$\langle \Delta I_1 \Delta I_2 \rangle = (1/2)(\langle \Delta I^2 \rangle - \langle \Delta I_1^2 \rangle - \langle \Delta I_2^2 \rangle). \tag{7.86}$$

Assume first that the incident light has a constant intensity (radiation of a single-mode laser), then $\Delta I_n = 0$ and, according to (7.86), the correlation is equal to zero. Then, $\langle I_1 I_2 \rangle = \langle I_1 \rangle \langle I_2 \rangle$. Now, suppose that usual light from a thermal or luminescent source be incident on the interferometer; then, according to (7.51), $\langle \Delta I^2 \rangle = \langle I \rangle^2$. Suppose that similar relations hold for the secondary beams as well, $\langle \Delta I_n^2 \rangle = \langle I_n \rangle^2$. Hence, with the help of (7.86) we find (Fig. 7.12) that

$$\begin{gathered}\langle \Delta I_1 \Delta I_2 \rangle = (1/2)(\langle I \rangle^2 - \langle I_1 \rangle^2 - \langle I_2 \rangle^2) = \langle I_1 \rangle \langle I_2 \rangle, \\ g_{12}^{(2)} \equiv \langle I_1 I_2 \rangle / \langle I_1 \rangle \langle I_2 \rangle = 2.\end{gathered} \tag{7.87}$$

This reasoning can be easily reproduced in the photon language by changing I_n to photon numbers N_n. In this case, the shot noise (Poissonian noise) is added to the variance, so that

$$\langle \Delta N^2 \rangle_{las} = \langle N \rangle, \;\; \langle \Delta N^2 \rangle_T = \langle N \rangle + \langle N \rangle^2. \tag{7.88}$$

In fact, the result is the same: *in thermal light, there is photon correlation caused by photon bunching* or, in other words, by the presence of excess noise in addition to shot noise.

Further, consider field with a fixed number N of photons. These N photons will be randomly split by the semi-transparent mirror between the two channels with the probabilities p and $q = 1 - p$. This picture corresponds to the well-known Bernoulli's probability model [Rytov (1976)], which gives the binomial distribution for the probability of N_1 photons going to channel 1,

$$P(N_1) = C_N^{N_1} p^{N_1} q^{N-N_1}. \tag{7.89}$$

The moments of this distribution have the form

$$\begin{aligned}\langle N_1\rangle &= pN, \quad \langle N_1^2\rangle = p^2N^2 + pqN,\\ \langle N_2\rangle &= qN, \quad \langle N_2^2\rangle = q^2N^2 + pqN.\end{aligned} \tag{7.90}$$

Hence, we find

$$\langle N_1N_2\rangle = (1/2)(N^2 - \langle N_1^2\rangle - \langle N_2^2\rangle) = pqN(N-1). \tag{7.91}$$

It is noteworthy that now the correlation is negative,

$$\langle \Delta N_1 \Delta N_2\rangle = \langle N_1N_2\rangle - \langle N_1\rangle\langle N_2\rangle = -pqN.$$

Then (see Fig. 7.12),

$$g_{12}^{(2)} = 1 - 1/N. \tag{7.92}$$

Thus, *photon anti-bunching in the initial beam leads to photon 'anti-correlation' in the two output beams.*

Consider, finally, the general case of a field with an arbitrary statistics. Then, (7.91) should be additionally averaged over the photon-number distribution $P(N)$ in the incident field. As a result,

$$\langle N_1N_2\rangle = pq\langle : N^2 :\rangle, \quad g_{12}^{(2)} = \langle : N^2 :\rangle/\langle N\rangle^2 \equiv g^{(2)}, \tag{7.93}$$

where $\langle : N^2 :\rangle \equiv \langle N(N-1)\rangle$ is the *normalized* (*factorial*) moment and the angular brackets denote averaging over the distribution $P(N)$. Thus, the *relative correlation of photon numbers at two field points,* $g_{12}^{(2)} - 1$, *is determined by the normalized factorial moment of the field* $g^{(2)}$. The same result follows from the rigorous quantum-theory approach (see (7.331)).

7.2.10 *Second-order coherence (added by the Editors)*

Unlike the first-order coherence parameters, which are defined unambiguously as the widths of the first-order correlation function (CF) in time and space, and can be measured as shown in Figs. 7.8–7.10, second-order coherence time and radius allow different definitions. This is because the second-order CF $G^{(2)}(t_1, \boldsymbol{r}_1, t_2, \boldsymbol{r}_2)$ has two time arguments and two space arguments, and there are several ways to define its width. In the stationary case, the second-order correlation function depends only on the difference of its time arguments; the same relates to the space arguments in the spatially homogeneous case: $G^{(2)}(t_1, \boldsymbol{r}_1, t_2, \boldsymbol{r}_2) \equiv G^{(2)}(\tau, \boldsymbol{\rho})$. *Then, the second-order coherence parameters can be defined as the widths of* $G^{(2)}(\tau, \boldsymbol{\rho})$ *in* τ *and* ρ. *In the non-stationary (spatially inhomogeneous) case, they can be*

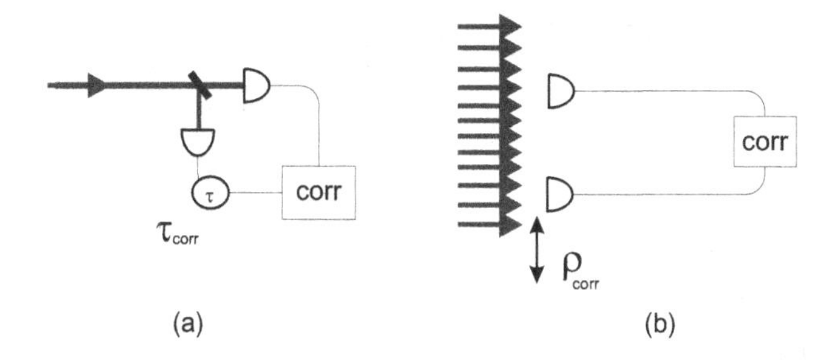

Fig. 7.14 Measurement of the temporal (a) and spatial (b) widths of the second-order correlation function.

introduced by analogy, as the widths of $G^{(2)}(t_1, \boldsymbol{r}_1, t_2, \boldsymbol{r}_2)$ *in* $t_1 - t_2$ *or* $\boldsymbol{r}_1 - \boldsymbol{r}_2$ *('conditional' width of the CF).* These parameters can be measured using two simple setups shown in Fig. 7.14. Intensities at two space-time points are measured by two detectors, usually photon-counting ones, and the coincidences of their counts are registered with the help of a coincidence (correlation) circuit. Time delay τ is usually introduced electronically, by delaying the output pulses of one of the detectors (Fig. 7.14(a)).[m] Space delay is introduced by displacing one of the detectors (Fig. 7.14(b)). Such setups (see, for instance, [Mandel (2004)]) allow one to measure the second-order correlation function and, in particular, its widths in space and time.

However, there is another way to introduce second-order coherence parameters. By analogy with the experimental schemes of measuring the first-order coherence time and radius (Figs. 7.8–7.10), which are based on the first-order interference, one can define the second-order coherence parameters using various experimental schemes for observing second-order interference (intensity interference). In particular, intensity interference can be observed using Michelson's or Young's interferometers (Fig. 7.15). One measures the coincidence counting rate between two detectors, which register the intensities of the field at two points (using a beamsplitter if necessary, as in Fig. 7.15(a)). The interference phase can be varied by moving the mirror in the Michelson interferometer (a) or by displacing one of the slits in the Young interferometer (b). The interference pattern will be formed by the dependence of the coincidence counting rate on the phase, provided that the first-order interference is absent. *The second-order coherence time can be introduced then as the delay in the Michelson interferometer at which the visibility of the intensity interference decays by a factor of two. Similarly, the second-order coherence radius can be defined as the distance between the slits in the Young's*

[m] In the case of SPDC, the correlation time is usually less than the resolution of the detectors; in such cases, other techniques should be used.

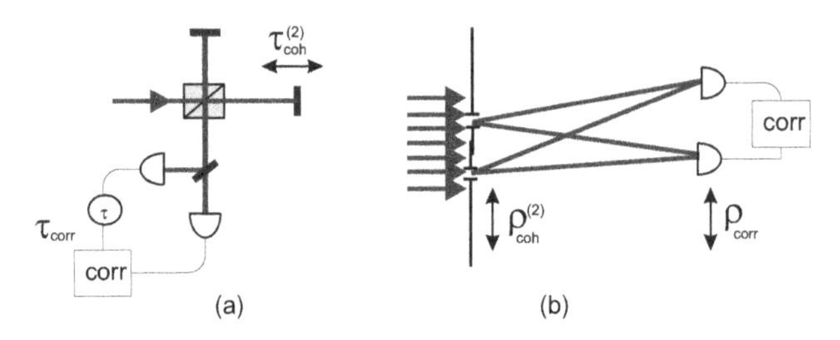

Fig. 7.15 Intensity interference observed using (a) Michelson's interferometer and (b) Young's interferometer. Coincidences of photocounts are observed for two detectors registering photon pairs emitted by sources separated in time (a) or space (b).

experiment at which the visibility of intensity interference decays by a factor of two.

The principal difference between this definition and the previous one is that they relate the second-order coherence parameters to the radiation emitters (or different contributions into radiation) and detectors, respectively. Note that it is the second definition that is meant at the beginning of this section (paragraph after Eq. (7.82)), leading to the conclusion that for two-photon light, $\rho^{(1)}_{coh} \ll \rho^{(2)}_{coh}$. In order to distinguish between these two different definitions, it is convenient to use the terms '*correlation time/correlation radius*' for the widths of $G^{(2)}(t_1, \boldsymbol{r}_1, t_2, \boldsymbol{r}_2)$ in $\tau \equiv t_1 - t_2$ and $\rho \equiv |\boldsymbol{r}_1 - \boldsymbol{r}_2|$, respectively (Figs. 7.14(a) and 7.14(b)), and the terms '*coherence time/coherence radius*' for the widths of the intensity interference patterns (Figs. 7.15(a) and 7.15(b)).

As an example, consider the two alternative definitions in the case of two-photon light generated via SPDC. Let us discuss only temporal coherence, and suppose that the pump is a sequence of relatively long Fourier-limited pulses of duration T_p. The correlation time will then be determined by the length of the nonlinear crystal[n] and be close to the first-order coherence time, which is given by the inverse spectral width $\Delta\omega$ of SPDC radiation [Goodman (1985)]: $\tau_{corr} \sim \tau^{(1)}_{coh} = 2\pi/\Delta\omega$. At the same time, according to the second definition, $\tau^{(2)}_{coh}$ can be measured experimentally by observing two-photon interference with SPDC radiation fed into a Michelson interferometer (Fig. 7.15(a)). Numerous experiments on two-photon interference [Mandel (2004)] show that the visibility will be high as long as the time delay does not exceed the pump coherence time, which in the case considered here coincides with T_p. Thus, for this example $\tau^{(2)}_{coh} \approx T_p$ and $\tau^{(1)}_{coh} \sim \tau_{corr} \ll \tau^{(2)}_{coh}$.

[n]Strictly speaking, it is the minimal value of the correlation time that is determined by the crystal length; the correlation time can then be increased due to the propagation of light through a dispersive medium.

The same reasoning will be valid for spatial coherence parameters, $\rho_{coh}^{(1,2)}$.

7.3 Hamiltonian form of Maxwell's equations

In this section, we will show that Maxwell's equations for the transverse part of the field, i.e., for the radiation field, can be reduced to a system of independent equations for harmonic oscillators. These equations can be easily represented in the form of classical Hamilton's equations. This enables one to use the quantization algorithm which defines the commutator of two operators in terms of Poisson's brackets for the corresponding classical values.

7.3.1 *Maxwell's equations in the* **k**, *t representation*

Suppose that we are interested in the evolution of a radiation field within a certain bounded space domain during a limited time interval T. Imagine that all space is divided in identical cubic cells with the linear size $L > cT$,[o], so that one of the cells contains all the field that is measured during the time T.

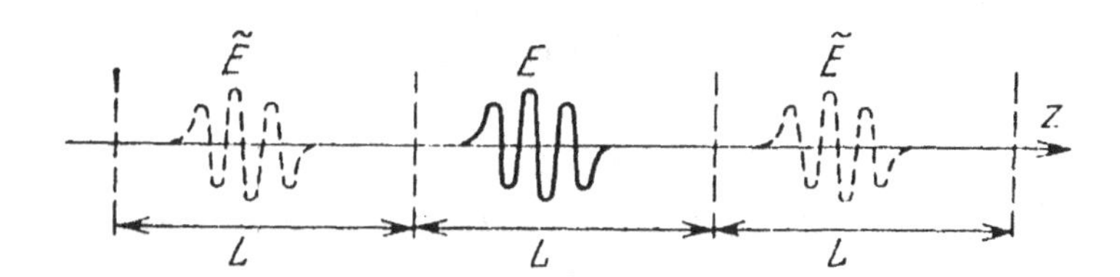

Fig. 7.16 To the definition of the quantization length L: $E(z)$ is the real field and $\tilde{E}(z)$ is the fictitious field that is periodic in space.

Consider the dependence of some field component on one of the coordinates, for instance, $E_x(z)$, at a fixed time moment (Fig. 7.16). Let us define a spatially periodic function $\tilde{E}_x(z)$ by the condition $\tilde{E}_x(z + nL) \equiv E_x(z)$, where $-L/2 < z < L/2$, $n = 0, \pm 1, \ldots$. Within the observation interval, $(-L/2, L/2)$, the fictitious periodic field $\tilde{E}$ coincides with the real one, E; therefore, $\tilde{E}$ and E are physically equivalent, and in future we will omit the tilde.

[o]This condition is assumed to be well satisfied, so that the field at the boundaries stays equal to zero during all observation time.

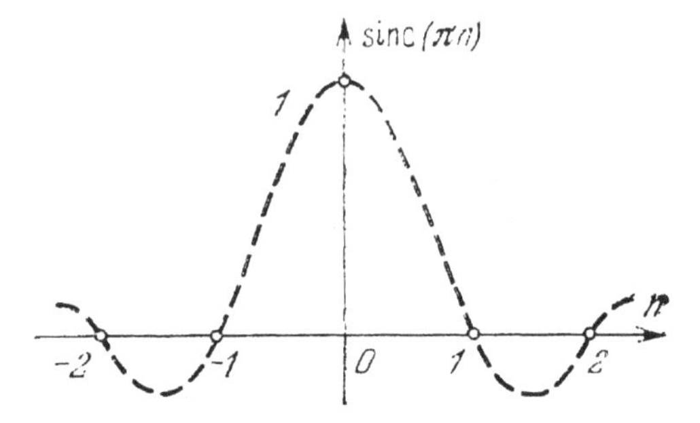

Fig. 7.17 The sinc(x) function in the cases of a discrete (points) and continuous (dashed line) arguments.

A field that is periodic in z can be represented as a sum of *spatial harmonics* (we omit the x index),

$$E(z) = \sum_{m=-\infty}^{\infty} E_m \exp(ik_m z),$$
$$k_m \equiv 2\pi m/L, \ m = 0, \pm 1, \pm 2, \dots \tag{7.94}$$

For finding the amplitude of the mth harmonic, we act on (7.94) from the left by the operator

$$\int_{-L/2}^{L/2} dz \exp(-ik_n z),$$

then

$$\int_{-L/2}^{L/2} dz \exp(-ik_n z) E(z) = \sum_m E_m L \mathrm{sinc}[\pi(m-n)] = E_n L. \tag{7.95}$$

The last inequality follows from the fact that the function of an integer value,

$$\int_{-L/2}^{L/2} dz \exp[-i(k_m - k_n)z] = L \mathrm{sinc}[\pi(m-n)] = L\delta_{mn}, \tag{7.96}$$

differs from zero only at a single point $m = n$, where it is equal to unity (Fig. 7.17). Since $E(z)$ is real, $E_{-m} = E_m^*$.

At the final stages of the calculation, the *quantization length* L can be usually assumed to be infinite; in this case, the Fourier series (7.94) becomes an integral,

$$E(z) = (L/2\pi) \int_{-\infty}^{\infty} dk E(k) e^{ikz}, \tag{7.97}$$

where

$$E(k) = L^{-1} \int_{-\infty}^{\infty} dz E(z) e^{-ikz}. \tag{7.98}$$

In the derivation of (7.98), the following representation of the delta function was used:

$$\int_{-\infty}^{\infty} dz e^{ikz} = \lim_{L\to\infty} L\mathrm{sinc}(kL/2) = 2\pi\delta(k). \tag{7.99}$$

The factor $L/2\pi$ is added to (7.97) for making the discrete Fourier components coincide at $L \to \infty$ with the continuous ones, $E_m \to E(k_m)$. Comparing (7.94) and (7.97), we find the rule for passing from summation to integration,

$$\sum_m \cdots \to (L/2\pi) \int dk \ldots \tag{7.100}$$

The value $L/2\pi$ is called the density of modes (in the one-dimensional case with a single polarization state). The inverse value, $2\pi/L$, is equal to the distance between the neighboring modes $m, m+1$ on the k axis.

By repeating this procedure for other components of the field E_y, E_z and for the dependencies on x, y, we obtain a three-dimensional Fourier series for the field,

$$\boldsymbol{E}(\boldsymbol{r}, t) = \sum_{lmn} \boldsymbol{E}_{lmn}(t) e^{2\pi i(lx+my+nz)/L} \equiv \sum_{\boldsymbol{k}} \boldsymbol{E}_{\boldsymbol{k}}(t) e^{i\boldsymbol{k}\cdot\boldsymbol{r}}, \tag{7.101}$$

$$\boldsymbol{E}_{\boldsymbol{k}}(t) = L^{-3} \int_{L^3} d^3 r \boldsymbol{E}(\boldsymbol{r}, t) e^{-i\boldsymbol{k}\cdot\boldsymbol{r}} = \boldsymbol{E}^*_{-\boldsymbol{k}}(t), \tag{7.102}$$

$$\boldsymbol{k} \equiv (2\pi/L)\{l, m, n\}, \; l, m, n = 0, \pm 1, \pm 2, \ldots \tag{7.103}$$

A similar series expansion can be written for the magnetic field. Due to representing an arbitrary field (7.101) as a sum of plane waves, a continuous spatial distribution $E_\alpha(\boldsymbol{r})$ is determined by a countable set of complex numbers $E_{k\alpha}$. The 'allowed' vectors $\boldsymbol{k}$ form a grating in the $\boldsymbol{k}$ space, which is divided in cells of volume $(2\pi/L)^3$ due to the periodicity condition. Note that because of the relation $\boldsymbol{E}_{-\boldsymbol{k}} = \boldsymbol{E}^*_{\boldsymbol{k}}$, not all numbers $E_{k\alpha}$ are independent. The sum (7.101) can be written in the following equivalent forms:

$$\boldsymbol{E} = \mathrm{Re} \sum_{\boldsymbol{k}} \boldsymbol{E}_{\boldsymbol{k}} e^{i\boldsymbol{k}\cdot\boldsymbol{r}} = \mathrm{Re}(\boldsymbol{E} + i\boldsymbol{F}) = \mathrm{Re} \sum_{\boldsymbol{k}} (\boldsymbol{E}_{\boldsymbol{k}} + i\boldsymbol{F}_{\boldsymbol{k}}) e^{i\boldsymbol{k}\cdot\boldsymbol{r}}, \tag{7.104}$$

where $\boldsymbol{F}(\boldsymbol{r}, t)$ is an arbitrary real field and $\boldsymbol{F}_{\boldsymbol{k}} = \boldsymbol{F}^*_{-\boldsymbol{k}}$ are its harmonics.

Expansion in spatial harmonics (7.101) enables the field $\boldsymbol{E}(\boldsymbol{r})$ to be unambiguously separated in two components: the transverse $\boldsymbol{E}_\perp(\boldsymbol{r})$ and the longitudinal $\boldsymbol{E}_\parallel(\boldsymbol{r})$ (the argument t is omitted),

$$\begin{aligned} \boldsymbol{E}_\perp(\boldsymbol{r}) &= \sum_{\boldsymbol{k}} \sum_{\nu=1,2} \boldsymbol{e}_{\boldsymbol{k}\nu} E_{\boldsymbol{k}\nu} e^{i\boldsymbol{k}\cdot\boldsymbol{r}}, \\ \boldsymbol{E}_\parallel(\boldsymbol{r}) &= \sum_{\boldsymbol{k}} \boldsymbol{e}_{\boldsymbol{k}3} E_{\boldsymbol{k}3} e^{i\boldsymbol{k}\cdot\boldsymbol{r}}, \end{aligned} \tag{7.105}$$

where the unity orthogonal vectors $\boldsymbol{e}_{k\nu}$ form a right-hand triple, and $\boldsymbol{e}_{k3} \equiv \hat{\boldsymbol{k}} \equiv \boldsymbol{k}/k$. It follows from (7.105) that

$$\begin{aligned} \mathrm{div}\boldsymbol{E} &= \sum_k i\boldsymbol{k} \cdot E_k e^{i\boldsymbol{k}\cdot\boldsymbol{r}} = \mathrm{div}\boldsymbol{E}_{\parallel}, \\ \mathrm{rot}\boldsymbol{E} &= \sum_k i\boldsymbol{k} \times E_k e^{i\boldsymbol{k}\cdot\boldsymbol{r}} = \mathrm{rot}\boldsymbol{E}_{\perp}, \end{aligned} \tag{7.106}$$

and similar relations are valid for the magnetic field. By substituting (7.106) into Maxwell's equations (4.9)–(4.12) and assuming $\epsilon = 1$, we obtain

$$c\mathrm{rot}\boldsymbol{H}_{\perp} - \dot{\boldsymbol{E}}_{\perp} = 4\pi \boldsymbol{j}_{\perp}, \quad c\mathrm{rot}\boldsymbol{E}_{\perp} + \dot{\boldsymbol{H}}_{\perp} = 0, \tag{7.107}$$

$$-\dot{\boldsymbol{E}}_{\parallel} = 4\pi \boldsymbol{j}_{\parallel}, \quad \dot{\boldsymbol{H}}_{\parallel} = 0, \tag{7.108}$$

$$\mathrm{div}\boldsymbol{E}_{\parallel} = 4\pi\rho, \quad \mathrm{div}\dot{\boldsymbol{H}}_{\parallel} = 0. \tag{7.109}$$

Hence, the longitudinal part of an alternating magnetic field is equal to zero, while the longitudinal part of the electric field is determined by the positions of the charges at the same time moment, without retardation. Therefore, *the radiation field in the vacuum*, which is of interest for optics, *is transverse, and it is determined by the dynamical equations (7.107) through the transverse part of given (external) currents* $\boldsymbol{j}_{\perp} \equiv \boldsymbol{j}$. (Hereafter, we omit the $\perp$ index.)

By substituting (7.105) into (7.107), we find the equations of motion for the spatial harmonics,

$$\dot{\boldsymbol{E}}_k - ic\boldsymbol{k} \times \boldsymbol{H}_k = -4\pi \boldsymbol{j}_k, \tag{7.110}$$

$$\dot{\boldsymbol{H}}_k + ic\boldsymbol{k} \times \boldsymbol{E}_k = 0, \tag{7.111}$$

which, after excluding $\boldsymbol{H}_k$, yields

$$\ddot{\boldsymbol{E}}_k + \omega_k^2 \boldsymbol{E}_k = -4\pi \dot{\boldsymbol{j}}_k. \tag{7.112}$$

Here, $\omega_k \equiv ck$ and

$$\boldsymbol{j}_k(t) \equiv L^{-3} \int_{L^3} d^3r \boldsymbol{\Pi}_k \cdot \boldsymbol{j}(\boldsymbol{r}, t) e^{-i\boldsymbol{k}\cdot\boldsymbol{r}} = \boldsymbol{j}^*_{-k}(t), \tag{7.113}$$

with $\boldsymbol{\Pi}_k$ being the projection tensor (see (4.20)). Thus, *Maxwell's equations for the transverse field in the* $\boldsymbol{k}, t$ *representation are reduced to a system on inhomogeneous equations for independent harmonic oscillators*. Note that the harmonics $\boldsymbol{E}_k$ and $\boldsymbol{E}_{-k}$ are always excited simultaneously since $\boldsymbol{E}_k = \boldsymbol{E}^*_{-k}$.

In the case of a free-space field, i.e., in the absence of currents in L^3, the spatial harmonics, according to (7.110)–(7.112), oscillate without damping with the eigenfrequencies of the modes ω_k,

$$\begin{aligned} \boldsymbol{E}_k(t)_{free} &= (\boldsymbol{E}_{k0}e^{-i\omega_k t} + \boldsymbol{E}'_{k0}e^{i\omega_k t})/2, \\ \boldsymbol{H}_k(t)_{free} &= \hat{\boldsymbol{k}} \times (\boldsymbol{E}_{k0}e^{-i\omega_k t} - \boldsymbol{E}'_{k0}e^{i\omega_k t})/2. \end{aligned} \tag{7.114}$$

Here, $\boldsymbol{E}_{k0}$ is the initial amplitude of the plane wave propagating in the $+\boldsymbol{k}$ direction, while $\boldsymbol{E}'_{k0}$ is the independent amplitude of the backward wave, propagating in the $-\boldsymbol{k}$ direction. The condition $\boldsymbol{E}_k = \boldsymbol{E}^*_{-k}$ yields $\boldsymbol{E}'_{k0} = \boldsymbol{E}^*_{-k0}$. Hence, by summing over all $\boldsymbol{k}$, we obtain

$$\begin{aligned} \boldsymbol{E}(\boldsymbol{r},t)_{free} &= \mathrm{Re}\sum_k \boldsymbol{E}_{k0}e^{i(\boldsymbol{k}\cdot\boldsymbol{r}-\omega_k t)}, \\ \boldsymbol{H}(\boldsymbol{r},t)_{free} &= \mathrm{Re}\sum_k \hat{\boldsymbol{k}} \times \boldsymbol{E}_{k0}e^{i(\boldsymbol{k}\cdot\boldsymbol{r}-\omega_k t)}. \end{aligned} \tag{7.115}$$

Thus, the state of a free-space field at an arbitrary point $\boldsymbol{r}, t$ is given by a set of complex vectors $\boldsymbol{E}_{k0}$.

In the presence of external currents in L^3, induced field is added to the free one. The induced field is determined by the $\boldsymbol{j}_k(t)$ functions according to the inhomogeneous equation (7.112). For instance, a monochromatic plane wave of the current will 'excite' the induced field with its frequency ω, which may differ from ω_k (compare (4.23)). In the general case, $\boldsymbol{E}_k(t)$ is certainly not a harmonic function. The induced field can be also searched in the form (7.115), by assuming $\boldsymbol{E}_{k0}$ to be slow functions of the coordinates in the case of stationary currents (see Chapter 6 where we used the notation $\boldsymbol{E}_{k0} = \boldsymbol{E}^{(+)}_k(z)$) or functions of the time in the case of non-stationary problems, typical for the quantum mechanics.

Sometimes, it is convenient to describe the field using the vector potential $\boldsymbol{A}(\boldsymbol{r},t)$. In the case of the *Coulomb gauge*, the field $\boldsymbol{A}$ is assumed to be transverse, and it is unambiguously defined by the relations

$$\mathrm{rot}\boldsymbol{A} \equiv \boldsymbol{H}, \ \ \mathrm{div}\boldsymbol{A} = 0. \tag{7.116}$$

By substituting $\mathrm{rot}\dot{\boldsymbol{A}}$ for $\dot{\boldsymbol{H}}$ into (7.107), we obtain $\mathrm{rot}(c\boldsymbol{E} + \dot{\boldsymbol{A}}) = 0$, i.e.,

$$\boldsymbol{E} = -\dot{\boldsymbol{A}}/c. \tag{7.117}$$

Hence, we find the relations between the spatial harmonics of a real field and its potential,

$$\dot{\boldsymbol{A}}_k = -c\boldsymbol{E}_k, \ \ \boldsymbol{A}_k = i\boldsymbol{k} \times \boldsymbol{H}_k/k^2, \ \ \boldsymbol{H}_k = i\boldsymbol{k} \times \boldsymbol{A}_k. \tag{7.118}$$

7.3.2 *Canonical field variables*

Equations (7.110)–(7.112) for $\boldsymbol{H}_k$ and $\boldsymbol{E}_k$ resemble the Hamilton equations for the canonical coordinates and momenta of a system of particles, q_i, p_i. The currents $\boldsymbol{j}_k$ play the role of generalized forces. However, in experiment, one usually observes traveling waves in the far-field zone of the source, having a certain propagation direction, for instance, along $+\boldsymbol{k}$; therefore, it is preferable that the canonical variables with the index $\boldsymbol{k}$ relate only to the 'forward' wave.

The set of the four numbers $\{l, m, n, \nu\} \equiv \{\boldsymbol{k}, \nu\} \equiv k$ defines a plane wave or a *mode* (oscillation type) in free space. (In what follows, we will numerate the modes by a single index k.) In the presence of currents, the instantaneous state of the field in two modes k and $\bar{k} \equiv \{-\boldsymbol{k}, \nu\}$ with the same linear polarizations $\boldsymbol{e}_{k\nu}$ is given by two complex scalars or four real ones,

$$E_k \equiv \boldsymbol{E}_k \cdot \boldsymbol{e}_{k\nu} \equiv E'_k + iE''_k, \quad H_k \equiv (\hat{\boldsymbol{k}} \times \boldsymbol{e}_{k\nu}) \cdot \boldsymbol{H}_k \equiv H'_k + iH''_k,$$
$$E_{\bar{k}} \equiv \boldsymbol{E}_{-k} \cdot \boldsymbol{e}_{k\nu} = E'_k - iE''_k, \quad H_{\bar{k}} \equiv (-\hat{\boldsymbol{k}} \times \boldsymbol{e}_{k\nu}) \cdot \boldsymbol{H}_{-k} \equiv -H'_k + iH''_k. \quad (7.119)$$

Instead of the magnetic field, one can use the vector potential. According to (7.118) and (7.119),

$$A_k \equiv \boldsymbol{A}_k \cdot \boldsymbol{e}_{k\nu} = -iH_k/k = (H''_k - iH'_k)/k,$$
$$A_{\bar{k}} = (H''_k + iH'_k)/k,$$

where k denotes simultaneously the absolute value of the $\boldsymbol{k}$ vector and the mode index.

Let us form linear combinations,

$$q_k \equiv (L^3/4\pi\omega_k^2)^{1/2}(E''_k + H''_k),$$
$$p_k \equiv -(L^3/4\pi)^{1/2}(E'_k + H'_k). \quad (7.120)$$

(the choice of the coefficients will be explained below, from (7.134).) With the help of (7.119), we see that the variables $q_{\bar{k}}, p_{\bar{k}}$ for the backward mode are independent of q_k, p_k,

$$q_{\bar{k}} \sim E''_{\bar{k}} + H''_{\bar{k}} = -E''_k + H''_k,$$
$$p_{\bar{k}} \sim -E'_{\bar{k}} - H'_{\bar{k}} = -E'_k + H'_k.$$

It is convenient to join the real 'coordinate' q_k and 'momentum' p_k of a mode in a single complex dimensionless variable,

$$a_k \equiv (2\hbar\omega_k)^{-1/2}(\omega_k q_k + ip_k) = (E_k + H_k)/2ic_k,$$
$$c_k \equiv (2\pi\hbar\omega_k/L^3)^{1/2}. \quad (7.121)$$

The inverse transformations can be easily found as well,

$$\begin{aligned} E_k &= ic_k(a_k - a^*_{\bar{k}}) = (\pi/L^3)^{1/2}[-p_k - p_{\bar{k}} + i\omega_k(q_k - q_{\bar{k}})], \\ H_k &= ic_k(a_k + a^*_{\bar{k}}) = (\pi/L^3)^{1/2}[-p_k + p_{\bar{k}} + i\omega_k(q_k + q_{\bar{k}})]. \end{aligned} \tag{7.122}$$

Note that in the case of a standing plane wave, $a_k = a_{\bar{k}}$, and therefore the variables q_k, p_k scale as the magnetic and electric fields, respectively,

$$\begin{aligned} E(\boldsymbol{r}, t) &= -4(\pi/L^3)^{1/2} p_k(t)\cos(\boldsymbol{k}\cdot\boldsymbol{r}), \\ H(\boldsymbol{r}, t) &= -4(\pi/L^3)^{1/2} \omega_k q_k(t)\sin(\boldsymbol{k}\cdot\boldsymbol{r}). \end{aligned} \tag{7.123}$$

In the new variables, the plane-wave expansion takes the form

$$\begin{aligned} \boldsymbol{E}(\boldsymbol{r}, t) &= i\sum_{\boldsymbol{k}} c_k \boldsymbol{a}_k(t) e^{i\boldsymbol{k}\cdot\boldsymbol{r}} + \text{c.c.}, \\ \boldsymbol{H}(\boldsymbol{r}, t) &= i\sum_{\boldsymbol{k}} c_k \hat{\boldsymbol{k}} \times \boldsymbol{a}_k(t) e^{i\boldsymbol{k}\cdot\boldsymbol{r}} + \text{c.c.}, \\ \boldsymbol{A}(\boldsymbol{r}, t) &= i\sum_{\boldsymbol{k}} (c_k/k) \boldsymbol{a}_k(t) e^{i\boldsymbol{k}\cdot\boldsymbol{r}} + \text{c.c.}, \end{aligned} \tag{7.124}$$

where complex vectors $\boldsymbol{a}_k \equiv \sum_\nu \boldsymbol{e}_{k\nu} a_{k\nu}$ have been introduced.

After substituting (7.122) into (7.110)–(7.112), we find the equations for the new variables,

$$\dot{q}_k = p_k - (4\pi L^3/\omega_k)^{1/2} j''_k, \tag{7.125}$$

$$\dot{p}_k = -\omega_k^2 q_k + (4\pi L^3)^{1/2} j'_k, \tag{7.126}$$

$$\dot{a}_k = -i\omega_k a_k + (2\pi i/c_k) j_k. \tag{7.127}$$

The general solution to the last equation has the form

$$a_k(t) = a_k(0)e^{-i\omega_k t} + \frac{2\pi i}{c_k}\int_0^t dt'\, e^{i\omega_k(t-t')} j_k(t'). \tag{7.128}$$

Expanding $a_k(t)$ and $j_k(t)$ in Fourier frequency integrals, from (7.127) we immediately find the induced part of the field in the $\boldsymbol{k}, \omega$ representation (compare with (4.23)),

$$a_k(\omega)_{ind} = \frac{2\pi c_k^{-1}}{\omega_k - \omega - i\gamma_k} j_k(\omega),$$

where we have added the damping, $\gamma_k > 0$. Hence, it is clear that at $\gamma_k \ll \omega_k$, the spectrum $a_k(\omega)$ of the amplitude $a_k(t)$ mainly contains only *positive* frequencies close to the eigenfrequency ω_k. If we neglect the negative-frequency part of $a_k(t)$, i.e., assume that

$$a_k(t) \approx a_k^{(+)}(t) \equiv \int_0^\infty d\omega e^{-i\omega t} a_k(\omega), \tag{7.129}$$

then every term in the sum (7.124) describes a plane wave propagating in the $+\boldsymbol{k}$ direction (in contrast to the sum (7.101)).

In a free-space field, this approximation, according to (7.128), is at $j_k = 0$ valid rigorously,

$$a_k(t)_{free} = a_k(0)e^{-i\omega_k t}. \tag{7.130}$$

By comparing (7.114) and (7.122), we find the relation $E_{k0} = 2ic_k a_k(0)$.

Hence, *the positive-frequency part of the field is determined by $\boldsymbol{a}_k(t)$ functions while the negative-frequency part, by $\boldsymbol{a}_k^*(t)$ functions,*

$$\begin{aligned} \boldsymbol{E}^{(+)}(\boldsymbol{r}, t) &= i \sum_k c_k \boldsymbol{a}_k(t) e^{i\boldsymbol{k}\cdot\boldsymbol{r}}, \\ \boldsymbol{E}^{(-)}(\boldsymbol{r}, t) &= -i \sum_k c_k \boldsymbol{a}_k^*(t) e^{-i\boldsymbol{k}\cdot\boldsymbol{r}}. \end{aligned} \tag{7.131}$$

In the quantum theory, these functions become photon creation and annihilation operators, $\boldsymbol{a}_k \to \hat{\boldsymbol{a}}_k$, $\boldsymbol{a}_k^* \to \hat{\boldsymbol{a}}_k^\dagger$.

7.3.3 °*Hamiltonian of the field and the matter*

From Maxwell's equations, it follows (see, for instance, (4.27)) that the instantaneous energy of the field is

$$\mathcal{E}(t) = (1/8\pi) \int_{L^3} d^3r(E^2 + H^2) \equiv \mathcal{H}_0. \tag{7.132}$$

Let us accept this expression as the Hamilton function of the free-field transverse part. By substituting here the plane-wave expansion (7.101) and taking into account the orthogonality condition (7.96), we obtain a diagonal quadratic form,

$$\mathcal{H}_0 = (L^3/8\pi) \sum_k (|\boldsymbol{E}_k|^2 + |\boldsymbol{H}_k|^2). \tag{7.133}$$

According to (7.114), at $j = 0$, $\mathcal{H}_0$ does not depend on t. With the help of (7.124), we find that

$$\mathcal{H}_0 = \sum_k (p_k^2 + \omega_k^2 q_k^2)/2 = \hbar \sum_k \omega_k |a_k|^2. \tag{7.134}$$

One can easily verify that the Hamilton equations,

$$\dot{q}_k = \partial\mathcal{H}/\partial p_k, \;\; \dot{p}_k = -\partial\mathcal{H}/\partial q_k, \tag{7.135}$$

at $\mathcal{H} = \mathcal{H}_0$ lead to the oscillator equations for q_k, p_k, a_k following from (7.125–7.127) at $j = 0$. This confirms that (7.132) was chosen correctly.

The joint Hamiltonian of the field and the system of charged particles inside the volume L^3 in the non-relativistic case (see Ref. [Landau (1973)]) is

$$\begin{gathered}\mathcal{H} = \mathcal{H}_0 + \sum_i (\boldsymbol{P}_i - e_i\boldsymbol{A}_i/c)^2/2m_i + \mathcal{H}_c, \\ \boldsymbol{A}_i \equiv \boldsymbol{A}(\boldsymbol{R}_i(t), t), \;\; \mathcal{H}_c \equiv \sum_{i<j} e_i e_j/|\boldsymbol{R}_i - \boldsymbol{R}_j|,\end{gathered} \tag{7.136}$$

where $\mathcal{H}_c$ is the energy of the Coulomb (longitudinal-field) interaction between the particles and $\boldsymbol{R}_i, \boldsymbol{P}_i$ are the canonical variables of the ith particle whose charge and mass are e_i, m_i.

Eqs. (7.135) and (7.136) lead to the relation between the 'kinetic' and canonical momenta of the particles,

$$m_i\boldsymbol{V}_i = \boldsymbol{P}_i - e_i\boldsymbol{A}_i/c, \tag{7.137}$$

where $\boldsymbol{V}_i \equiv \dot{\boldsymbol{R}}_i$ is the i-th particle velocity. Hence, the Hamiltonian (7.136) can be represented in a simple form,

$$\mathcal{H} = \mathcal{H}_0 + \sum_i m_i V_i^2/2 + \mathcal{H}_c. \tag{7.138}$$

According to (7.136), the Hamiltonian of the interaction between the particles and the transverse field is

$$\mathcal{V} = \sum_i \left(-\frac{e_i}{m_i c}\boldsymbol{P}_i \cdot \boldsymbol{A}_i + \frac{e_i^2}{2m_i c^2}A_i^2 \right). \tag{7.139}$$

In the case of particles with internal magnetic moment $\boldsymbol{\mu}_i$, one should add the energy of the spin interaction, $-\boldsymbol{\mu} \cdot \boldsymbol{H}_i$.

Let us show that (7.135), (7.136) lead to the usual Newton's equations with the Lorentz force for the particles and to Maxwell's equations with the external currents (7.107) for the field. In order to obtain Newton's equations, let us differentiate (7.137) in time. Taking into account that, according to (7.135), (7.136),

$$\dot{P}_{i\alpha} = (e_i/c)V_{i\beta}\,\partial A_{i\beta}/\partial R_{i\alpha}, \tag{7.140}$$

we find that

$$m_i\ddot{R}_{i\alpha} = \dot{P}_{i\alpha} - \frac{e_i}{c}\left(\frac{\partial A_{i\alpha}}{\partial t} + \frac{\partial A_{i\alpha}}{\partial R_{i\beta}}V_{i\beta}\right) = e_iE_{i\alpha} + \frac{e_i}{c}[\boldsymbol{V}_i \times \boldsymbol{H}_i]_\alpha. \tag{7.141}$$

Recall that the fields are taken at the point of the particle location, therefore,

$$d\boldsymbol{A}_i/dt \neq \partial \boldsymbol{A}_i/\partial t = -c\boldsymbol{E}_i.$$

The field equations can be found by differentiating the total Hamiltonian (7.136) w.r.t. the canonical variables p_k, q_k of the field. Then the 'forces' acting on the field from the particles are determined by the second term in (7.138). Let us first differentiate it in $A_{i\alpha}$ with the help of (7.137):

$$\frac{\partial}{\partial A_{i\alpha}}\frac{m_i V_i^2}{2} = m_i V_{i\beta}\frac{\partial V_{i\beta}}{\partial A_{i\alpha}} = -\frac{e_i}{c}V_{i\alpha} = \frac{\partial}{\partial A_{i\alpha}}\left(-\frac{e_i}{c}\boldsymbol{V}_i\cdot\boldsymbol{A}_i\right). \tag{7.142}$$

It follows that the Hamiltonian of the interaction between the transverse field and non-relativistic spin-free particles can be represented, instead of (7.139), as

$$\mathcal{V}' \equiv -\sum_i \frac{e_i}{c}\boldsymbol{V}_i\cdot\boldsymbol{A}_i = -\frac{1}{c}\int d^3r\boldsymbol{j}\cdot\boldsymbol{A}, \tag{7.143}$$

where $\boldsymbol{j}(\boldsymbol{r},t)$ is the external current density determined by the coordinates and velocities of the particles,

$$\boldsymbol{j}(\boldsymbol{r},t) = \sum_i e_i\boldsymbol{V}_i(t)\delta^{(3)}(\boldsymbol{r}-\boldsymbol{R}_i(t)), \tag{7.144}$$

and the prime reminds that, according to (7.142), the Hamiltonian (7.143) provides an exact description only for the perturbation of the field by the particles and not vice versa.

Hamilton's equations (7.135) immediately lead to the equations of motion for an arbitrary function of canonical coordinates, $f(q_k, p_k, t)$,

$$df/dt = \partial f/\partial t + \{f, \mathcal{H}\}, \tag{7.145}$$

$$\{f, g\} \equiv \sum_k\left(\frac{\partial f}{\partial q_k}\frac{\partial g}{\partial p_k} - \frac{\partial g}{\partial q_k}\frac{\partial f}{\partial p_k}\right). \tag{7.146}$$

One can easily verify that after the linear transformation (7.121) from q_k, p_k to the new independent variables a_k, a_k^*, the *Poisson bracket* (7.146) takes the form

$$\{f, g\} = \frac{1}{i\hbar}\sum_k\left(\frac{\partial f}{\partial a_k}\frac{\partial g}{\partial a_k^*} - \frac{\partial g}{\partial a_k}\frac{\partial f}{\partial a_k^*}\right). \tag{7.147}$$

Assuming $f \equiv a_k$, with the help of (7.134) and (7.143) we obtain

$$\dot{a}_k = \frac{1}{i\hbar}\frac{\partial H}{\partial a_k^*} = -i\omega_k a_k + \frac{i}{\hbar c}\int d^3r\boldsymbol{j}\cdot\frac{\partial \boldsymbol{A}_i}{\partial a_k^*}. \tag{7.148}$$

With an account for (7.124) and (7.144), this equation coincides with Eq. (7.127), which was obtained from Maxwell's equations with external currents.

Thus, we have written the equations for the field and the matter in the canonical form (7.135) with the Hamiltonian (7.136). Before using this result for the quantization of the field equations, let us consider the interaction Hamiltonian in the dipole approximation.

7.3.4 °*Dipole approximation*

In quantum electronics, it is often possible to use approximations instead of the exact expressions for the perturbation energy (7.139), (7.143). In the case of a plane free monochromatic wave, $H = E$. Therefore, in the first order in V_i/c, one can neglect the magnetic part of the Lorentz force in the Newton equation,

$$m\ddot{\boldsymbol{R}}_i \approx e_i \boldsymbol{E}(\boldsymbol{R}_i, t). \tag{7.149}$$

Further, let the particles occupy a restricted space area with the linear size a much less than the scale of the field variation, $\lambda\!\!\!^- = c/\omega$. Then the field can be written as a series expansion in $\boldsymbol{R}_i$ and only the first few terms should be taken into account. In this case, (7.139) leads to the multi-field expansion of the perturbation Hamiltonian for the particles in powers of $R_i/\lambda\!\!\!^-$. In the zeroth (dipole) approximation, $\boldsymbol{A}(\boldsymbol{R}_i) \approx \boldsymbol{A}(\boldsymbol{r}_0) \equiv \boldsymbol{A}_0$, where $\boldsymbol{r}_0$ is some fixed point inside the system of particles (for instance, the center of mass). Then, according to (7.140), $\dot{\boldsymbol{P}}_i = 0$, and (7.141) takes the form (compare with (7.149))

$$m_i\ddot{\boldsymbol{R}}_i = e_i \boldsymbol{E}_0, \tag{7.150}$$

where $\boldsymbol{E}_0 \equiv \boldsymbol{E}(\boldsymbol{r}_0, t)$. This equation, according to (7.135), follows from the interaction Hamiltonian of the form

$$\mathcal{V}_{dip} \equiv -\boldsymbol{d}(t) \cdot \boldsymbol{E}_0, \tag{7.151}$$

where

$$\boldsymbol{d}(t) \equiv \sum_i e_i(\boldsymbol{R}_i(t) - \boldsymbol{r}_0), \quad |\boldsymbol{R}_i - \boldsymbol{r}_0| \ll \lambda\!\!\!^-. \tag{7.152}$$

Here, in contrast to (7.139), the field is a given external parameter. Note also that the dipole moment of a neutral system does not depend on the choice of $\boldsymbol{r}_0$.

Let the matter consist of N separate motionless molecules with the dipole moments $\boldsymbol{d}_j$ and centers at $\boldsymbol{r}_j$. Then the energy of the matter in the external field, according to (7.151) is (compare with (4.28))

$$\mathcal{V}_{dip} = -\sum_{j=1}^{N} \boldsymbol{d}_j(t) \cdot \boldsymbol{E}(\boldsymbol{r}_j, t) = -\int d^3r \boldsymbol{P}(\boldsymbol{r}, t) \cdot \boldsymbol{E}(\boldsymbol{r}, t), \tag{7.153}$$

$$\boldsymbol{P}(\boldsymbol{r}, t) = \sum_{j=1}^{N} \boldsymbol{d}_j(t)\delta^{(3)}(\boldsymbol{r} - \boldsymbol{r}_j). \tag{7.154}$$

The interaction Hamiltonian *for the field* in the dipole approximation follows from (7.143) after replacing $\boldsymbol{A}_i$ by $\boldsymbol{A}_0$,

$$\mathcal{V}'_{dip} = -\frac{1}{c}\dot{\boldsymbol{d}} \cdot \boldsymbol{A}_0 = -\boldsymbol{d} \cdot \boldsymbol{E}_0 - \frac{1}{c}\frac{d}{dt}(\boldsymbol{d} \cdot \boldsymbol{A}_0). \tag{7.155}$$

If we restrict the consideration to the case of quasi-monochromatic currents and fields, then $\boldsymbol{d} \cdot \boldsymbol{A}_0$ contains two components: a constant one and one oscillating with the double frequency. As a result, only the first term in (7.155), coinciding with (7.151), provides an accumulating interaction,

$$\mathcal{V}'_{dip} \approx \mathcal{V}_{dip} = -\boldsymbol{d} \cdot \boldsymbol{E}_0. \tag{7.156}$$

Thus, *the dipole Hamiltonian $\mathcal{V}_{dip}$ can be also used for calculating the emitted field in 'single-frequency' problems.* It follows from (7.156) that

$$\dot{a}_k + i\omega_k a_k = \frac{1}{i\hbar}\frac{\partial \mathcal{V}_{dip}}{\partial a_k^*} = \frac{c_k}{\hbar} d_k \exp(-i\boldsymbol{k} \cdot \boldsymbol{r}_0), \tag{7.157}$$

where $d_k \equiv \boldsymbol{d} \cdot \boldsymbol{e}_k$. The same result can be obtained from the exact equation (7.127) in the case of a neutral system, after taking into account (7.144), replacing $\exp(-i\boldsymbol{k} \cdot \boldsymbol{R}_i)$ by $\exp(-i\boldsymbol{k} \cdot \boldsymbol{r}_0)$, and replacing $\boldsymbol{V}_i$ by $-i\omega_k \boldsymbol{R}_i$.

Often, instead of (7.139), one uses the approximation

$$\mathcal{V} \approx -\sum_i e_i \boldsymbol{P}_i \cdot \boldsymbol{A}_i / m_i c, \tag{7.158}$$

i.e., neglects the term that is quadratic in eA. (Note that in the case of a single electron in a harmonic field, this term is on the order of αE_i^2 where $\alpha = \lambda\!\!\!^{-2} r_e$ is the polarisability of a free electron (6.36).) This approximation is only valid in the first order of the perturbation theory, i.e., in the calculation of single-quantum effects. Eq. (7.158) also follows from (7.143) if the canonical and kinetic momentums are assumed to be the same. Eq. (7.158) leads to the following equation of motion for the particle:

$$m_i \ddot{\boldsymbol{R}}_i = -\frac{e_i}{c}\frac{dA_i}{dt}, \tag{7.159}$$

which at $a \ll \lambda\!\!\!^{-}$ coincides with (7.150).

For bonded electrons in atoms and small molecules, $a \sim 10^{-8}$ cm, and the condition for the validity of the dipole approximation (7.151), (7.155) is satisfied up to the X-ray range. Recall that magnetic moments related to spin and orbital motion are on the order of the *Bohr magneton* μ_0, which is two orders of magnitude as small as one Debye,

$$2\mu_0 \equiv e\hbar/mc = e\lambda\!\!\!^{-}_c \sim ea_0/137. \tag{7.160}$$

However, despite the relatively small value of multi-field effects, their manifestation in the optical range is important and can be easily observed: for instance, in the effect of the optical activity (polarization rotation) and in the appearance of forbidden lines in spectra.

A *free electron* in a harmonic field, according to (7.149), oscillates with the amplitude $a_1 = eE_1/m\omega^2$ and the velocity $a_1\omega$; therefore, the conditions $a \ll \lambda\!\!\!^{-}$ and $V \ll c$ have the same form,

$$E_1 \ll mc^2/e\lambda\!\!\!^{-} \sim 10^8\,\text{G}. \tag{7.161}$$

This estimate was made for $\lambda = 1\,\mu$ and corresponds to a practically impossible intensity 10^{18} W/cm^2. Nevertheless, by taking into account the magnetic-field effect in a light wave (Sec. 6.2) one can calculate the quadratic polarisability of a free electron and observable nonlinear effects.

7.4 Quantization of the field

Thus, we have represented the field equations in the form of Hamilton's equations for the spatial harmonicsq $E_k(t), H_k(t)$ (or their linear combinations q_k, p_k, a_k). Now, we can pass to the main stage of the quantum description, which is finding the commutation rules for the dynamical field variables.

7.4.1 *Commutation relations*

After passing to the quantum description, all canonical variables q_k, p_k and their functions $f(q_k, p_k)$ become linear operators $\hat{q}_k, \hat{p}_k, \hat{f}_k$ acting according to certain rules on the state vector of the system. The difference between the actions of the operator products fg and gf[p] can be defined in terms of the Poisson brackets (7.146),

$$fg - gf \equiv [f, g] = i\hbar\{f, g\}. \tag{7.162}$$

Here, all variables are considered at the same time moment. In particular, we find that

$$[q_k, p_{k'}] = i\hbar\delta_{k,k'}, \quad [q_k, q_{k'}] = [p_k, p_{k'}] = 0, \tag{7.163}$$

[p]The 'hat' sign over the operators will be only used where necessary.

$$[a_k, a^\dagger_{k'}] = \delta_{k,k'}, \ \ [a_k, a_{k'}] = [a^\dagger_k, a^\dagger_{k'}] = 0, \tag{7.164}$$

$$[f, a^\dagger_k] = \partial f/\partial a_k, \ \ [a_k, f] = \partial f/\partial a^\dagger_k, \tag{7.165}$$

$$[E_k, H_{k'}] = -2c_k^2 \delta_{k,\bar{k}'}, \ \ [E_k, E_{k'}] = [H_k, H_{k'}] = 0, \tag{7.166}$$

where

$$c_k^2 = 2\pi\hbar\omega_k/L^3, \ \ k \equiv \{\boldsymbol{k}, \nu\}, \ \ \bar{k} \equiv \{-\boldsymbol{k}, \nu\}. \tag{7.167}$$

From (7.166), using the linear relations (7.124), one can also find the commutators for the fields $\boldsymbol{E}(\boldsymbol{r}, t)$, $\boldsymbol{H}(\boldsymbol{r}, t)$.

The classical complex variable a_k^* is put into correspondence with the *photon creation operator* $a^\dagger_k$, which is Hermite conjugated to the *photon annihilation operator* a_k. The Hermitian operator $a^\dagger_k a_k = (a^\dagger_k a_k)^\dagger \equiv N_k$ is called the *photon-number operator*, and most often, it is this operator that corresponds to optical observables. For instance, the spectral brightness can be expressed in terms of N_k as

$$I_{\omega\Omega}(\boldsymbol{k}, \nu) = \hbar c \lambda^{-3} \langle N_k \rangle. \tag{7.168}$$

This expression, in contrast to (7.19), contains averaging over the wave function or the density operator ρ.

Often, in addition to N_k, other operators can be of interest. For instance, the mean value of $a^\dagger_k a_{k'}$ characterizes the statistical coupling between modes k and k'. One can show that the rate of an m-quantum stimulated transition scales as the mean value of the operator $a_k^{\dagger m} a_k^m \equiv : N_k^m :$. Here, colons denote *normal ordering*, i.e., putting all a_k operators on the right of $a^\dagger_k$. The mean values of these operators,

$$G_k^{(m)} \equiv \langle : N_k^m : \rangle = \langle a_k^{\dagger m} a_k^m \rangle, \tag{7.169}$$

are called *normal* (*normally ordered, factorial*) moments of order m for mode k. The relation between the factorial moments $G^{(m)}$ and the usual ones $\langle N^m \rangle$ can be easily found from the operator identities following from (7.162) or (7.164),

$$[a^m, N] = m a^m, \ \ [N, a^{\dagger m}] = m a^{\dagger m}.$$

Hence,

$$: N^m := N(N-1)\dots(N-m+1). \tag{7.170}$$

(Hereafter, we omit the subscript k whenever a single mode is considered.)

In the Heisenberg picture, the wave function and the density matrix are constant, while the operators depend on time according to the Heisenberg equations, which can be obtained from (7.145) and (7.162),

$$df/dt = \partial f/\partial t + [f, \mathcal{H}]/i\hbar. \tag{7.171}$$

For instance, assuming $f \equiv a_k$ and using (7.124), (7.134), (7.143) and (7.165), we obtain the Heisenberg equation for the annihilation operator in the form (7.127). Similarly, all other relations from Sec. 7.3 remain valid after changing the classical dynamical variables to operators in the Heisenberg representation. It is important that the operator products are written in the symmetrized form, for instance,

$$|a|^2 \to (a^\dagger a + aa^\dagger)/2 = a^\dagger a + 1/2 = aa^\dagger - 1/2, \tag{7.172}$$

where we used (7.165) for obtaining the last equalities. Note that in the operator identities similar to (7.172), 1/2 is understood as $\hat{I}/2$, where $\hat{I}$ is the *unity*, or *identity*, operator, $\hat{I}\Psi = \Psi$.

Spatial harmonics of a free field depend on time harmonically with the frequency $\omega_k = ck$. Hence, we find the *two-time* commutators,

$$[a_k(t), a_{k'}^\dagger(t')]_{free} = \delta_{kk'} \exp[-i\omega_k(t - t')], \tag{7.173}$$

and similar relations for other field variables. In the presence of external currents, (7.173) can be replaced by a more complicated dependence on t and t', but it should turn into (7.164) at $t = t'$. This conservation of commutation relations follows from the unitarity of the operators time evolution.

7.4.2 *Quantization of macroscopic field in matter*

Macroscopic field in a non-magnetic material is described by Maxwell's equations with the phenomenological dielectric function ϵ (in the linear approximation). In the transparency windows, $\epsilon''(\omega) \approx 0$, and the energy of the free field is preserved, so that we can again use the Hamiltonian formalism and quantize the field variables. If we also neglect the dispersion $\epsilon'(\omega)$, then the procedure will be similar to the one of Sec. 7.3, with the only changes in the speed of light ($c \to c/n$) and in the orientation of the polarization unit vectors $\boldsymbol{e}_k$ (in the case of an anisotropic medium).

One can show [Klyshko (1980)] that with an account for linear dispersion, the relation between the macroscopic field $\boldsymbol{E}(\boldsymbol{r}, t)$ and the photon creation and annihilation operators $a_k^\dagger, a_k$ in a transparent medium has the form (7.124) provided that the c_k coefficients are multiplied by a factor of

$$\xi_k \equiv \left[2\omega \left(\frac{\partial}{\partial \omega} \omega^2 \boldsymbol{e} \cdot \boldsymbol{\epsilon} \cdot \boldsymbol{e} \right)^{-1} \right]_k^{1/2} = \left(\frac{uv}{c^2 \cos\rho} \right)_k^{1/2} \approx \frac{1}{n_k}, \tag{7.174}$$

where $v_k \equiv \omega_k/k = c/n_k$ and $u_k \equiv \partial\omega_k/\partial k \cos\rho_k$ are the phase and group velocities, respectively and ρ_k is the angle between the ray and wave vectors. The spatial harmonics of the magnetic field in (7.124) should be then multiplied by n_k.

7.4.3 *Quantization of the field in a cavity*

The field in a closed cavity with ideally reflecting walls can be represented as a sum of real orthogonal eigenfunctions $\boldsymbol{u}_k(\boldsymbol{r})$, $\boldsymbol{v}_k(\boldsymbol{r})$ of the corresponding boundary problem,

$$\boldsymbol{E} = \sum_k p_k(t)\boldsymbol{u}_k(\boldsymbol{r}), \ \boldsymbol{H} = \sum_k \omega_k q_k(t)\boldsymbol{v}_k(\boldsymbol{r}), \tag{7.175}$$

where ω_k are the cavity eigenfrequencies.

For instance, free field in a rectangular cavity is a superposition of standing plane waves. (The field is not transverse in this case, see [Landau (1982)]) The allowed values of the wavevector are determined by the cavity dimensions L_α, $\alpha = x, y, z$ (compare with (7.103)),

$$k = \pi\left\{\frac{l}{L_x}, \frac{m}{L_y}, \frac{n}{L_z}\right\}, \ l, m, n = 0, 1, 2, \ldots \tag{7.176}$$

A standing plane wave is a superposition of two counter-propagating waves with the equal amplitudes $a_k = a_{-k}$; in this case, according to (7.123), $E \sim p_k \cos kz$ and $H \sim \omega_k q_k \sin kz$. (The scaling coefficient can be found from (7.134) at $L^3 \equiv L_x L_y L_z$ and is on the order of $(16\pi/L^3)^{1/2}$.) Hence, taking into account the uncertainty relation $\Delta q \Delta p \geqslant \hbar/2$, one comes to the conclusion that the accuracy of a simultaneous measurement of $\boldsymbol{E}(\boldsymbol{r}, t)$ and $\boldsymbol{H}(\boldsymbol{r}, t)$ inside the cavity is limited.

Sometimes, the free-space field is also expanded in standing waves of the form $\cos \boldsymbol{k}\cdot\boldsymbol{r}$, $\sin \boldsymbol{k}\cdot\boldsymbol{r}$, but then the amplitudes of the standing waves do not have a direct relation to the values observed in experiment. Indeed, for selecting the plane wave $+\boldsymbol{k}$, the detector should be placed in the far-field zone of the emitter, where the $-\boldsymbol{k}$ wave is absent. The relation between the far field and the a_k operators is considered in [Klyshko (1980)].

7.5 °States of the field and their properties

Next, we consider the various states of the field, both pure and mixed ones, and their properties, as well as the mean values and distributions of the observables in these states. It is convenient to use some *basis* set of wave functions, so that an arbitrary state can be represented as an expansion over this basis. This procedure is similar to expanding an arbitrary vector over the set of unit vectors of some frame of reference in real space. We will consider the basis sets generated by various operators: energy $\mathcal{H}$, coordinate q, momentum p, photon annihilation $a \sim \omega q + ip$, as well as relations between these bases. In this consideration, we will use compact Dirac's notation, which will be briefly described in the next subsection.

7.5.1 *Dirac's notation*

An arbitrary instantaneous state of a quantum system is defined by the wave function $\psi(x) \equiv \langle x|\psi\rangle$, where x is some set of variables, discrete or continuous, that are sufficient for the *complete description* of the system. A complete description of a 'one-dimensional' spinless particle, in particular, of an oscillator or a field mode, is given by a single variable: the coordinate ($x \equiv q$), the momentum ($x \equiv p$), or the energy ($x \equiv \mathcal{E}$).[q]

The function $\langle x|\psi\rangle$ is called the x-representation of the system. The state itself, without specifying the representation, is denoted by $|\psi\rangle$, or $|\ \rangle$, or $|t\rangle$. The complex conjugated function, $\psi^*(x)$, is denoted by $\langle\psi|x\rangle = \langle x|\psi\rangle^*$, i.e., one can write $|\ \rangle^* = \langle\ |, \langle\ |^* = |\ \rangle$.

In the x-representation, the state $|\ \rangle$ is given by a set (discrete or continuous) of numbers $\langle x_1|\ \rangle \equiv c_1, \langle x_2|\ \rangle \equiv c_2, \ldots$, which can be naturally considered as the components of some vector in a multi-dimensional space. Then, $\langle x_n|\ \rangle$ is an analogue of the inner product of the unit vector $\langle x_n|$ and the state vector $|\ \rangle$, i.e., the projection of $|\ \rangle$ onto the n-th axis. Any vector can be represented as a sum of the unit vectors multiplied by the c_n coefficients,

$$|\ \rangle = \sum_n c_n|x_n\rangle \equiv \sum_n |x_n\rangle\langle x_n|\ \rangle, \tag{7.177}$$

or, in more compact notation,

$$|\ \rangle = \sum_n |n\rangle\langle n|\ \rangle. \tag{7.178}$$

Similarly,

$$\langle\ | = \sum_n \langle\ |n\rangle\langle n|. \tag{7.179}$$

In the case of a continuous variable, summation in (7.177) and other similar equations is replaced by integration,

$$|\ \rangle = \int dx|x\rangle\langle x|\ \rangle. \tag{7.180}$$

The vectors $\langle\ |$ and $|\ \rangle$ are called, respectively, the *bra*- and *ket*-vectors (being parts of a *bracket*).

[q]Recall that in classical mechanics, the state is given by the *numbers* q, p, while in quantum mechanics, by the function $\psi(q)$ (or $\psi(p), \psi(\mathcal{E}), \ldots$).

Projections of different unity vectors onto one another are equal to zero (in the case of an orthogonal system),

$$\langle n|n'\rangle = \delta_{nn'}, \quad \langle x|x'\rangle = \delta(x - x'). \tag{7.181}$$

One says that a frame of reference (a basis) is complete if any vector can be represented in the form (7.177). The completeness can be expressed in the form of a tensor equation

$$\hat{I} = \sum_n |n\rangle\langle n|. \tag{7.182}$$

Here, $\hat{I}$ is a unit tensor ($\hat{I}|\ \rangle = |\ \rangle$), and $|a\rangle\langle b|$ denotes a dyadic tensor, or the outer product of the vectors $|a\rangle$ and $\langle b|$. The action of a dyadic on the vectors is obvious from its notation,

$$\begin{aligned} \{|a\rangle\langle b|\}|\ \rangle &\equiv |a\rangle\langle b|\ \rangle = \langle b|\ \rangle|a\rangle, \\ \langle\ |\{|a\rangle\langle b|\} &\equiv \langle\ |a\rangle\langle b| = \langle b|\langle\ |a\rangle. \end{aligned} \tag{7.183}$$

The tensor $|a\rangle\langle a| \equiv \hat{P}_a$ with $\langle a|a\rangle = 1$ is called a *projector* since its action on a vector $|\ \rangle$ selects the component of this vector along $|a\rangle$: $\hat{P}_a|\ \rangle = |a\rangle\langle a|\ \rangle \equiv c_a|a\rangle$.

The *expansion of the unity* (7.182) provides an easy way for forming various *representations* of scalars $\langle a|b\rangle$, vectors $|\ \rangle$, tensors (operators) f:

$$\langle a|b\rangle = \langle a|I|b\rangle = \sum_n \langle a|n\rangle\langle n|b\rangle, \tag{7.184}$$

$$|\ \rangle = I|\ \rangle = \sum_n |n\rangle\langle n|\ \rangle, \tag{7.185}$$

$$f = IfI = \sum_{nn'} f_{nn'}|n\rangle\langle n'|, \tag{7.186}$$

with $f_{nn'} \equiv \langle n|f|n'\rangle$ in the last equation.

An operator f can act on the right on a ket vector and on the left on a bra vector, creating new vectors, $f|a\rangle \equiv |b\rangle$ and $\langle a|f \equiv \langle c|$, with other directions and lengths. (The length or, more precisely, the *norm* of a vector $|a\rangle$ is defined as the number $\langle a|a\rangle^{1/2}$.) If it is only the length of a vector that is changed, the vector is called the eigenvector (right or left) for this operator. It is convenient to denote an operator and its eigenvectors and eigenvalues by the same character,

$$\hat{f}|f_n\rangle = f_n|f_n\rangle. \tag{7.187}$$

The set of eigenvectors $|f_n\rangle$ usually forms a basis, not necessarily orthogonal. The operator $f^\dagger$, Hermite conjugated to f, is defined by the equation

$$f^\dagger|a\rangle = \{\langle a|f\}^*, \tag{7.188}$$

or $(f^\dagger)_{ab} = (f_{ba})^*$. If $f^\dagger = f$, f is an *Hermitian operator*, and $f_{ab} = f^*_{ba}$, $f_n = f^*_n$, $\langle f_n|f_{n'}\rangle = 0$ (at $f_n \neq f_{n'}$).

In quantum mechanics, it is postulated that the probability distribution $P(f|t)$ for the observable f at a time t for an ensemble of systems being in the same quantum state $|t\rangle$ is determined by the projections of $|t\rangle$ onto the eigenvectors $|f\rangle$ of the operator $\hat{f}$,

$$P(f|t) = C|\langle f|t\rangle|^2 = C\langle t|\hat{P}_f|t\rangle, \tag{7.189}$$

where C^{-1} is the normalization sum or integral ($C = 1$ for normalized vectors). If the observable f has a continuous spectrum, then $P(f)$ has dimensionality $1/f$ and is the probability distribution density. The distribution (7.189) is defined in terms of the Schrödinger variables. In the Heisenberg representation, it has the form

$$P(f|t) = C\langle t_0|P_f(t)|t_0\rangle, \tag{7.190}$$

$$\hat{P}_f(t) \equiv |f(t)\rangle\langle f(t)|. \tag{7.191}$$

In the case of a mixed state, the operator $\hat{P}_f$ should be averaged with the density operator. For instance, in the Schrödinger representation,

$$P(f|t) = \mathrm{Tr}\{|f\rangle\langle f|\rho(t)\} = \langle f|\rho(t)|f\rangle, \tag{7.192}$$

i.e., *the distribution of an observable f is defined by the diagonal elements of the density matrix in the f-representation (the populations).*

According the postulate of the wave function *reduction*, when some value (for instance, energy) is measured by means of a classical device, the measurement brings the system from the initial state $|\ \rangle$ into the state $|\mathcal{E}_1\rangle$, where $\mathcal{E}_1$ is determined by the reading of the detector. Thus, *a measurement is simultaneously the preparation of a system with a known wave function.* In order to prepare a system in a given state $|\mathcal{E}_1\rangle$, one has to measure the energy of a sufficiently large number of systems in different initial states until a necessary reading $\mathcal{E}_1$ is achieved. When a classical device shows a reading f this means it has brought the system into the state $\hat{P}_f|\ \rangle = |f\rangle$; the *back-action* of the device on the state is described by the projector $\hat{P}_f \equiv |f\rangle\langle f|$.

7.5.2 *Energy states*

Usually, it is the energy that is measured in quantum optics; correspondingly, as the basis one usually chooses the set of the eigenstates of the energy operator for separate modes of the free field, i.e., the harmonic-oscillator Hamiltonian,

$$\mathcal{H}_0 = (p^2 + \omega^2 q^2)/2 = \hbar\omega(N + I/2), \tag{7.193}$$

where $N \equiv a^\dagger a$ is the photon-number operator for the chosen mode. The k index will be omitted, as a rule, whenever we consider only a single mode. By definition, the eigenvalues and eigenfunctions satisfy the equality

$$\mathcal{H}_0|N\rangle = \mathcal{E}_N|N\rangle. \tag{7.194}$$

According to (7.193), the energy states $|N\rangle$ or, in short, the *N-states*, also known as *Fock states*, are also eigenstates for the $a^\dagger a$ operator, $(\hat{N} - N)|N\rangle = 0$, where $N \equiv \mathcal{E}/\hbar\omega - 1/2$.

The $\mathcal{H}_0$ operator is Hermitian; therefore, the vectors $|N\rangle$ form a complete orthogonal normalized 'frame of reference',

$$\langle N|N'\rangle = \delta_{NN'}, \quad \sum_N |N\rangle\langle N| = I, \tag{7.195}$$

$$|\ \rangle = \sum_N |N\rangle\langle N|\ \rangle, \quad f = \sum_{NN'} f_{NN'}|N\rangle\langle N'|. \tag{7.196}$$

Using the commutation rule $[a, a^\dagger] = I$, it is not difficult to show (see [Klyshko (1980)]) that N are integers,

$$\mathcal{E}_N = (N + 1/2)\hbar\omega, \quad N = 0, 1, 2, \ldots \tag{7.197}$$

Thus, *the energy of a single mode can only take a set of discrete equidistant values differing by $\hbar\omega$, the energy of a photon.* A mode has the smallest possible energy if $N = 0$, which corresponds to the vacuum state $|0\rangle$, while the states with $N > 0$ are called *N-photon states*.

It is also easy to find out how the $a, a^\dagger$ operators act on $N-$ *states*,

$$a|N\rangle \equiv N^{1/2}|N - 1\rangle, \tag{7.198}$$

$$a^\dagger|N\rangle \equiv (N + 1)^{1/2}|N + 1\rangle. \tag{7.199}$$

These relations explain why $a, a^\dagger$ are called photon annihilation and creation operators; they also show that N-states are not eigenstates for the operators $q, p, a, a^\dagger$. Hence, *if the field is in some N-state (including the vacuum state), the measurement of the electric field will reveal quantum fluctuations.* This conclusion immediately follows from the fact that $\mathcal{H}_0$ does not commute with q, p.

According to (7.199), the N-state can be obtained by acting N times on the vacuum state by the operator $a^\dagger$,

$$|N\rangle = (N!)^{-1/2}(a^\dagger)^N|0\rangle. \tag{7.200}$$

However, in practice it is very difficult to bring some free-field or cavity mode into a pure N state (except the vacuum one), especially for $N \geqslant 2$. Usually, the actual state of a mode is an incoherent mixture of a few first N states, while the state of an ideal laser is a coherent mixture of many N-states.

All considerations given above related to a fixed time moment. The time dependence of the state vector of a single free-field mode is described in the Schrödinger representation by the equation

$$i\hbar d|t\rangle/dt = \mathcal{H}_0|t\rangle, \tag{7.201}$$

and if at time $t = 0$ the mode was in an N-photon state, then, according to (7.194),

$$|t\rangle_N = |N\rangle e^{-iN\omega t} \equiv |N, t\rangle, \tag{7.202}$$

where $\omega = ck$. *Hence, in an N-state, the mean values, moments, and distributions of all observables, including q, p, are stationary.* An arbitrary pure state of a mode depends on the time as

$$|t\rangle = \sum_N c_N|N, t\rangle = c_0|0\rangle + c_1 e^{-i\omega t}|1\rangle + \ldots, \tag{7.203}$$

where $c_N \equiv \langle N|t_0\rangle$. In the presence of currents, the coefficients in this expansion are time-dependent, although this dependence is usually slow compared to the one of $\exp(-i\omega t)$.

So far, we have been discussing the state of a single mode. In the case of independent modes, the energy wave function of the total field is obtained by simply multiplying the energy wavefunctions for the modes with definite photon numbers,

$$|\{N_k\}\rangle \equiv \Pi_k|N_k\rangle_k \equiv |N_1, N_2, \ldots\rangle, \tag{7.204}$$

$$|\{N_k\}, t\rangle = |\{N_k\}\rangle \exp(-i\mathcal{E}t/\hbar),\ \mathcal{E} \equiv \hbar \sum_k N_k\omega_k. \tag{7.205}$$

Thus, an energy state is fixed by giving the photon numbers $\{N_k\}$ in all modes (*occupation numbers*), while an arbitrary state can be represented as a superposition of states with all possible combinations $\{N\}$,

$$|t\rangle = \sum_{\{N_k\}} c(\{N_k\})|\{N_k\}, T\rangle. \tag{7.206}$$

In the presence of external currents, the state amplitudes $c(\{N_k\})$ become time-dependent, and can be only determined by applying the perturbation theory (compare with Sec. 2.1).

7.5.3 *Coherent states*

As it was shown by Glauber, another convenient basis is formed by the eigenstates of the non-Hermitian photon annihilation operator,

$$a|z\rangle = z|z\rangle. \tag{7.207}$$

The states $|z\rangle$ are called *coherent states*. Since $a \sim \omega q + ip$, one can expect that the spectrum of a is continuous and complex, i.e., $z = z' + iz''$ is an arbitrary complex number.

From the definition (7.207), it follows that the action of an arbitrary operator function $f(\hat{a})$ on the vector $|z\rangle$ is reduced to a simple multiplication of this vector by a usual (*c-number*) function $f(z)$,

$$f(\hat{a})|z\rangle = f(z)|z\rangle. \tag{7.208}$$

The equation conjugated to (7.207) has the form

$$\langle z|a^\dagger = z^*\langle z|, \tag{7.209}$$

i.e., ket-vectors $\langle z|$ are left-hand eigenvectors for the $a^\dagger$ operator.

Using (7.207), (7.208), (7.209), we immediately find the mean number of photons in a coherent state $|z\rangle$,

$$\langle N\rangle_z \equiv \langle z|N|z\rangle = |z|^2. \tag{7.210}$$

In the case of a coherent state, all factorial moments of the photon number are also calculated in a simple way (see (7.170)),

$$G_z^{(m)} \equiv \langle z| : N^m : |z\rangle = |z|^{2m} = \langle N\rangle_z^m. \tag{7.211}$$

This equality shows that the moments are *factorable*.

In (7.211), we have used the notation $: N^m :\equiv: a^\dagger a \dots a^\dagger a :\equiv a^{\dagger m}a^m$. Generally, the colons denote the operation of *normal ordering*, which means placing all a operators on the right of all $a^\dagger$ operators. This operation ignores the noncommutativity of the operators, i.e., between the colons the operators can be written in any order. Note that $: \cdots :$ is a nonlinear operation; for instance, the operator $: aa^\dagger :=: a^\dagger a + I := a^\dagger a$ is not equal to the operator $: a^\dagger a : +I = a^\dagger a + I$.

From (7.207), (7.209), (7.208), it follows that

$$\langle : f(a^\dagger, a) :\rangle_z = f(z^*, z). \tag{7.212}$$

For finding the mean value of an arbitrary operator $f(a^\dagger, a)$ in a coherent state, it is sufficient to represent this operator, using the equality $aa^\dagger = a^\dagger a + I$, as a sum of normally ordered operators, and then replace $a^\dagger$ by z^* and a by z. Often, normal ordering can be performed through the series expansion,

$$f(a^\dagger, a) = \sum_{mn} c_{mn} a^{\dagger m} a^n,$$
$$\langle f(a^\dagger, a)\rangle_z = \sum_{mn} c_{mn} z^{*m} z^n. \tag{7.213}$$

For instance,

$$\langle aa^\dagger\rangle_z = \langle a^\dagger a + I\rangle_z = |z|^2 + 1,$$
$$N^2 \equiv a^\dagger a a^\dagger a = a^\dagger(a^\dagger a + I)a =: N^2 : +N, \tag{7.214}$$
$$\langle N^2\rangle_z = |z|^4 + |z|^2.$$

Using the last equality, one can write the second moment and the variance $\langle \Delta N^2\rangle \equiv \langle N^2\rangle - \langle N\rangle^2$ in terms of the first moment,

$$\langle N^2\rangle_z = \langle N\rangle_z(\langle N\rangle_z + 1), \ \langle \Delta N^2\rangle_z = \langle N\rangle_z. \tag{7.215}$$

This relation is typical for a *Poissonian* random variable (see below).

In the general case, 'normal ordering' of an arbitrary operator is not a simple task (see examples in Ref. [Louisell (1964)]). Sometimes, the following operator identity is helpful [Klauder (1968)]:

$$\exp(\mu a^\dagger + \eta a) = C\exp(\mu a^\dagger)\exp(\eta a) = C^{-1}\exp(\eta a)\exp(\mu a^\dagger), \tag{7.216}$$

where $C \equiv \exp(\mu\eta/2)$. At $\eta = -\mu^*$, the operator in (7.216) is called the *displacement operator* and is denoted as $D(\mu)$,

$$D(\mu) \equiv \exp(\mu a^\dagger - \mu^* a). \tag{7.217}$$

From (7.200), (7.217), and the equality obtained below, (7.222), it follows that

$$D(z)|0\rangle = |z\rangle. \tag{7.218}$$

One can show [Bloembergen (1965)] that *a classical external current j_k converts a mode from a vacuum state, $|0\rangle$, into a coherent one, $|z\rangle$, i.e., its action can be described by the displacement operator*. The amplitude z coincides in this case with the classical amplitude found from (7.128).

Let us show that an oscillator in a coherent state will indeed have a Poissonian distribution of the energy, i.e., that in a coherent state photons behave, in a sense, like a chaotic flow of sand grains. For this, we will find the transformation matrix $\langle N|z\rangle$ relating the bases $|N\rangle$ and $|z\rangle$. Let us left-multiply (7.207) by $\langle N|$, which yields

$$\langle N|a|z\rangle = z\langle N|z\rangle. \tag{7.219}$$

Hence, taking into account (7.188) and (7.199), we obtain

$$(N+1)^{1/2}\langle N+1|z\rangle = z\langle N|z\rangle.$$

From this recurrent relation, we find

$$\langle N|z\rangle = (N!)^{-1/2} z^N \langle 0|z\rangle. \tag{7.220}$$

The remaining unknown factor $\langle 0|z\rangle$ can be assumed to be real; then it can be found from the normalization condition,

$$\langle z|z\rangle = \sum_N |\langle N|z\rangle|^2 = \langle 0|z\rangle^2 \exp|z|^2 = 1. \tag{7.221}$$

Hence, the expansion of a coherent state over Fock states follows,

$$|z\rangle = \sum_N |N\rangle\langle N|z\rangle, \tag{7.222}$$

$$\langle N|z\rangle = (N!)^{-1/2} z^N \exp(-|z|^2/2). \tag{7.223}$$

Recall that $\langle N|z\rangle$ is the N-representation of a coherent wave function, while $\langle z|N\rangle = \langle N|z\rangle^*$ is the z-representation of a number-state wave function. Therefore, the probability of measuring N photons per mode in a coherent state is determined by a Poissonian distribution with the parameter $\langle N\rangle = |z|^2$,

$$P(N|z) = \langle N\rangle^N \exp(-\langle N\rangle)/N! \tag{7.224}$$

With the help of (7.223), one can easily verify that different z-vectors are not orthogonal to each other (in contrast to N-vectors),

$$\langle z_1|z_2\rangle = \sum_N \langle z_1|N\rangle\langle N|z_2\rangle = \exp[-|z_1 - z_2|^2/2 + i\mathrm{Im}(z_1^* z_2)], \tag{7.225}$$

$$|\langle z_1|z_2\rangle|^2 = \exp(-|z_1 - z_2|^2). \tag{7.226}$$

This fact does not cancel the completeness property (as in the case of a usual oblique-coordinate reference system): any arbitrary vector or diadic tensor can be still expanded over the set of $|z\rangle$ vectors or $|z_1\rangle\langle z_2|$ diadic tensors, respectively. Indeed, it follows from (7.222) that

$$|z\rangle\langle z| = e^{-|z|^2} \sum_{MN} \frac{z^{*M} z^N}{(M!N!)^{1/2}} |M\rangle\langle N|. \tag{7.227}$$

Let us sum these diadic tensors over all $z \equiv \rho e^{i\varphi}$,

$$\int d^2z |z\rangle\langle z| = \sum_{MN} |M\rangle\langle N| (M!N!)^{-1/2} \int_0^\infty d\rho e^{-\rho^2} \rho^{M+N+1} \int_0^{2\pi} d\varphi e^{i(N-M)\varphi}, \tag{7.228}$$

where $d^2z = dz'dz'' = \rho\, d\rho\, d\varphi$. The integral in φ yields $2\pi\delta_{NM}$, while the integral in ρ is equal to $N!/2$. Hence, taking into account (7.195), we obtain the z-diadic expansion of the operator unity, i.e., the completeness condition, in the form

$$I = \pi^{-1} \int d^2z |z\rangle\langle z|. \tag{7.229}$$

Hence, an arbitrary vector can be represented as

$$|\ \rangle = \pi^{-1} \int d^2z |z\rangle\langle z|\ \rangle. \tag{7.230}$$

In particular,

$$|z_1\rangle = \pi^{-1} \int d^2z |z\rangle\langle z|z_1\rangle. \tag{7.231}$$

This relation, together with (7.225), shows that the basis vectors $|z\rangle$ can be expressed in terms of each other, i.e., the z basis is *over-complete*. Roughly speaking, it means that the number of coordinates exceeds the dimensionality of the space.

The time dependence of a coherent state can be easily found by substituting (7.202) into (7.222),

$$|z, t\rangle = \sum |N\rangle\langle N|z\rangle e^{-iN\omega t} = |ze^{-i\omega t}\rangle \equiv |z(t)\rangle. \tag{7.232}$$

Thus, *free evolution does not turn a coherent state into some other type of state, similarly to the case of an energy state (see (7.202))* and in contrast to the cases of q- and p-states (see below).

If all modes are in coherent states, then the state vector of the field will be

$$|\{z_k\}\rangle = |z_1\rangle_1 |z_2\rangle_2 \cdots \equiv |z_1, z_2, \ldots\rangle. \tag{7.233}$$

According to (7.131), this vector is the eigenvector for the positive-frequency field operator $\hat{\boldsymbol{E}}^{(+)}(\boldsymbol{r}, t)$ with the eigenvalue

$$\boldsymbol{E}^{(+)}(\boldsymbol{r}, t) \equiv i \sum_k c_k \boldsymbol{e}_k z_k \exp(i\boldsymbol{k} \cdot \boldsymbol{r} - i\omega_k t). \tag{7.234}$$

The mean value of the field in a coherent state is equal to the real part of this expression,

$$\langle\{z_k\}|\hat{\boldsymbol{E}}(\boldsymbol{r}, t)|\{z_k\}\rangle = 2\mathrm{Re}\boldsymbol{E}^{(+)}(\boldsymbol{r}, t). \tag{7.235}$$

Further, according to (7.212), all normally ordered field moments (correlation functions),

$$G^{(n)}_{1\ldots 2n} \equiv \langle \hat{E}^{(-)}_1 \ldots \hat{E}^{(-)}_n \hat{E}^{(+)}_{n+1} \ldots \hat{E}^{(+)}_{2n}\rangle, \tag{7.236}$$

which usually determine the readings of optical detectors, are *factorable* in the case of a coherent state, i.e., can be expressed in terms of the products of the first moments,

$$G^{(n)}_{1\ldots 2n}(\{z_k\}) = E^{(-)}_1 \ldots E^{(+)}_{2n}. \tag{7.237}$$

Here, the eigenvalues of the fields $E^{(\pm)}_i \equiv E_{\alpha_i}(\boldsymbol{r}_i, t_i)$ are determined by the set $\{z_k\}$ according to Eq. (7.234).

7.5.4 *Coordinate and momentum states*

By definition, the eigenvectors $|q\rangle$ of the coordinate operator $\hat{q}$ satisfy the condition

$$\hat{q}|q\rangle = q|q\rangle. \tag{7.238}$$

From $\hat{q}^\dagger = \hat{q}$, it follows that $q^* = q$ and $\langle q|\hat{q} = q\langle q|$. Similarly, one can define the eigenvectors $|p\rangle$ of the momentum operator, $\hat{p}|p\rangle = p|p\rangle$. The spectrum of q is continuous; therefore, the q-representation of an arbitrary state vector, the orthonormality condition and the completeness condition have the form

$$|\,\rangle = \int dq|q\rangle\langle q|\,\rangle, \tag{7.239}$$

$$\langle q|q'\rangle = \delta(q - q'), \tag{7.240}$$

$$\hat{I} = \int dq|q\rangle\langle q|. \tag{7.241}$$

According to (7.240), q-vectors have infinite norm, $\langle q|q\rangle = \infty$, which leads to certain difficulties. As an example, let us find, using the general rule (7.189), the probability density for the coordinate of a system in a state $|\,\rangle = |q_1\rangle$:

$$P(q|q_1) = C|\langle q|q_1\rangle|^2 = C[\delta(q - q_1)]^2, \tag{7.242}$$

$$C^{-1} = \int dq|\langle q|q_1\rangle|^2 = \int dq[\delta(q - q_1)]^2. \tag{7.243}$$

The squared delta-function has a meaning provided that one of its representations with a finite width Δq is used. In this case, $\delta(0) = 1/\Delta q$ (see (6.139)), and one can replace $\delta(q)^2$ by $\delta(q)/\Delta q$. The width Δq is chosen from the physical considerations: it should be much less than the interval on which the functions that are multiplied by the delta function before the integration vary considerably. In this example, however, Δq is canceled since $C = \Delta q$,

$$P(q|q_1) = \delta(q - q_1). \tag{7.244}$$

Sometimes, it is more convenient to use discrete q and p representations with the usual normalization, $\langle q_m|q_n\rangle = \langle p_m|p_n\rangle = \delta_{mn}$. For passing to discrete spectra $\{q_n\}$ and $\{p_n\}$, the wave functions $\psi(q) \equiv \langle q|\ \rangle$ or their Fourier transforms $\psi(p) \equiv \langle p|\ \rangle$ should be considered either periodic or differing from zero only within finite intervals, L and $\hbar K$, respectively (compare with the discretization procedure for wavenumbers in Sec. 7.3). This restriction is equivalent to the assumption that $\psi(q), \psi(p)$ vary very little on the intervals $\Delta q \equiv 1/K, \Delta p \equiv \hbar/L$, and is always valid for sufficiently large L and K.

Further, let us find the transformation function $\langle q|p\rangle$, i.e., the q-representation of a p-state, by assuming

$$\langle q|\hat{p} = -i\hbar\frac{d}{dq}\langle q|. \tag{7.245}$$

In this relation, q is a continuous parameter of the vector $\langle q|$, and the differentiation operator acts on this parameter. Multiplying (7.245) by $|p\rangle$, we obtain the equation

$$-i\hbar\frac{\partial}{\partial q}\langle q|p\rangle = p\langle q|p\rangle, \tag{7.246}$$

whose solution, evidently, is

$$\langle q|p\rangle = (2\pi\hbar)^{-1/2}e^{ipq/\hbar}. \tag{7.247}$$

The normalization constant here was found by substituting into (7.240) the diadic expansion of a unity,

$$\int dp\langle q|p\rangle\langle p|q'\rangle = \delta(q-q').$$

According to (7.247), *in a q-state, all momentums are equally probable, while in a p-state, all coordinates are equally probable,*

$$\begin{aligned} P(p|q) &\sim |\langle p|q\rangle|^2 = 1/2\pi\hbar, \\ P(q|p) &\sim |\langle q|p\rangle|^2 = 1/2\pi\hbar. \end{aligned} \tag{7.248}$$

Similarly, one can find a function of two variables $\langle q|z\rangle \equiv \psi_z(q)$, whose square determines the probability density distribution $P(q|z)$ for the coordinate in a z state. (Note that the symbol $P(z|q)$ is meaningless since the non-Hermitian operator a does not correspond to any physical observable.) From (7.121) and (7.245), we obtain that

$$\langle q|a = (\tilde{q} + d/d\tilde{q})\langle q|/\sqrt{2}, \tag{7.249}$$

where $\tilde{q} \equiv (m\omega/\hbar)^{1/2}q$ and we have introduced the mass m of an equivalent oscillator. Multiplication by $|z\rangle$ with an account for (7.207), (7.209) yields

$$(\partial/\partial\tilde{q} + \tilde{q} - \sqrt{2}z)\langle q|z\rangle = 0. \tag{7.250}$$

This equation is satisfied by the function

$$\langle q|z\rangle = C_1(z)\exp[-(\tilde{q} - \sqrt{2}z)^2/2]. \tag{7.251}$$

Similarly,

$$\langle p|z\rangle = C_2(z)\exp[-(\tilde{p} + i\sqrt{2}z)^2/2], \tag{7.252}$$

where $\tilde{p} \equiv p/(\hbar\omega m)^{1/2}$. The normalization constants are defined here only up to the phase factors,

$$\begin{aligned} C_1 &= (m\omega/\pi\hbar)^{1/4}\exp[-z''^2 + i\varphi_1(z)], \\ C_2 &= (\pi\hbar\omega m)^{-1/4}\exp[-z'^2 + i\varphi_2(z)]. \end{aligned} \tag{7.253}$$

According to the definition (7.121),

$$\tilde{q} = (a + a^\dagger)/\sqrt{2},\ \tilde{p} = (a - a^\dagger)/i\sqrt{2}, \tag{7.254}$$

hence it follows from (7.207), (7.209) that

$$\langle\tilde{q}\rangle_z \equiv \langle z|\tilde{q}|z\rangle = \sqrt{2}z',\ \langle\tilde{p}\rangle_z = \sqrt{2}z''. \tag{7.255}$$

As a result, the distributions following from (7.251), (7.252) can be represented in the form

$$\begin{aligned} P(\tilde{q}|z) &= \pi^{-1/2}\exp[-(\tilde{q} - \langle\tilde{q}\rangle)^2], \\ P(\tilde{p}|z) &= \pi^{-1/2}\exp[-(\tilde{p} - \langle\tilde{p}\rangle)^2]. \end{aligned} \tag{7.256}$$

Thus, *for an oscillator in a coherent state, the coordinate and amplitude have Gaussian distributions with the variances*

$$\langle\Delta q^2\rangle = \hbar/2m\omega,\ \langle\Delta p^2\rangle = \hbar\omega m/2,\ \langle\Delta\tilde{q}^2\rangle = \langle\Delta\tilde{p}^2\rangle = 1/2 \tag{7.257}$$

and the minimal possible product of uncertainties,[r]

$$\Delta q\Delta p = \hbar/2. \tag{7.258}$$

The distributions of the coordinate and momentum at $z \neq 0$ differ from the vacuum one only by the displacement of the center of reference by $\sqrt{2}z'$ and $\sqrt{2}z''$; their variances are not increased, in contrast to the energy variance (see (7.215)). The relative widths of the distributions, $\Delta q/\langle q\rangle$ and $\Delta p/\langle p\rangle$, are inversely proportional to z' and z''.

It follows from (7.232) that the evolution of an oscillator in a coherent state is described by changing z to $z_0e^{-i\omega t}$. Let $z_0 = z_0^* \equiv q_0/\sqrt{2}$, then, according to (7.255), one should set in (7.256)

$$\langle\tilde{q}\rangle = q_0\cos\omega t,\ \langle\tilde{p}\rangle = -q_0\sin\omega t. \tag{7.259}$$

[r]Here, as usual, Δx denotes two different values, the operator $x - \langle x\rangle$ and the number $\langle(x - \langle x\rangle)^2\rangle^{1/2}$.

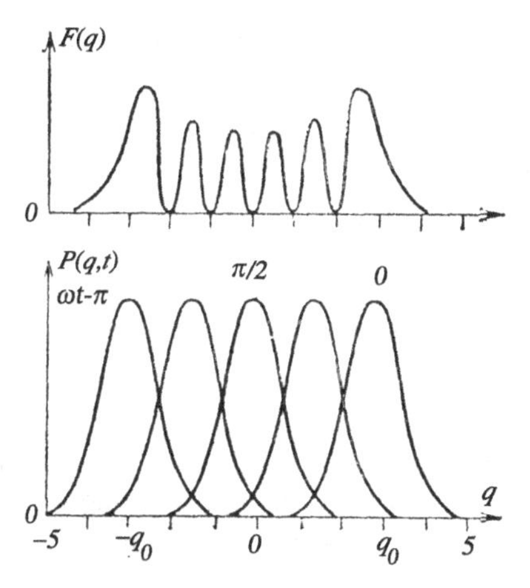

Fig. 7.18 Distribution of the coordinate for an oscillator in a Fock state (top) and a coherent state (bottom) with the same mean photon numbers, equal to 5. The coordinate q is in $\sqrt{\hbar/m\omega}$ units.

As a result, the distributions shift without changing their shapes (Fig. 7.18),

$$P(\tilde{q}|z,t) = \pi^{-1/2}\exp[-(\tilde{q} - q_0\cos\omega t)^2],$$
$$P(\tilde{p}|z,t) = \pi^{-1/2}\exp[-(\tilde{p} + q_0\sin\omega t)^2]. \tag{7.260}$$

Thus, the mean coordinate and momentum in the case of a coherent state depend on time the same way as the corresponding values of a classical oscillator. As $|z_0|$ increases, the relative fluctuations are reduced and a quantum oscillator becomes more and more similar to a classical one.

Note that the wave function (7.251) with the substitution $z_t \equiv z(t) = z_0e^{i\omega t}$ should satisfy the Schrödinger equation in the q representation,

$$\left(\frac{2i}{\omega}\frac{\partial}{\partial t} + \frac{\partial^2}{\partial q^2} - q^2\right)\psi_z(q,t) = 0,$$
$$\psi_z(q,t) \equiv \langle q|z_t\rangle. \tag{7.261}$$

(Here and below, we use dimensionless variables $q \equiv \tilde{q}$, $p \equiv \tilde{p}$.) This condition allows one to find the phase in (7.253). As a result, the 'coherent' wave function can be represented as

$$\psi_z(q,t) = \pi^{-1/4}\exp[-(q - \sqrt{2}z_t)^2/2 - z_t''^2 + i(z_t'z_t'' - \omega t/2)]$$
$$= \pi^{-1/4}\exp\{-(q - \langle q_t\rangle)^2/2 + i[(q - \langle q_t\rangle/2)\langle p_t\rangle - \omega t/2]\}. \tag{7.262}$$

Above, we have found the matrices of transitions from the q-representation to p- and z-representations. Similarly, one can find the functions $\langle N|q\rangle \equiv \psi_N(q)$

determining the coordinate distribution in Fock states and the photon-number distribution in q-states. These functions satisfy Eq. 7.261 with $\partial\psi/\partial t$ replaced by $-iN\omega\psi$; they are equal to the Hermite polynomials multiplied by the vacuum function $\langle 0|q\rangle = \exp(-q^2/2)$. They are obtained if one multiplies (7.200) by $\langle q|$ and replaces $\langle q|a^\dagger$ with $2^{-1/2}(q - d/dq)\langle q|$,

$$\langle q|N\rangle = (2^N N!\pi^{1/2})^{-1/2}(q - d/dq)^N \exp(-q^2/2). \tag{7.263}$$

7.5.5 *Squeezed states*

One should keep in mind that in the case of all states except the energy (Fock) ones, the distributions and moments are time-dependent (see (7.259), (7.260) for a coherent state). For instance, one can show that q-states periodically become p-states and vice versa (Fig. 7.20).

Consider the evolution of variances for the coordinate and momentum in the case of an arbitrary initial state. Solutions to the Heisenberg equations for the coordinate and momentum operators have the 'classical' form,

$$\hat{q}(t) = \hat{q}\cos\tau + \hat{p}\sin\tau, \;\; \hat{p}(t) = \hat{p}\cos\tau - \hat{q}\sin\tau, \tag{7.264}$$

where $\tau \equiv \omega t, \;\; \hat{q} \equiv \hat{q}(0), \;\; \hat{p} \equiv \hat{p}(0)$.

In the general case, from (7.264) we can find the time dependence of the coordinate and momentum variances,

$$D_q(\tau) = D_p(\tau - \pi/2) = D_q\cos^2\tau + D_p\sin^2\tau + D_{qp}\sin 2\tau. \tag{7.265}$$

Here, the following notation was introduced:

$$\begin{aligned} D_x(t) &\equiv \langle[\Delta x(t)]^2\rangle, \;\; \Delta x(t) \equiv x(t) - \langle x(t)\rangle, \;\; D_x \equiv D_x(0), \\ D_{qp}(t) &\equiv \langle\Delta q(t)\Delta p(t) + \Delta p(t)\Delta q(t)\rangle/2 \\ &= \langle a(t)^2 - a^\dagger(t)^2\rangle/2i - \langle q(t)\rangle\langle p(t)\rangle; \end{aligned}$$

the averaging runs over the initial state of the oscillator $|t_0\rangle$. Thus, *the variances of the coordinate and momentum oscillate anti-phased with the frequency* 2ω, *and their sum is an integral of motion*,

$$D_q(t) + D_p(\tau) = D_q + D_p = 2(\langle N\rangle - \langle a^\dagger\rangle\langle a\rangle) + 1 = 2D_{aa^\dagger}. \tag{7.266}$$

According to (7.265), the variances are constant only under the condition that $D_q = D_p$ and $D_{qp} = 0$. Using (7.254), one can verify that this is the case for Fock states ($D_q = D_p = N + 1/2$) and coherent states ($D_q = D_p = 1/2$).

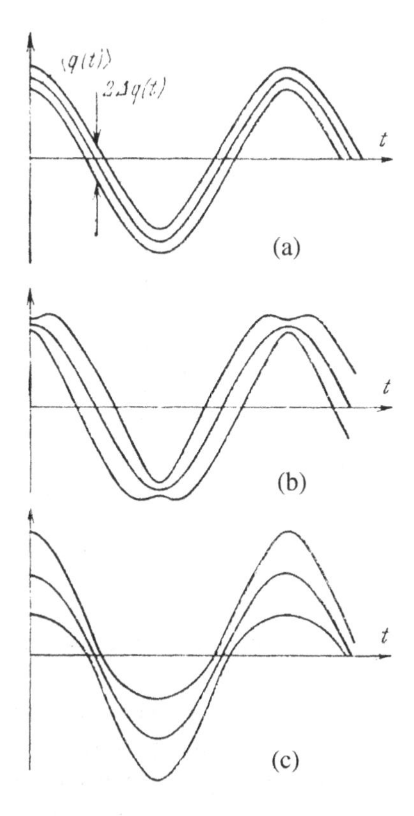

Fig. 7.19 Coherent (a) and squeezed (b,c) states of an oscillator. The figure shows the time dependencies of the mean coordinate and the coordinate uncertainty. The dependencies are calculated using Eq. (7.265) with $q_0 = 5$, $D_{qp} = 0$, $\Delta q \Delta p = 1/2$ and $\Delta q = 1/\sqrt{2}$ (a), 0.2 (b), 2.5 (c).

Recently, it was proposed to generate and measure the so-called *squeezed states* [Walls (1983)], for which $D_q \ll 1/2$ (or $D_p \ll 1/2$) and $D_q D_p = 1/4$ (Fig. 7.19).[s] For a mechanical oscillator or electromagnetic field in such a state, repeated stroboscopic measurements with an appropriate phase will reveal fluctuations reduced with respect to the zero-point vacuum ones, $\sqrt{\hbar/2m\omega}$. Thus, in principle, *zero-point vacuum fluctuations do not restrict the limiting accuracy of the coordinate or momentum measurement.* Squeezed states can be of interest for the information transmission and for the measurement of tiny forces caused, for instance, by gravitational waves [Braginsky (1980)]. Note that up to now, attention

[s]Editors' note: About the same time as the book was published, squeezed states were produced in experiment, first through four-wave mixing and then via parametric amplification, see [Bachor (2004); Walls (1994)]. Amplitude squeezing of photocurrent was also observed using a negative feedback loop [Yamamoto (1999)].

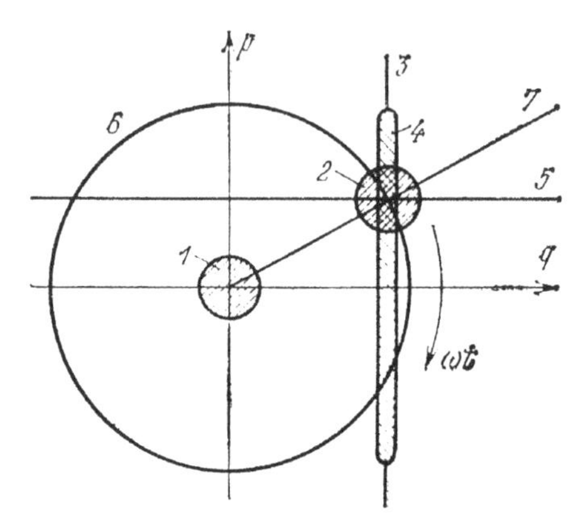

Fig. 7.20 Various states of a quantum oscillator shown on the phase plane. The horizontal and vertical sizes of the figures correspond to the uncertainties in the coordinate and momentum, respectively; the sizes along the radius and the azimuth correspond, respectively, to the amplitude and phase uncertainties. 1, the vacuum state; 2, a coherent state; 3, a coordinate state; 4, a squeezed state; 5, a momentum state; 6, a photon-number state; 7, a phase state.

is attracted to certain questions in the quantum measurement theory [Braginsky (1975, 1980)].[t]

We have considered four types of states generated by the operators p^2+q^2, $p+iq$, p, q, as well as the relations between these states. Similarly, one can construct a lot of other states. Worth mentioning are the eigenstates of the phase operator [Fain (1972); Loudon (2000)], which are most close to a classical oscillation with a fixed phase.

It is convenient to show various states on a phase diagram $(\tilde{q}, \tilde{p})$ as figures of different shapes whose linear dimensions are equal to the uncertainties $\Delta\tilde{q}, \Delta\tilde{p}$ in these states (Fig. 7.20). The area of each figure cannot be less than about a unity. A classical state of an oscillator is shown by a point $(\tilde{q}_1, \tilde{p}_1)$, a coherent state, by a circle of a unity diameter and the center at point $(\tilde{q}_1, \tilde{p}_1)$; an N-photon state, by a thin circle with the diameter $(N+1/2)^{1/2}$ centered at the origin; a q-state, by a thin

[t]Editors' note: nowadays, squeezed states still attract a great attention in connection with different applications. We refer to the following ones:

- optical communication and optical measurements, due to the fact that weaker signals can be transmitted with the same signal-to-noise ratio and the same light power;
- gravitational-wave detection;
- precise measurement of spatial displacements with multimode squeezed states; super-resolution;
- quantum imaging with surpassing the quantum noise limit;
- one-way quantum computing with cluster states based on squeezed states;
- quantum memory based on squeezed-light interaction with atomic ensembles and single atoms;
- research in fundamental quantum physics.

vertical straight line; a p-state, by a thin horizontal straight line.[u] However, one should keep in mind that these figures have only qualitative meaning; strictly, they do not correspond to any joint distributions $P(q, p)$, which do not exist in quantum mechanics.

Similar to the states themselves, the figures depicting them on the phase plane change in time due to the natural evolution (described by the Schrödinger equation) or due to the *reduction* as a result of the measurement back-action. For instance, after an accurate measurement of q, a coherent circle in Fig. 7.20 will turn into a vertical line. The evolution due to free oscillations is described by the counter-clockwise rotation of the figure around the origin, with the angular rate ω, or, alternatively, the clockwise rotation of the frame of reference $\tilde{q}, \tilde{p}$.

7.5.6 *Mixed states*

If the field is interacting (or has interacted) with another quantum object, for instance, with an atom, then, by definition, there are no separate wave functions for the field, $\psi(x_E, t)$, and the atom, $\psi(x_A, t)$; one can only speak of the joint wave function, $\psi(x_E, x_A, t)$. Similarly, one cannot speak of the state vector $|\ \rangle_k$ of a given mode k if it is coupled with another mode k', or several modes. For instance, a classical point-like source with the frequency ω excites a spherical wave, in which case all plane waves with $|k| = \omega/c$ are coupled. Similar coupling of modes with the same frequency ('transverse' modes) occurs due to the diffraction. Modes with different frequencies ('longitudinal' modes) can be coupled due to the matter anharmonicity [Klyshko (1980)].[v]

In all cases where the system is described by a non-complete set of variables, one says that it is in a *mixed* state (Sec. 3.1). Then, instead of a state vector $|\ \rangle$, the system is characterized by a certain *operator* ρ, called the *density operator*. In the special case of a pure state, the density operator ρ is a projector, $\rho_{pure} = |\ \rangle\langle\ |$, while for a mixed state, ρ is given by a sum of projectors (see (7.186)).

The density operator, like any other field operator, can be written in various bases (representations), in terms of the density matrices of the form $\rho_{NN'} \equiv$

[u]Editors' note: Probably one should add squeezed vacuum, a state generated at the output of an unseeded optical parametric amplifier. In the diagram, it will be shown by an 'ellipse' at the origin, with the area being the same as for a coherent state [Bachor (2004); Walls (1994)].

[v]Editors' note: Note that the description of a composite system consisting of two or several interacting sub-systems (or sub-systems having interacted in the past) directly relates to the concept of entangled states ([Peres (1993); Nielsen (2000); Bouwmeester (2000)], Sec. 7.5.7). This family of states was not considered in the original book at all, partly due to the fact that in 1986, entangled states were not as popular as nowadays. At the same time the author put a lot of efforts to avoid using vague terms while describing physical phenomena ([Klyshko (1994)]).

$\langle N|\rho|N'\rangle$, $\rho_{qq'}$, $\rho_{zz'}$ and so on. In most cases, the density operator of the field is written in the N-basis, but in some cases the z-basis is more convenient. For a single mode, ρ can be written in terms of N- and z-projectors as (the mode index k is omitted)

$$\rho = \sum_{NN'} |N\rangle\langle N'| \tag{7.267}$$

$$= \int d^2z P(z)|z\rangle\langle z|. \tag{7.268}$$

Note that here, the *diagonal* z-representation is used, which is possible in many cases and is provided by the over-completeness of the z-basis. Equation (7.268) is called the *Glauber-Sudarshan representation* or the P-representation. The normalization and Hermiticity conditions of ρ are

$$\sum_N \rho_{NN} = 1, \quad \rho_{NN'} = \rho^*_{N'N},$$

$$\int d^2z\, P(z) = 1, \quad P(z) = P^*(z).$$

The mean value of any field operator f can be expressed in terms of ρ according to the relation $\langle f\rangle = \mathrm{Tr}(\rho f)$ (Sec. 3.2), or according to (7.267), (7.268),

$$\langle f\rangle = \sum_{NN'} \rho_{NN'} f_{N'N} = \int d^2z P(z)\langle z|f|z\rangle. \tag{7.269}$$

Hence, with the help of (7.212), we find

$$\langle : f(a^\dagger, a) :\rangle = \int d^2z P(z) f(z^*, z). \tag{7.270}$$

Thus, the $P(z)$ function provides an easy way to calculate the mean values of normally ordered operators. In particular, the normally ordered moments can be found as

$$G^{(n)} = \int d^2z P(z)|z|^{2n}. \tag{7.271}$$

Equations (7.269), (7.270) show that the weighting function $P(z)$ plays the role of the probability for the oscillator to have a complex amplitude z, i.e., $\tilde{q} = \sqrt{2}z'$, $\tilde{p} = \sqrt{2}z''$. However, $P(z)$ can take negative values; besides, even with $P(z) = \delta^{(2)}(z - z_1)$, i.e., in the case of a pure coherent state, $\tilde{q}$ and $\tilde{p}$ have zero-point fluctuations.[w] Therefore, $P(z)$ is called a *quasi-probability*.

[w] Editors' note: also called shot-noise fluctuations.

Moreover, the quasi-probability of a coherent state $P(z)$ allows one to find the probability distribution $P(f)$ of an arbitrary observable f. For this, one should replace f in (7.269) by the projector $P(f) = |f\rangle\langle f|$,

$$P(f) = \int d^2zP(z)|\langle z|f\rangle|^2 = \int d^2zP(f|z)P(z). \quad (7.272)$$

The two-dimensional Fourier-transform of the quasi-probability $P(z)$ is called the normally ordered *characteristic function*,

$$\chi(\mu, \mu^*) \equiv \langle e^{\mu a^\dagger} e^{-\mu^* a}\rangle = \int d^2zP(z)e^{\mu z^* - \mu^* z}. \quad (7.273)$$

This definition yields a usual (not generalized) function for all states, in contrast to $P(z)$.[x] From the definition of χ, it follows that $G^{(n)}$ can be calculated by means of differentiation, instead of more complicated integration,

$$G^{(n)} = \left(-\frac{\partial}{\partial\mu}\frac{\partial}{\partial\mu^*}\right)^n \chi(\mu, \mu^*) \,|_{\mu=0} \,. \quad (7.274)$$

Thus, *a mixed state of a mode can be described by the* $\rho_{NN'}$ *matrix, or one of the functions* $P(z)$, $\chi(\mu, \mu^*)$.

A mixed state of a multi-mode field is given by the density matrix

$$\langle N_1, N_2, \ldots |\rho| N_1', N_2', \ldots\rangle = \langle\{N_k\}|\rho|\{N_k'\}\rangle,$$

or by the quasi-probability $P(\{z_k\})$, or by its Fourier transform, $\chi(\{\mu_k, \mu_k^*\})$. In the case of independent modes, these values are factorable, 'reducible'. It should be stressed that with the help of the quasi-probability function, the operation of quantum averaging of normally ordered operators (which are usually of interest) takes the 'classical' form (7.270), which is also maintained in the case of a multi-mode field. For instance, the correlation functions (7.236) are found by averaging their mean values in a coherent state (7.237) with the quasi-probability,

$$G^{(n)}_{1\ldots 2n} = \int \ldots \int P(\{z_k\})G^{(n)}_{1\ldots 2n}(\{z_k\}) \prod_k d^2z_k. \quad (7.275)$$

Recall that here, $z_k \equiv z(\boldsymbol{k}, \nu)$ has the meaning of the amplitude (in $2c_k$ units) of the plane wave E_{k0} propagating in the $\boldsymbol{k}$ direction and having polarization $\boldsymbol{e}_\nu$.

Further, let us consider some examples of mixed states of the field.

In the case of a *stationary* field, the density operator does not depend on time, i.e., $[\rho, \mathcal{H}] = 0$, which in the N-representation yields

$$\langle\{N_k\}|\rho|\{N_k'\}\rangle \sum_k (N_k' - N_k)\omega_k = 0.$$

[x]For instance, $P(z)$ for N states contains the $2N$th-order derivative of the delta function [Perina (1972)].

Hence, *a stationary density matrix is diagonal in the occupation numbers of modes with different frequencies.* The function $P(\{z_k\})$ depends then only on the absolute values $|z_k|$, since $z_k(t) \sim e^{-i\omega t}$.

In a stationary state, all simultaneous moments, $\langle f^m(t)\rangle$, and correlation functions, $\langle f(t)g(t+\tau)\ldots\rangle$, are independent of t. Usually, one assumes that the system is *ergodic*, i.e., that the ensemble means $\langle f\rangle$ coincide with the values measured in experiment, $f_{exp}(t)$, and averaged over time, and that the observed variations (fluctuations) of $f_{exp}(t)$ in time are caused by the uncertainty of $\hat{f}$ in a pure quantum or mixed ensemble. An important role is also played by the *periodically non-stationary states*, which include, in particular, coherent states.

Often, one can use the approximation of statistically independent modes, $\rho = \prod_k \rho_k$. Then, the diagonal element $\langle N_k|\rho_k|N_k\rangle$ has the meaning of the population (occupation number) of the kth mode N-photon state. In the case of a stationary field with independent modes, populations fully characterize the properties of the field. In particular, the mean photon number per mode, which determines the main photometry parameter, the spectral brightness, is

$$\langle N_k\rangle = \mathrm{Tr}(a_k^\dagger a_k \rho_k) = \sum_{N=0}^{\infty} N\langle N|\rho_k|N\rangle. \tag{7.276}$$

In an *equilibrium state*, the density operator is determined by the Gibbs distribution, $\rho_k \sim \exp(-\mathcal{H}_k/\kappa T)$, where T is the thermostat temperature (Sec. 3.2). In an equilibrium state, the population of an N-photon level, i.e., the probability to find simultaneously N photons in one mode, or the probability for the mode energy to take the value $\hbar\omega(N + 1/2)$, depends on N exponentially,

$$P_T(N) = Ce^{-Nx}, \tag{7.277}$$

where

$$C = P(0) = 1 - e^{-x}, \;\; x \equiv \hbar\omega/\kappa T.$$

Equation (7.277) is called the *Planck's*, or *geometric, distribution*, as it forms a geometric sequence, or the *Bose-Einstein distribution*. By substituting (7.277) into (7.276), we find the mean number of photons per mode, also called the degeneracy factor of the photon 'gas',

$$\langle N\rangle_T = (e^x - 1)^{-1} \equiv \mathcal{N} \equiv \delta. \tag{7.278}$$

This equality enables δ to be used instead of x as the parameter of the distribution; then (7.277) takes the form

$$P_T(N) = P(0)/(1 + 1/\delta)^N, \;\; P(0) = 1/(1 + \delta). \tag{7.279}$$

In a non-equilibrium field, the modes can also have exponential energy distribution, provided that they are chaotically excited by many independent sources. This is the case for thermal radiation, fluorescence, or superfluorescence (in the linear regime, see Sec. 7.1). Then, the dependence of $\langle N_k \rangle$ on $|\boldsymbol{k}| = \omega/c$ determines the frequency spectrum of the radiation, while the dependence on the direction, $\boldsymbol{k}/k$, determines the angular spectrum and the direction of incoherent radiation. Recall that in non-laser light, usually all $\langle N_k \rangle \ll 1$. For instance, for the green part of the sunlight spectrum, $\langle N_k \rangle \sim 10^{-2}$, hence the probabilities to find 0, 1, and 2 photons in one mode are approximately equal to 0.99, 10^{-2}, and 10^{-4}.

One can show [Glauber (1965)] that the quasi-probability of a chaotically excited mode is a two-dimensional Gaussian function with the variance $\langle N \rangle/2$,

$$P_T(z) = \exp(-|z|^2/\langle N \rangle)/\pi\langle N \rangle. \tag{7.280}$$

Hence, taking into account (7.273),

$$\chi_T(\mu, \mu^*) = \exp(-\mu\mu^*\langle N \rangle) \tag{7.281}$$

and, according to (7.274), only even symmetric moments are nonzero,

$$G_T^{(m)} \equiv \langle : N^m : \rangle_T = m!\langle N \rangle^m. \tag{7.282}$$

It follows from (7.282) and (7.211) that *an m-quantum transition is m! times as probable in a thermal field than in a coherent field with the same* $\langle N \rangle$ (see (6.212)), which is due to the long 'tail' of the thermal distribution. At $m = 2$, (7.282) describes *photon bunching* (Secs. 7.2, 7.6).

By substituting (7.280) into (7.272), one can see that the distribution of the coordinate and momentum for a thermal state are Gaussian as well, with zero mean values and the variances determined by the relation $\langle N \rangle + 1/2 = \langle \tilde{p}^2 + \tilde{q}^2 \rangle/2$, i.e.,

$$\langle \Delta\tilde{q}^2 \rangle_T = \langle \Delta\tilde{p}^2 \rangle_T = \langle N \rangle_T + 1/2 = (1/2)\coth(x/2). \tag{7.283}$$

Note that additive multi-mode parameters of the field, such as, for instance, the electric field amplitude E_α at point $(\boldsymbol{r}, t)$, will have Gaussian distribution regardless of the states of separate modes (provided that they are independent), by virtue of the central limit theorem.

As we have already mentioned, a pure coherent state does not belong to the class of stationary states, since $z(t) = \rho\exp(-i\omega t + i\varphi)$. However, one can construct a 'stationary coherent state' by forming a mixture of coherent states with the same amplitudes ρ and random phases φ. Such a state, apparently, is described by the quasi-probability of the form [Glauber (1965)]

$$P(z) = \delta(|z| - \rho)/2\pi\rho = \delta(|z|^2 - \rho^2)/\pi, \tag{7.284}$$

which describes an ensemble of ideal lasers with uncertain phases. It is easy to check, with the help of (7.272), (7.224), that the energy distribution for such an ensemble will be still Poissonian, with $\langle N\rangle = \rho^2$.

7.5.7 *Entangled states (added by the Editors)*

There is an important class of quantum states, named *entangled states*, which at the time when the book was published were only starting to be discussed, mainly in connection with the experiments on Bell tests. Now, these states form the base for such important branches of modern science as quantum information and quantum communications. The details can be found in many beautiful books and review articles [Peres (1993); Scully (1997); Nielsen (2000); Bouwmeester (2000); Bachor (2004); Bruss (2002)]. Below, we present just a brief review of their properties.

For the first time the entangled state was introduced by Schrödinger in 1935, who mentioned that '*the best possible knowledge of the whole does not include the best possible knowledge of its parts*' [Schrödinger (1935)]. One can interpret this statement as follows: it is impossible to describe the subsystems $A, B, \ldots,$ forming a composite system, in terms of wave functions while the whole state does possess a wave function. *A pure state of a bipartite system is called separable if and only if it can be written as*

$$|\Psi_\Sigma\rangle = |\Psi_A\rangle \otimes |\Psi_B\rangle; \tag{7.285}$$

otherwise it is entangled.

In the simplest case of two qubits, $|\Psi_1\rangle = \alpha_1|0_1\rangle + \beta_1|1_1\rangle$ and $|\Psi_2\rangle = \alpha_2|0_2\rangle + \beta_2|1_2\rangle$, their joint state belongs to the $2\times 2 = 4$-dimensional Hilbert space and, in the general case, is not separable,

$$|\Psi_{12}\rangle = c_1|0_1\rangle|0_2\rangle + c_2|0_1\rangle|1_2\rangle + c_3|1_1\rangle|0_2\rangle + c_4|1_1\rangle|1_2\rangle,\ \sum_{j=1}^{4}|c_j|^2 = 1; \tag{7.286}$$

apparently, $|\Psi_{12}\rangle \neq |\Psi_1\rangle \otimes |\Psi_2\rangle$.

However, definition (7.285) does not tell us whether a given state is more entangled or less entangled. For *quantifying entanglement*, there is a simple parameter introduced by Wootters, the *concurrence* C, which indicates how much entanglement is stored in a composite state of two qubits:

$$0 \leq C \equiv 2|c_1c_4 - c_2c_3| \leq 1. \tag{7.287}$$

If the state is separable, then $C = 0$. An example of maximally entangled two-qubit states ($C = 1$) are the *Bell states*, which form a complete basis for two-qubit systems:

$$\Psi_{12}^{(\pm)} \equiv \frac{1}{\sqrt{2}}[|0_1\rangle|1_2\rangle \pm |1_1\rangle|0_2\rangle], \tag{7.288}$$

$$\Phi_{12}^{(\pm)} \equiv \frac{1}{\sqrt{2}}[|0_1\rangle|0_2\rangle \pm |1_1\rangle|1_2\rangle]. \tag{7.289}$$

The state $\Psi_{12}^{(-)}$ is called *the singlet state* and plays a special role due to its remarkable symmetry properties. The other three states form the *triplet*. For instance the singlet state is invariant with respect to choice of basis. The Bell states are an important tool in modern quantum optics but probably they are most famous for violating the Bell inequality [Klyshko (1998); Grynberg (2010)]

Another way to quantify entanglement is through the so-called *Schmidt decomposition*, which exists for any bipartite composite system:

$$\Psi_{AB} = \sum_{j=1}^{D} \sqrt{\lambda_j}|u_j\rangle|v_j\rangle, \sum_{j=1}^{D} = 1, \tag{7.290}$$

where $|u_j\rangle$ and $|v_j\rangle$ are bases for subsystems A and B, both having dimensionality D. The degree of entanglement can be estimated with the Schmidt number,

$$1 \leq K \equiv \left(\sum_{j=1}^{D} \lambda_j^2\right)^{-1} \leq D, \tag{7.291}$$

which can be interpreted as the number of nonzero Schmidt modes in the expansion (7.290). It is easy to calculate the Schmidt number for the Bell states (7.288), (7.289): $K = 2$. A state is separable if and only if $K = 1$. The Schmidt decomposition is a very useful approach from the physical viewpoint as it allows one to interpret the natural eigenmodes $|u_j\rangle$ and $|v_j\rangle$ of the system under study (called the *Schmidt modes*) in terms of entanglement [Mandel (2004)].

Another closely related measure of entanglement is the von Neumann entropy,

$$S \equiv -\sum_{j=1}^{D} \lambda_j \log_2 \lambda_j = S(\rho_A) = S(\rho_B), \tag{7.292}$$

where $S(\rho_{A,B}) \equiv \mathrm{Tr}_{B,A}\rho_{AB}$ is the reduced density matrix for subsystem A (B). This definition returns us to the initial meaning of entanglement introduced by Schrödinger. Indeed, the notion of the entropy relates to the uncertainty in the (sub)system. For a pure state the von Neumann entropy equals zero, while for completely mixed states it takes the maximal value.

The definition (7.285) can be also formulated for mixed states. Namely, *a mixed state is called separable if and only if it can be written as*

$$\rho_\Sigma = \sum_j p_j |\Psi_A^j\rangle\langle\Psi_A^j\rangle \otimes |\Psi_B^j\rangle\langle\Psi_B^j\rangle, \tag{7.293}$$

where $\sum_j p_j = 1,\ 0 \leq p_j \leq 1$. *Otherwise, the state* ρ_Σ *is entangled.*

As an example let us consider the so-called *Werner states*,

$$\rho_W \equiv x|\Psi^-\rangle\langle\Psi^-| + \{1-x\}\left(|\Psi^-\rangle\langle\Psi^-| + |\Psi^+\rangle\langle\Psi^+| + |\Phi^+\rangle\langle\Phi^+| + |\Phi^+\rangle\langle\Phi^+|\right). \tag{7.294}$$

It turns out that the state (7.294) is separable for $x < 1/3$.

Let us mention that sometimes one falsely associates entanglement with the violation of the Bell inequalities. In this connection, the Werner states (7.294) represent an important example: at $1/3 < x < 1/\sqrt{2}$ they are not separable (and hence are entangled) but do not violate the Bell inequalities.

It is worth noting that the concept of entangled states was actively developed in quantum information science and formally relates to systems of quantum bits (*qubits*) or quantum dits (*qudits*) which are rather abstract notions. Nowadays there are several physical systems that play the role of qubits, such as polarization states of single photons, two-level atoms or ions in traps, etc. Real physical qubits need to be carefully specified to avoid a contradiction with the formal description. For instance, the well-known permutation property of identical particles leads to entanglement since their wave function should be symmetrized, hence the singlet state of two spin-1/2 particles takes the form

$$\Psi^- = \frac{1}{\sqrt{2}}[|\uparrow\rangle|\downarrow\rangle - |\downarrow\rangle|\uparrow\rangle]. \tag{7.295}$$

However, this sort of states can not serve as a resource of quantum information since it is impossible to perform local operations over the subsystems and there is no possibility to change the coefficients in the coherent superposition (7.295). This fact was discussed by Zanardi and Peres [Peres (1993)] but using systems of identical particles in the protocols of quantum information and quantum communication seems to be still an open question.

Starting with the pioneering experiments performed by Fry and independently by Aspect [Grynberg (2010)] with two-photon fluorescence of atoms, entangled states were actively studied in quantum optics. In particular, the most popular object was the two-photon state (*biphoton*) [Klyshko (1998)] created via spontaneous parametric down conversion (see Secs. 6.5 and 7.6). Several types of bipartite entangled states were suggested depending on the available degrees of freedom under consideration: entanglement between polarization and momentum, polarization and frequency, energy and time, frequency and momentum. The particular type of entanglement is determined by the phase-matching conditions and/or further transformations performed over the biphotons. For instance, polarization-momentum entangled state can be achieved by choosing the non-collinear degenerate regime of SPDC when two photons with the same frequencies propagate along different directions θ, θ' and carry orthogonal polarizations H (horizontal) or V (vertical):

$$\Psi = \frac{1}{\sqrt{2}}[|H_\theta\rangle|V_{\theta'}\rangle + e^{i\varphi}|V_\theta\rangle|H_{\theta'}\rangle]. \tag{7.296}$$

Sometimes the entangled state can be created by means of post-selection, i.e. by taking into account only part of the state, for instance using a coincidence circuit, which measures only events corresponding to incoming photons coinciding in time. Such a scheme exploiting energy-time entanglement was suggested by Franson. The scheme contained unbalanced Mach-Zehnder interferometers in both signal and idler channels, which introduced delays exceeding the coherence length (see Sec. 7.2) of each photon,

$$l_{coh} = c\tau_{coh} = c\left(\frac{d^2k}{d\omega^2}|_{\omega_p/2}\, l\right)^{1/2}, \tag{7.297}$$

where l is the length of the crystal generating biphotons and ω_p is the pump laser frequency. Starting with a separable state of two photons (the same frequency but different angles of propagations),

$$\Psi_{s,i} = \frac{1}{\sqrt{2}}[|S\rangle_s + e^{i\varphi_s}|L\rangle_s] \otimes \frac{1}{\sqrt{2}}[|S\rangle_i + e^{i\varphi_i}|L\rangle_i] = \Psi_s \otimes \Psi_i, \tag{7.298}$$

and using post-selection technique by picking up only photons passing through the long (L) or short (S) paths simultaneously, the final state (after the renormalization caused by the non-unitary operation of post-selection) becomes an entangled one,

$$\Psi = \frac{1}{\sqrt{2}}[|S\rangle_s|S\rangle_i + e^{i(\varphi_s+\varphi_i)}|L\rangle_s|L\rangle_i], \tag{7.299}$$

where $\varphi_{s,i}$ are the phase delays introduced in the signal and idler channels. By varying the phase delays one can observe the second-order interference in the coincidence counts, while there is no modulation in the intensities both for the signal and idler channels. Note that sign "+" in the phase of (7.299) is typical for two-photon interference experiments (compare with the typical sign "–" observed for usual classical interference experiments).

Many other types of entangled states have been studied. We would like to specially mention some important cases, namely:

- bipartite multidimensional systems and the so-called Fedorov's ratio, which serves as an operational entanglement quantifier;

- bright-light entanglement and, in particular, quadrature and polarization entangled states [Bachor (2004)];
- multi-particle entanglement like *GHZ* and *W-states*.

A special class of entangled states, *cluster states*, attracts a great attention due to one-way quantum computation scheme suggested by Raussendort, Browne, and Briegel in 2003. Also, a special class of *graph states* seems to be useful for quantum computation, quantum error correction, and other applications of multi-particle entangled states.

Perhaps one of the most stupendous applications of entangled states relates to *quantum key distribution*, where entangled photon pairs are used for establishing a secret key between remote parties both in free space and using single-mode fibres. Another field under an unquenchable interest is *testing the foundations of the quantum theory*, where the use of multidimensional entangled systems gives unexpected results with respect to the case of bipartite systems.

7.6 °Statistics of photons and photoelectrons

Let us consider in more detail the statistics of the photon number N (see also Ref. [Loudon (2000)]). For simplicity, we will mostly consider a single mode. Below, we will show that the distribution $P(N) \equiv \rho_{NN}$ or the moments $\langle N^m \rangle \equiv \sum N^m P(N)$ can be experimentally obtained from the statistics of the number of photocounts, i.e., the number of electrons released by light from the photocathode of a PMT during some sample time T. Such methods form the base for the optical mixing spectroscopy [Cummins (1974)] where the traditional *spectral analysis of light* is replaced by the *statistical analysis of photocurrent* at the PMT output.

7.6.1 *Photon statistics*

The distribution $P(N)$ at a fixed time moment for an arbitrary instantaneous state of the mode is determined by the general rule,

$$P(N) = \mathrm{Tr}\{\rho |N\rangle\langle N|\}, \tag{7.300}$$

where $|N\rangle\langle N| \equiv \hat{P}(N)$ is the projection operator. Assuming in (7.272) $f = \hat{N} = a^\dagger a$, we obtain the z-representation for ρ_{NN},

$$P(N) = \int d^2 z P(z) |z|^{2N} e^{-|z|^2} / N! \tag{7.301}$$

Further, using (7.270), one can write $P(N)$ as an infinite series of normally ordered moments of orders $m \geqslant N$,

$$P(N) = \frac{1}{N!}\langle : \hat{N}^N e^{-\hat{N}} : \rangle = \frac{1}{N!}\sum_{k=0}^{\infty}\frac{(-1)^k}{k!}G^{(N+k)}. \qquad (7.302)$$

Thus, *the photon-number distribution has the form of a Poissonian one, $\mu^N e^{-\mu}/N!$, with the random parameter $\mu \equiv \hat{N}$, which requires additional quantum averaging with normal ordering.*

Above, we have considered two examples of $P(N)$ distributions: the Poissonian one, (7.224), for a coherent state, and the geometric one, (7.279), for a chaotic (thermal) state. These are single-parameter distributions: they are fully characterized by, for instance, the first moment, $\langle N\rangle$. One can show (see, for instance, Ref. [Loudon (2000)]) that for a single-mode laser much above the oscillation threshold, $P(N)$ is Poissonian and for a laser below the threshold, it is geometric. Near the threshold, the distribution has a shape that is intermediate between these two limiting cases. This distribution, in the simplest models, is determined by two parameters, for instance, $\langle N\rangle$ and the excess inversion above the threshold.

It is useful to consider another type of states, namely, an incoherent mixture of the vacuum, $|0\rangle$, and the K-photon state, $|K\rangle$. Then, $P(N)$ differs from zero only at two points,

$$P(N) = P(0)\delta_{N0} + P(K)\delta_{NK}. \qquad (7.303)$$

Hence, $\langle N\rangle = KP(K)$, and all values can be expressed in terms of a single parameter, the mean photon number,

$$P(K) = \langle N\rangle/K, \;\; P(0) = 1 - P(K), \qquad (7.304)$$

$$\langle N^m\rangle = K^m P(K) = K^{m-1}\langle N\rangle, \; 0 \leqslant N \leqslant K. \qquad (7.305)$$

In the limiting case $\langle N\rangle = K$, we obtain a pure energy state with K photons; however, it is interesting to consider the more realistic case $\langle N\rangle \ll K$. Such states can be generated via a K-photon decay of a single excited atom into one mode. Repetition of this process in time leads to the appearance of 'K-photon light', radiation consisting of K-photon groups. Two-photon light can be also generated via the spontaneous ($\langle N\rangle \ll 1$) parametric down-conversion of 'usual' light, in which photons have Poissonian or Bose-Einstein distribution.

Figure 7.21 shows the plots of all three distributions we have considered. Below, it will be shown that the termination of the K-photon distribution at $N > K$ leads to the effect of *photon anti-bunching* for $\langle N\rangle > K - 1$, while at $K > 1, \langle N\rangle \ll K - 1$, it leads to the effect of super-bunching. (Conventionally, the Poissonian distribution is considered as having no bunching.)

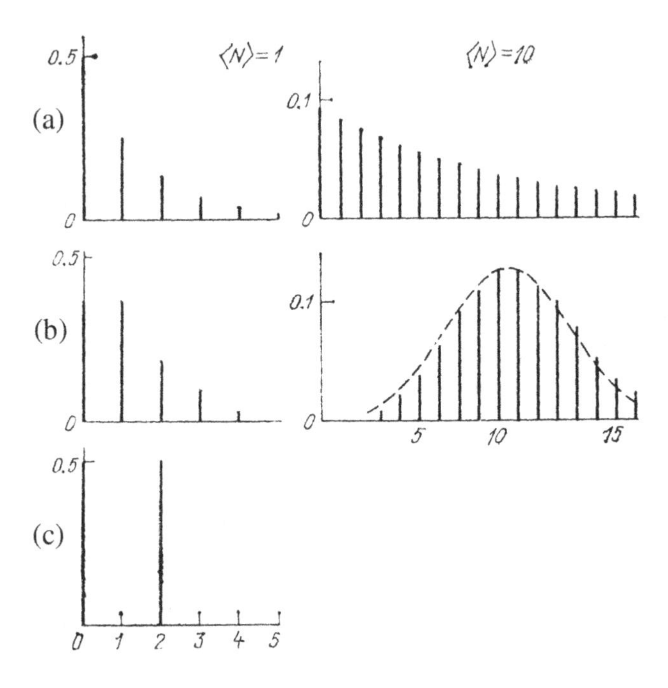

Fig. 7.21 Typical photon-number distributions: (a) geometric; (b) Poissonian; (c) two-photon mixed. The dashed line shows a Gaussian function with the mean value 9.5 (see (7.320)).

Often, calculations and interpretation of the experiments are much simplified by passing, from usual moments $\langle N^m \rangle$, to *factorial moments*,

$$G^{(m)} \equiv \langle : N^m : \rangle = \langle N(N-1)\ldots(N-m+1)\rangle, \qquad (7.306)$$

and their generating functions,

$$Q(x) \equiv \sum_{N=0}^{\infty}(1+x)^N P(N) = \langle : e^{xN} : \rangle. \qquad (7.307)$$

The last equality follows from (7.302). Note that $Q(x)$ differs from the $\langle e^{xN}\rangle$ function, whose derivatives yield usual moments. If $Q(x)$ is known, the factorial moments and the distribution $P(N)$ can be easily found by differentiating it at points $x = 0$ and $x = -1$. Indeed, it follows from the definitions that

$$G^{(m)} = Q^{(m)}(0), \qquad (7.308)$$

$$P(N) = Q^{(N)}(-1)/N!, \qquad (7.309)$$

where

$$Q^{(m)}(x) \equiv d^m Q(x)/dx^m. \qquad (7.310)$$

By substituting into (7.307) one-by-one the distributions (7.224), (7.279), (7.303), we find the generating functions for the three considered types of oscillator states,

$$Q_z(x) = e^{x\langle N\rangle}, \tag{7.311}$$

$$Q_T(x) = 1/(1 - x\langle N\rangle), \tag{7.312}$$

$$Q_K(x) = 1 + \langle N\rangle[(1+x)^K - 1]/K. \tag{7.313}$$

Hence, according to the rule (7.308), we immediately obtain

$$\begin{aligned} G_z^{(m)} &= \langle N\rangle^m, \ G_T^{(m)} = m!\langle N\rangle^m, \\ G_K^{(m)} &= \langle N\rangle(K-1)\ldots(K-m+1), \ 1 < m \leqslant K. \end{aligned} \tag{7.314}$$

An important property of the generating functions (GFs) follows from the definition (7.307): the GF $Q(x)$ of a sum $N = \sum N_i$ of independent integers N_i is equal to the product of the GFs for these integers,

$$Q(x) = \prod_i Q_i(x). \tag{7.315}$$

For instance, let a set of M similar atoms independently[y] emit K-photon light; then, using (7.313) and (7.315), we find

$$Q(x) = \prod_{i=1}^{M} \{1 + \langle N_i\rangle[(1+x)^K - 1]/K\} \approx \exp\{\langle N\rangle[(1+x)^K - 1]/K\}, \tag{7.316}$$

where we assumed $M \to \infty$ and $\langle N_i\rangle \to 0$ with a finite $\langle N\rangle \equiv \sum\langle N_i\rangle$. In the case $K = 1$, (7.316) becomes (7.311), i.e., weak single-photon radiation from a large number of independent atoms results in a Poissonian distribution (in the absence of interference). At $K > 1$, (7.316) describes Poissonian distribution for groups of K photons. For instance, in the case of two-photon radiation,

$$\begin{aligned} Q(x) &= \exp[\mu(2x + x^2)], \ \ \mu \equiv \langle N\rangle/2, \\ P(2N) &= \mu^N e^{-\mu}/N!, \ \ P(2N+1) = 0. \end{aligned} \tag{7.317}$$

Apparently, this example also includes the case where M is the number of independent modes and N is the total number of photons in these modes.

Consider another example of applying the composition rule (7.315). Let us focus on the total photon number N in M independent modes with the geometric distributions and the same mean photon numbers, $\langle N_k\rangle = \langle N\rangle/M \equiv \delta$ (the

[y]Here we neglect interference, which one can do in the case of multi-mode detectors with a large detection volume (Sec. 7.2).

degeneracy factor). The generating function, according to (7.312) and (7.315), is

$$Q(x) = \prod_{k=1}^{M} 1/(1 - x\langle N_k\rangle) = 1/(1 - x\delta)^M. \tag{7.318}$$

Hence, at $M \gg 1$, $x\delta \ll 1$, with finite $\langle N\rangle$ we once again obtain a Poissonian distribution,

$$Q(x) \approx e^{x\langle N\rangle} \; (M \gg 1). \tag{7.319}$$

Thus, *the total number of photons in a large number of independent modes with geometric distributions tends to a Poissonian distribution.* Note that at $\langle N\rangle \gg 1$, a discrete Poissonian distribution can be approximated by a normal distribution (Fig. 7.21(b)) with the first moment and variance coinciding,

$$P(N) \approx \exp\{-(N - \langle N\rangle)^2/2\langle N\rangle\}/\sqrt{2\pi\langle N\rangle}. \tag{7.320}$$

Relative fluctuations become small at $N \gg 1$,

$$\Delta N/\langle N\rangle = \langle N\rangle^{-1/2} = (\hbar\omega/\langle\mathcal{E}\rangle)^{1/2} \ll 1. \tag{7.321}$$

Thus, *for high mean energy and large mode number we have obtained a Gaussian distribution, according to the central limit theorem and classical thermodynamics.*

In the general case, (7.318) leads to the Pascal distribution,

$$\begin{aligned} G^{(k)} &= \frac{(k + M - 1)!}{(M - 1)!}\delta^k = \langle N\rangle^k\left(1 + \frac{1}{M}\right)\ldots\left(1 + \frac{k-1}{M}\right), \\ P(N) &= \frac{(N + M - 1)!}{N!(M - 1)!}\frac{\delta^N}{(1 + \delta)^{M+N}}. \end{aligned} \tag{7.322}$$

Hence, assuming $k = 2$, we find that the relative fluctuations decrease with the growth of the number of modes,

$$\frac{\Delta N}{\langle N\rangle} = \left(\frac{1}{\langle N\rangle} + \frac{1}{M}\right)^{1/2}. \tag{7.323}$$

7.6.2 *Photon bunching and anti-bunching*

Among various types of instantaneous states of a quantum oscillator, the coherent ones are special because, in a certain sense, they are most close to the state of a classical oscillator with fixed coordinate and velocity. According to (7.224), the distribution of the photon number N for a coherent state is Poissonian, so that the variance coincides with the mean photon number,

$$\langle\Delta N^2\rangle_z = \langle N\rangle = |z|^2. \tag{7.324}$$

Thus, the N distribution for a coherent quantum ensemble coincides with the distribution of chaotically scattered classical particles over a set of cells, either in space or in time. For an arbitrary state, the variance $\langle \Delta N^2 \rangle$ can certainly differ from $\langle N \rangle$; in this case, for $\langle \Delta N^2 \rangle > \langle N \rangle$ one speaks of *photon bunching*, while for $\langle \Delta N^2 \rangle < \langle N \rangle$, of *photon anti-bunching*. These effects can be quantitatively described in terms of the normalized second factorial moment,

$$g \equiv \frac{\langle : N^2 : \rangle}{\langle N \rangle^2} = \frac{\langle N^2 \rangle - \langle N \rangle}{\langle N \rangle^2} = 1 + \frac{\langle \Delta N^2 \rangle - \langle N \rangle}{\langle N \rangle^2}. \tag{7.325}$$

For a Poissonian distribution, $g = 1$, in the case of bunching $g > 1$, and in the case of anti-bunching $g < 1$. Because the rate of a stimulated two-photon transition scales as $\langle : N^2 : \rangle$, g determines the ratio of *two-photon efficiencies* for a given field and for a coherent field with the same mean energy. Note that if we ignore the normal ordering in the definition of g (7.325), then g will be always greater than a unity,

$$g_{class} \equiv \langle N^2 \rangle / \langle N \rangle^2 = 1 + \langle \Delta N^2 \rangle / \langle N \rangle^2 \geqslant 1. \tag{7.326}$$

In a chaotic (thermal) state, there is always bunching, since, according to (7.314), for the geometric distribution,

$$\langle \Delta N^2 \rangle_T = \langle N \rangle (1 + \langle N \rangle), \tag{7.327}$$

and $g_T = 2$. The variance is above $\langle N \rangle$ due to the relatively slow decay of the geometric ('exponential') distribution (Fig. 7.21), which makes 'groups' of N photons with $N \neq \langle N \rangle$ occur more often than in the case of a Poissonian distribution. As a result, relative fluctuations of the photon number in a thermal state tend at $\langle N \rangle \gg 1$ to the unity (and not to zero, as for a coherent state),

$$(\Delta N / \langle N \rangle)_T = (1/\langle N \rangle + 1)^{1/2} \to 1. \tag{7.328}$$

Sometimes one says that the first and the second terms in (7.327) and (7.328) correspond, respectively, to the corpuscular and wave sides in the wave-particle duality of a photon, and that the bunching effect (i.e., the second term) confirms that the photons have a tendency to joining in groups. This terminology masks the fact that the variance of some observable f characterizes the state, and not the properties of f. This approach also ignores the presence of states with anti-bunching, i.e., with the relative fluctuations less than $1/\sqrt{\langle N \rangle}$ and with two-photon efficiency less than the one for a coherent state with the same mean energy.

An evident example of an anti-bunched state are photon-number states. In a K-photon state, the fluctuations of N are absent, $P(N) = \delta_{NK}$, so that $\langle N^m \rangle = \langle N \rangle^m$ and $g = 1 - K^{-1} < 1$.

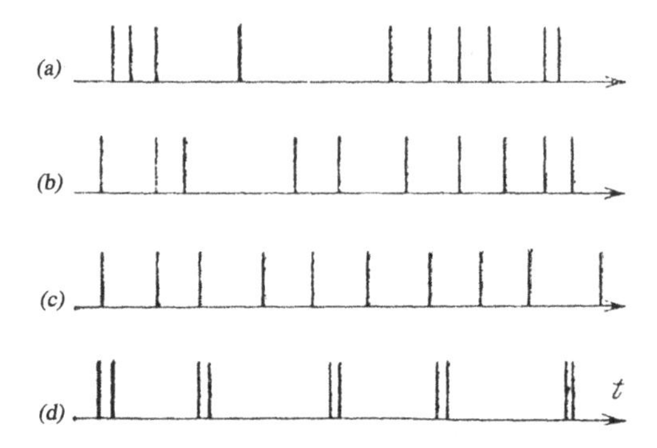

Fig. 7.22 A typical distribution of 10 photons in time in the cases of (a) chaotic light (bunching); (b) coherent light; (c) single-photon light (anti-bunching); (d) two-photon light (super-bunching).

The above-considered mixture of the vacuum and the K-photon state (7.303) also manifests anti-bunching under the condition $\langle N \rangle > K - 1$. Indeed, it follows from (7.305) that

$$g_K = (K - 1)/\langle N \rangle. \tag{7.329}$$

Hence, single-photon decay of separate atoms turns the field into an anti-bunched state, $g = 0$, which was observed in the resonance fluorescence (Sec. 5.2) of sodium atomic beam [Paul (1977)]. Certainly, in experiment the averaging runs not over the ensemble of experimental setups but over time.

This effect can be explicitly explained by the fact that the decay of a single atom cannot create two photons; therefore, the emitted photons are always separated by a certain time interval τ required for a second excitation and de-excitation of the same atom or the next atom in the beam (Fig. 7.22(c)). Recall that our 'single-mode' theory relates only to time intervals T much less than τ; therefore, $P(N) = 0$ for $N > 1$.

Radiation with anti-bunching ('single-photon light') can also emerge as a result of multi-photon absorption (Sec. 6.4) of 'usual' (chaotic or coherent) radiation. The reason is that K-photon absorption, apparently, influences only the 'tail' of the $P(N)$ distribution at $N \gg K$, which leads to the rarefication of photon groups and the reduction of fluctuations in the initial radiation. On the contrary, the saturation effect in single-photon absorption (Sec. 4.3) makes fluctuations more pronounced, i.e., leads to photon bunching. This forms the base for a method of obtaining very short (picosecond)[z] pulses in mode-locked lasers.

In the case of weak multi-photon light, $K > 1$, $\langle N \rangle \ll K - 1$, Eq. (7.329) yields $g \gg 1$ (Fig. 7.23). This effect can be called 'photon super-bunching'. It

[z]Editors' note: at present, tens of femtoseconds are achieved.

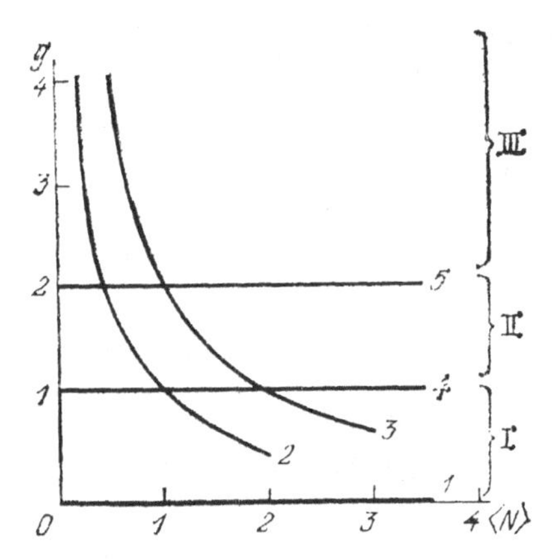

Fig. 7.23 The bunching parameter g as a function of the mean photon number $\langle N \rangle$ in the cases of single-photon light (1), two-photon light (2), three-photon light (3), coherent light (4), and chaotic light (5). In range I, there is photon anti-bunching, in range II, bunching, in range III, super-bunching.

can be explained by the existence of regular groups of K photons separated by large time intervals (Fig. 7.22(d)). A convenient way of obtaining directed two-photon light is parametric down-conversion (Sec. 6.5) in piezoelectric crystals. Below, it will be shown (see (7.356)) that super-bunching can be used for the absolute (reference-free) measurement of the quantum efficiency of photodetectors [Klyshko (1980)].

States of field with anti-bunching and the above-mentioned squeezed states attract much attention nowadays, similarly to the way the bunching effect was very popular soon after its discovery by Hanbury Brown and Twiss in 1955. The anti-bunching effect, observed in 1977, is considered as disproving the semi-classical radiation theory (see (7.326)). Indeed, according to this theory, a single-mode laser with stabilized intensity should create photocurrent with minimal possible fluctuations, which, similarly to the shot noise, should be Poissonian due to the equivalence of all time moments. Thus, it seems to be impossible to create a more uniform time distribution of photoelectrons than a Poissonian one. In contrary, from the photon viewpoint, a flux of photons that are equidistant in time (obtained, for instance, via a periodic excitation of a single atom) will cause an equidistant flux of photoelectrons, provided that the quantum efficiency η is high enough.

It is typical that in states with $g < 1$, for instance, K-photon ones, quasi-probabilities $P(z)$ have huge singularities or take negative values [Glauber (1965); Perina (1972)], which allows one to consider such states as essentially non-classical, i.e., having no classical analogues.

Consider now the quantum interpretation of interference experiments like the Young one and the Hanbury Brown–Twiss one (Sec. 7.2). While quantizing the field, it is now convenient to expand it not over plane waves but over orthogonal eigenfunctions $u_n(\boldsymbol{r})$ of the corresponding boundary problem, which takes into account the existence of screens with slits and semi-transparent mirrors. In the simplest case of a monochromatic field and single-mode detectors, all field can be represented as a single mode,

$$E(\boldsymbol{r}, t) = u(\boldsymbol{r})a(t) + u^*(\boldsymbol{r})a^\dagger(t) \equiv E^{(+)} + E^{(-)}, \tag{7.330}$$

where $a(t) = ae^{-i\omega t}$ is the photon annihilation operator in this mode.

The counting rate of a single-photon detector placed at point $\boldsymbol{r}_1$ scales as the mean intensity,

$$G_1^{(1)} \equiv \langle E_1^{(-)} E_1^{(+)} \rangle = |u_1|^2 \langle N \rangle, \tag{7.331}$$

where the subscript 1 substitutes the argument $\boldsymbol{r}_1$, $N \equiv a^\dagger a$, and the averaging runs over the initial state of the field. We stress that $u(\boldsymbol{r})$ is the solution to the classical wave equation, it describes the propagation of classical waves, their diffraction and interference. Thus, the *concepts of a photon and a wave by no means contradict each other*, and here we do not need to speak about the wave-particle duality. The operators a, $a^\dagger$, N do not depend on the coordinate; therefore, a photon is an elementary excitation of the *whole* field and the question 'which slit did the photon go through?' is meaningless. The space structure of the field is determined by the function $u(\boldsymbol{r})$, its square gives the probability to discover a photon at an arbitrary point $\boldsymbol{r}$; therefore, it plays the role of the wavefunction of a photon.

The corpuscular properties of a photon are only revealed through detection, when the energy of the whole field, $\hbar\omega$ (in the case $|t_0\rangle = |1\rangle$), 'gets focused' at a single 'point'. This is manifested most clearly while observing flashes on the screen of an image intensifier. Note that the duality of wavepackets, discussed quite often, is not typical for only quantum mechanics; the same property is possessed by classical waves.

Further, correlation of the counts of m single-photon detectors placed at points $\boldsymbol{r}_1, \ldots, \boldsymbol{r}_m$ scales as the normally ordered moment of order m,

$$G_{1\ldots m}^{(m)} \equiv \langle E_1^{(-)} \ldots E_m^{(-)} E_m^{(+)} \ldots E_1^{(+)} \rangle = |u_1 \ldots u_m|^2 \langle : N^m : \rangle. \tag{7.332}$$

The first factor here describes the influence of the spatial positions of the detectors. For the case $m = 2$, this result was obtained in Sec. 7.2 from a simple model assuming the binomial distribution of photons at points $\boldsymbol{r}_1, \boldsymbol{r}_2$.

Let, for instance, two-photon light be incident on an intensity interferometer (Fig. 7.13); then, according to (7.314), the readings of the correlator will scale as

$$G_{12}^{(2)} = |u_1 u_2|^2 \langle N \rangle,$$

while the mean counting rate will scale as $G_n^{(1)}$. Hence, the relative correlation is

$$g_{12}^{(2)} = G_{12}^{(2)}/G_1^{(1)}G_2^{(1)} = 1/\langle N\rangle. \qquad (7.333)$$

In the case of a pure two-photon state, $\langle N\rangle = 2$, and the so-called negative correlation, or anti-correlation, takes place, with $g_{12}^{(2)} < 1$ (Fig. 7.12).

Above, for simplicity we discussed the statistics of a single mode; however, many conclusions can be generalized to the case of a multi-mode field. Indeed, in spontaneous two-photon emission (Sec. 6.2) and in spontaneous parametric down-conversion (Sec. 6.5), photons in pairs usually belong to different modes, $\boldsymbol{k}_1$ and $\boldsymbol{k}_2$, which differ both in frequency and in direction. Then, a small contribution from the state $|1\rangle_1|1\rangle_2$ is added to the vacuum, which provides the equality between the moments,

$$\langle N_1 N_2\rangle = \langle N_1\rangle = \langle N_2\rangle.$$

As a result, at $\langle N_i\rangle \ll 1$, the probability of discovering two photons is much greater than the product of single-photon probabilities,

$$g(k_1, k_2) \equiv \langle N_1 N_2\rangle/\langle N_1\rangle\langle N_2\rangle = 1/\langle N_i\rangle \gg 1. \qquad (7.334)$$

This inequality, which can also be interpreted as the *super-bunching effect* (compare with (7.329) at $K = 2$), was experimentally confirmed by measuring the rate of coincidences between two PMTs.

7.6.3 *Statistics of photoelectrons*

Let $V_{det} \ll V_{coh}$, then the radiation incident on a PMT photocathode can be considered as single-mode (Sec. 7.2). Following Scully [Arecchi (1974)], from explicit combinatorics considerations we will now show (see also Refs. [Loudon (2000); Klauder (1968); Glauber (1965); Perina (1972)]) that the Mandel formula (7.71) maintains its form even in the framework of the quantum theory if αI is replaced by ηN, where η is the PMT quantum efficiency and N is the operator of the photon number in the detection volume (equal to the operator of the photon number in one mode times the factor V_{det}/V_{coh}).

Consider first a field in a pure Fock state with the photon number N incident on a photodetector (PMT). Let the probability of registering a single photon be η $(0 \leqslant \eta \leqslant 1)$, then the probability of registering any m photons out of their total number $N \geqslant m$, apparently, is determined by the Bernoulli binomial distribution (Fig. 7.24),

$$P(m|N) = C_N^m \eta^m (1-\eta)^{N-m}. \qquad (7.335)$$

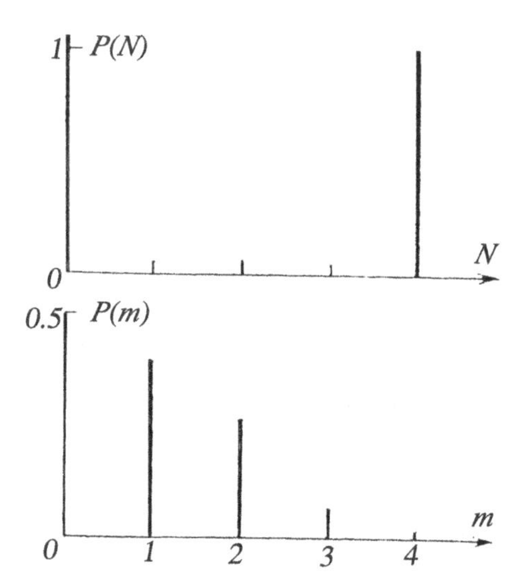

Fig. 7.24 Distributions of the photon number, $P(N)$, and the photoelectron number, $P(m)$, in the case of the detector efficiency 0.3 and the field being in a four-photon Fock state.

Hence, in the case of an arbitrary state of the field, we find the relation between the distribution of photons, $P(N) \equiv \rho_{NN}$, and the distribution of photoelectrons (or, in other words, *photocounts*),

$$P(m) = \sum_{N=m}^{\infty} P(m|N)P(N). \tag{7.336}$$

In the case $\eta = 1$, we have $P(m|N) = \delta_{mN}$, and the distributions of electrons and photons coincide. At $\eta < 1$, the binomial transformation (7.336) adds additional stochasticity, distorts $P(N)$ and complicates solving the inverse problem, which is the determination of the field statistics from the measured photocount statistics. (Using (7.346), one can show that $P(N)$ can be formally expressed in terms of $P(m)$ through a relation similar to (7.336) but with η replaced by $1/\eta$.)

Note that if η is understood as the probability of a photon 'survival' after light passing a layer of matter with single-photon absorption, then m has the meaning of the photon number at the output of the layer, while N is the input number of photons. Then, $\eta = e^{-\alpha l}$, where α is the absorption coefficient and l is the layer thickness. Thus, Eq. (7.336) and all its corollaries are also applicable to the transformation of the field statistics caused by linear absorption of the field by cold matter.

Let us pass from the N-representation to the z-representation using (7.301). As a result, (7.336) leads to the quantum analogue of the Mandel formula (7.71), which gives the photoelectron distribution in terms of the quasi-probability $P(z)$ of a z state,

$$P(m) = \int d^2zP(z)(\eta|z|^2)^m e^{-\eta|z|^2}/m! \quad (7.337)$$

This relation, after taking into account (7.270), can be represented in an invariant form,

$$P(m) = \langle : (\eta N)^m e^{-\eta N} : \rangle / m!, \quad (7.338)$$

where $N = a^\dagger a$. Thus, the distribution of photoelectrons is of a Poissonian form but has additional quantum averaging including the operation of normal ordering. Note that the obtained expression differs from the photon distribution (7.302) only by the replacement of N by ηN.

The integral transformation (7.337)can be easily realized for the cases of coherent and chaotic states. Then, it turns out that the functional forms of the distributions for photons and photoelectrons in these two cases coincide. Indeed, assuming in (7.337) $P(z) = \delta^{(2)}(z - z_1)$, we find

$$P_z(m) = \langle m \rangle^m e^{-\langle m \rangle}/m!; \quad (7.339)$$

further, it follows from (7.337) and (7.280) that

$$P_T(m) = \langle m \rangle^m / (1 + \langle m \rangle)^{m+1}, \quad (7.340)$$

where $\langle m \rangle = \eta \langle N \rangle$.

Recall that $P(m)$ can be also understood as the photon distribution at the output of a cold layer with single-photon absorption ($\eta = e^{-\alpha l}$). Thus, linear absorption at $T = 0$K does not change the form of the photon distribution in the cases of chaotic and coherent incident radiation (in contrast to N-photon absorption, see (**??**)). Absorption or amplification at $T \neq +0$ is accompanied by chaotic spontaneous emission, which changes the shape of the distribution.

The obtained relations (7.336), (7.337), (7.338) between the distributions of photons and photoelectrons are rather complicated. The relations between the generating functions and the factorial moments are much simpler,

$$m^{(k)} = \eta^k N^{(k)}, \quad (7.341)$$

where

$$m^{(k)} \equiv \langle m(m-1)\ldots(m-k+1) \rangle, \quad (7.342)$$

$$N^{(k)} \equiv \langle : N^k : \rangle = \langle N(N-1)\ldots(N-k+1) \rangle. \quad (7.343)$$

Here, the functions of m are averaged using the discrete distribution $P(m)$,

$$\langle f(m)\rangle \equiv \sum_{m=0}^{\infty} f(m)P(m). \tag{7.344}$$

The quantum mean value, $\langle f(N)\rangle$, has a similar form in the N representation.

Equation (7.341) can be easily obtained using the formalism of generating functions for photons, $Q_{phot}(x)$ (see (7.307)), and electrons,

$$Q_{el}(x) \equiv \sum_{m=0}^{\infty} (1+x)^m P(m). \tag{7.345}$$

By substituting (7.338), we find

$$Q_{el}(x) = Q_{phot}(\eta x) = \langle : e^{x\eta N} : \rangle. \tag{7.346}$$

Hence, with an account for (7.308), we come to (7.341).

From (7.341), we easily obtain relations between usual moments,

$$\langle m\rangle = \eta\langle N\rangle, \tag{7.347}$$

$$\langle m^2\rangle = \eta\langle N\rangle + \eta^2(\langle N^2\rangle - \langle N\rangle), \tag{7.348}$$

and between the variances (compare with (7.78)),

$$\Delta m^2 = \langle m\rangle(1-\eta) + \eta^2\langle \Delta N^2\rangle. \tag{7.349}$$

Thus, to the usual shot noise of the photocurrent, $\langle m\rangle$, 'photon noise', $\langle \Delta N^2\rangle$, is added with the weight η^2 (which seems natural), but simultaneously, the term $\eta^2\langle N\rangle$ is subtracted (which is surprising for semi-classical theory). In the case of photon anti-bunching, $\langle \Delta N^2\rangle < \langle N\rangle$, so that the photocurrent noise is less than the Poissonian, i.e., electrons appear with some regularity, 'repulse' each other.

Note that, according to (7.341), *the bunching parameters (as well as all normalized factorial moments) of photons and electrons coincide*,

$$g = m^{(2)}/\langle m\rangle^2 = N^{(2)}/\langle N\rangle^2.$$

If N is understood in these equations as the total number of photons in M independent modes, then the equations will describe the statistics of an M-mode detector. For instance, in (7.315) and (7.346) we find that

$$Q_{el}(x) = \prod_{k=1}^{M} Q_k(\eta x), \tag{7.350}$$

where $Q_k(x)$ is the generating function for the kth mode.

In the case of modes with a geometric (thermal) distribution of photons, (7.312) leads to (compare with (7.318))

$$Q_{el}(x) = 1/(1 - x\eta\delta)^M, \quad (7.351)$$

where $\delta = \langle N \rangle / M$ is the mean number of photons in one mode.

At $M \gg 1$ and $x\eta\delta \ll 1$, this function tends to (compare with (7.319))

$$Q_{el}(x) \approx e^{x\eta\delta M} = e^{x\langle m \rangle}. \quad (7.352)$$

Hence, *a multi-mode detector registering a thermal field produces photocounts with Poissonian statistics and does not manifest bunching* (similarly to a multi-mode field, see (7.319)).

For a finite number of modes M, from (7.351) we find the distribution and factorial moments that coincide with (7.322) after the replacement of δ by $\eta\delta$, or $\langle N \rangle$ by $\langle m \rangle$. In particular, from (7.323) we find relative fluctuations,

$$\Delta m / \langle m \rangle = (1/\langle m \rangle + 1/M)^{1/2}. \quad (7.353)$$

Similarly, replacing x by ηx in (7.317), we find the generating function for a multi-mode detector registering two-photon light,

$$Q_{el}(x) = \exp[\langle m \rangle (x + \eta x^2/2)]. \quad (7.354)$$

It is important that at $\eta \neq 1$, this function differs from (7.317), which opens an interesting possibility of absolute (reference-free) measurement of the PMT quantum efficiency η. Indeed, from (7.354) we find that

$$\langle m^2 \rangle = \langle m \rangle (\langle m \rangle + 1 + \eta), \quad (7.355)$$

or

$$\eta = \langle \Delta m^2 \rangle / \langle m \rangle - 1 = \langle m \rangle (g - 1). \quad (7.356)$$

Thus, *by measuring the mean value and variance (or the bunching parameter) of the photocounts one can find the quantum efficiency.*

7.7 °Interaction of an atom with quantized field

So far, we have been calculating the probabilities of quantum transitions in the framework of the semiclassical approach, where the spontaneous transitions were described not rigorously, with the help of additional rules or analogies introduced without sufficient justification. In the present section, we will fill this gap. It will be shown that the interaction between two stationary quantum systems can be conveniently described phenomenologically in terms of normally and anti-normally ordered correlation functions (CFs). These functions have more direct relation to the energy exchange than the symmetrized CFs that are commonly used. We will also consider the symmetry of CFs, the relations between different CFs, and the dependence of CFs on the microscopic parameters of the systems and their Green's functions, i.e., response to coherent perturbations.

7.7.1 *Absorption and emission probabilities*

Consider the interaction of a single atom with quantized field. Let the atom and the field be independent at the initial time moment, $|t_0\rangle = |m\rangle|i\rangle \equiv |mi\rangle$, where $|m\rangle$, $|i\rangle$ are the initial states of the atom and the field, respectively. In the first order of the perturbation theory, the amplitude $c_{nf}(t)$ of the transition into some state $|nf\rangle$ is determined by the matrix element of the interaction operator, $\langle nf|\mathcal{V}|mi\rangle$ (Sec. 2.1). In the dipole approximation,

$$c_{nf} = -\frac{1}{i\hbar}\int_{t_0}^{t} dt' \langle nf|\boldsymbol{d}(t')\cdot\boldsymbol{E}(t')|mi\rangle, \tag{7.357}$$

where the operators are considered in the interaction picture, i.e., without the account for the perturbation.

Let us split these operators into the positive- and negative-frequency parts. At $t - t_0 \equiv T \gg 1/\bar{\omega}$, fast oscillating (with approximately twice the mean frequency $\bar{\omega}$) products $\boldsymbol{d}^{(+)}\cdot\boldsymbol{E}^{(+)}$ and $\boldsymbol{d}^{(-)}\cdot\boldsymbol{E}^{(-)}$ have no contribution into the integral (7.357); therefore, one can write

$$-\mathcal{V}(t) \approx \boldsymbol{d}^{(-)}(t)\cdot\boldsymbol{E}^{(+)}(t) + \boldsymbol{d}^{(+)}(t)\cdot\boldsymbol{E}^{(-)}(t). \tag{7.358}$$

This is the rotating-wave approximation, which we already discussed in Sec. 2.2.

Further, assume that the initial and final states of the atom and/or the field are energy ones, then the first term in (7.358) gives a nonzero contribution only for a quantum transmitted from the field to the atom, while the second one, only for a quantum transmitted from the atom to the field. Hence, the probability of a transition with absorption is

$$P_{\uparrow}(nf|mi) = \hbar^{-2}\int_{t_0}^{t} dt'dt'' d_{mn}^{(+)}(t')d_{nm}^{(-)}(t'')E_{if}^{(-)}(t')E_{fi}^{(+)}(t''), \tag{7.359}$$

where we have used the equality $f_{mn}^{(+)} = (f_{nm}^{(-)})^*$ and assumed, for simplicity, that the vectors $\boldsymbol{d}$ and $\boldsymbol{E}$ are parallel. In the case of emission, apparently, the superscripts (+) and (−) in (7.359) should be interchanged. Note that the approximation (7.358) does not influence the transition probability in the case where one of the interacting systems is in an energy state, since the products of the form $f_{mn}^{(+)}f_{nm}^{(+)}$ are zero then.

If we are not interested in the final state of the system, (7.359) should be summed over all possible states $|nf\rangle$. These states form a complete set; therefore, the total probability of the transition 'up' will be (compare with (2.82))

$$P_{\uparrow} = \hbar^{-2} \int_{t_0}^{t} dt' dt'' F^{(-)}(t', t'') G^{(+)}(t', t''), \tag{7.360}$$

where

$$F^{(-)}(t', t'') \equiv \langle d^{(+)}(t') d^{(-)}(t'') \rangle, \quad G^{(+)}(t', t'') \equiv \langle E^{(-)}(t') E^{(+)}(t'') \rangle \tag{7.361}$$

are, respectively, the anti-normally ordered CF for the dipole moment of the atom and the normally ordered CF for the field at the initial state (which, apparently, can be mixed as well).

Similarly, the probability of the emission of a quantum is determined by the square of (7.357) with only the second term in (7.358) taken into account,

$$P_{\downarrow} = \hbar^{-2} \int_{t_0}^{t} dt' dt'' F^{(+)}(t', t'') G^{(-)}(t', t''). \tag{7.362}$$

Thus, *the probability of an energy quantum transfer from one quantum system to another one is determined by the product of the unperturbed correlation functions, the normal one for the emitting system and the anti-normal one for the absorbing system* (provided that the initial states are independent and at least one of them is not coherent). The last condition is to exclude coherent interactions, which depend on the phases of the states.

The mean variation of the total photon number within time T, in the second order of the perturbation theory, is equal to the difference between (7.362) and (7.360),

$$\Delta N = \hbar^{-2} \int_{t_0}^{t} dt' dt'' (F^{(+)} G^{(-)} - F^{(-)} G^{(+)}). \tag{7.363}$$

In the ground states, normal CFs are equal to zero; therefore, the spontaneous emission of an atom is determined by $F^{(+)}$ (this conclusion was already used in Chapter 5), while the probability of a cold detector registering a photon is determined by $G^{(+)}$ (Secs. 7.2, 7.6).

Note that the obtained result (7.363) is valid for any systems with the interaction energy of the form $\sum f_i g_i$.

7.7.2 *Spontaneous emission*

According to (7.362), the probability of a spontaneous transition is

$$P_{sp} = \hbar^{-2} \int_{t_0}^{t} dt' dt'' F^{(+)}) G^{vac}, \tag{7.364}$$

where G^{vac} is the anti-normally ordered CF of the field in the vacuum state. According to (7.131),

$$G^{vac}(t',t'') = \sum_{k'k''} c_{k'}c_{k''}\langle 0|a_{k'}(t')a^{\dagger}_{k''}(t'')|0\rangle = \sum_{k} c_k^2 e^{-i\omega_k(t'-t'')}, \tag{7.365}$$

where $c_k^2 = 2\pi\hbar\omega_k/L^3$. Note that regardless of the state of the field,

$$G^{(-)}(t',t'') = G^{(+)}(t'',t') + G^{vac}(t',t''). \tag{7.366}$$

If only two levels are taken into account (see (5.30)),

$$d^{(+)}(t) = d_0\sigma^{12}e^{-i\omega_0 t}, \ d^{(-)}(t) = d_0^*\sigma^{21}e^{i\omega_0 t}, \tag{7.367}$$

where $d_0 \equiv d_{12}$, $d_{11} = d_{22} \equiv 0$, $\omega_0 \equiv \omega_{21} > 0$, $\sigma^{12} = (\sigma^{21})^{\dagger} \equiv |1\rangle\langle 2|$. Since

$$\sigma^{kl}\sigma^{mn} = \sigma^{kn}\delta_{lm}, \ \langle\sigma^{mn}\rangle = \rho_{nm}, \tag{7.368}$$

it follows that

$$\begin{aligned} F^{(+)}(t',t'') &= |d_0|^2 e^{i\omega_0(t'-t'')}\rho_{22}, \\ F^{(-)}(t',t'') &= |d_0|^2 e^{-i\omega_0(t'-t'')}\rho_{11}. \end{aligned} \tag{7.369}$$

Thus, *the correlation functions of a two-level atom scale as the populations of the two levels: the normally ordered one, as the upper-level population, the anti-normally ordered one, as the lower-level population.*

With an account for (7.365) and (7.369), Eq. (7.364) takes the form

$$P_{sp} = \hbar^{-2}|d_0|^2\rho_{22}\sum_k c_k^2 \sin^2\vartheta_k \int dt'dt'' e^{i(\omega_k-\omega_0)(t''-t')},$$

where ϑ_k is the angle between $\boldsymbol{k}$ and $\boldsymbol{d}_0$. At $t, t_0 \to \pm\infty$, the integral yields

$$[2\pi\delta(\omega_k - \omega_0)]^2 = 2\pi T\delta(\omega_k - \omega_0). \tag{7.370}$$

Passing to integration over modes (see (7.100)), we obtain

$$W_{sp} \equiv \frac{P}{T} = \frac{|d_0|^2\rho_{22}}{2\pi\hbar}\int d^3k\omega_k \sin^2\vartheta_k\delta(\omega_k - \omega_0). \tag{7.371}$$

The integral of $\sin^2\vartheta_k$ over all directions equals $8\pi/3$; as a result, at $\rho_{11} = 1$ we again obtain the familiar equation for the probability of a spontaneous transition of an excited two-level atom,

$$W_{sp} = (4k_0^3/3\hbar)|d_0|^2, \tag{7.372}$$

where $k_0 \equiv \omega_0/c$.

Spontaneous emission from a three-level system has been already considered in Sec. 5.2, where it was shown that $P_{sp}(T)$ oscillates with the frequency ω_{32} corresponding to the splitting between the two upper levels (the *quantum beats effect* with the coherent initial state of the atom). Also, recall that the spontaneous emission of N two-level atoms leads to the *superradiance effect* in which W_{sp} is increased by a factor of N^2 (Sec. 5.3).

7.7.3 Interaction of stationary systems

If the initial unperturbed states of both interacting systems are stationary, then all CFs in (7.363) depend only on the difference of the integration variables, $t' - t'' \equiv \tau$. (This assumption excludes from consideration the quantum beats, see Sec. 5.2, and the coherent states, see Sec. 7.5.) If, in addition, the observation time $t - t_0 \equiv T$ is much greater than the correlation times of the atom and the field, then the double integral in (7.363) scales as T, and one can introduce time-independent rates of 'up' and 'down' transitions, $W \equiv P/T$.

Let us define the correlation functions of a single variable,

$$F_\tau^{(\pm)} \equiv F^{(\pm)}(t, t+\tau) = \langle d^{(\mp)}(0) d^{(\pm)}(\tau)\rangle = F_{-\tau}^{(\pm)*},$$
$$G_\tau^{(\pm)} \equiv G^{(\pm)}(t, t+\tau) = \langle E^{(\mp)}(0) E^{(\pm)}(\tau)\rangle = G_{-\tau}^{(\pm)*}, \tag{7.373}$$

then (7.363) becomes

$$\Delta N = \hbar^{-2} \int_0^T d\tau (T - \tau)[F_\tau^{(+)} G_\tau^{(-)} - F_\tau^{(-)} G_\tau^{(+)} + (\tau \to -\tau)]. \tag{7.374}$$

If T is much greater than the correlation time of the atom and the field, then it follows from (7.374) that

$$W = \hbar^{-2} \int_{-\infty}^{\infty} d\tau (F_\tau^{(+)} G_\tau^{(-)} - F_\tau^{(-)} G_\tau^{(+)}), \tag{7.375}$$

where $W \equiv (P_\downarrow - P_\uparrow)/T$ and the sign of W determines the direction of the quantum transfer (from the atom to the field at $W > 0$).

Our initial model describes the time evolution of the states of the atom and the field. Angle brackets in (7.373) denote averaging over the ensemble of experiments with the same initial conditions and different τ. If, however, we assume that both systems are in contact with their thermostats (Fig. 7.25), which continuously restore and maintain the initial stationary states, we will obtain an ergodic model. Then, (7.375) describes a continuous flux of energy quanta transmitted from one thermostat to the other one through the 'atom-field' system.

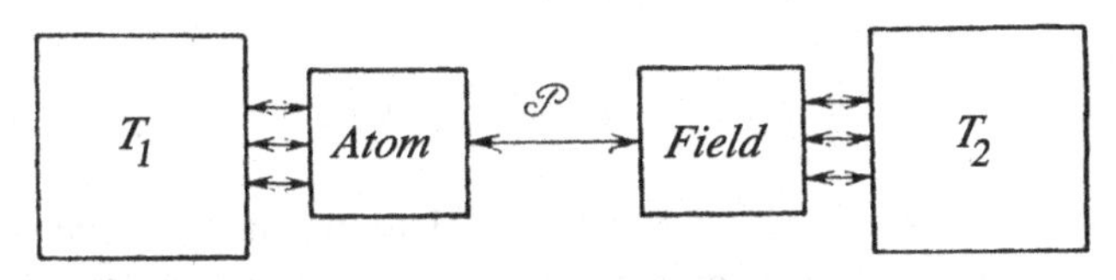

Fig. 7.25 Stationary interaction between an atom and a field. The atom-field coupling is often the 'bottleneck' for the energy exchange between the thermostats T_1, T_2.

Let us now write the CF (7.373) in terms of the microscopic parameters of the systems. Averaging with the help of stationary density matrices (not necessarily equilibrium ones) yields

$$\begin{aligned} G_\tau^{(+)} &= \sum_k c_k^2 \langle N_k \rangle e^{-i\omega_k \tau}, \\ G_\tau^{(-)} &= \sum_k c_k^2 (\langle N_k \rangle + 1) e^{i\omega_k \tau}, \\ F_\tau^{(+)} &= \sum_{m>n} |d_{mn}|^2 \rho_{mm} e^{-i\omega_{mn}\tau}, \\ F_\tau^{(-)} &= \sum_{m>n} |d_{mn}|^2 \rho_{nn} e^{i\omega_{mn}\tau}, \end{aligned} \tag{7.376}$$

where $\langle N \rangle_k \equiv \langle a_k^\dagger a_k \rangle$ are mean photon numbers in the modes and ρ_{nn} are relative populations of the atomic levels. After substituting (7.376) into (7.375), the rate of quantum exchange takes the form

$$W = 2\pi\hbar^{-2} \sum_{k,m>n} c_k^2 |d_{mn}|^2 \delta(\omega_k - \omega_{mn}) [(\langle N_k \rangle + 1)\rho_{mm} - \langle N_k \rangle \rho_{nn}]. \tag{7.377}$$

The three terms of the last factor correspond to the three types of transitions between each pair of levels according to Einstein (Chapter 2).

By inserting into (7.377) the energy of a quantum $\hbar\omega_k$, apparently, we find the power of emission or absorption,

$$\mathcal{P} = (2\pi)^2 L^{-3} \sum_{k,m>n} \omega_k^2 |d_{mn}|^2 \delta(\omega_k - \omega_{mn}) (\rho_{mm} - \Delta_{nm} \langle N_k \rangle), \tag{7.378}$$

where $\Delta_{nm} \equiv \rho_{nn} - \rho_{mm}$ is the population difference. This equation at $\langle N_k \rangle = 0$ describes the spontaneous emission of an atom into the vacuum and at $\rho_{mm} = 0$, the radiation heating of a cold atom.

Note that the normally and anti-normally ordered CFs introduced by (7.373) are not real and do not possess a certain parity (in contrast to classical CFs). However, one can construct even (symmetrized) and odd (anti-symmetrized) combinations,

$$\begin{aligned} F_\tau^{(s)} &\equiv \mathrm{Re} F_\tau = \frac{1}{2} \langle d_0 d_\tau + d_\tau d_0 \rangle = F_{-\tau}^{(s)}, \\ F_\tau^{(a)} &\equiv \mathrm{Im} F_\tau = \frac{1}{2i} \langle d_0 d_\tau - d_\tau d_0 \rangle = -F_{-\tau}^{(a)}, \end{aligned} \tag{7.379}$$

and similar functions $G_\tau^{(s)}$, $G_\tau^{(a)}$ for the field. Here, we have also introduced the 'total' CF F_τ, equal to the sum of normally and anti-normally ordered CFs,

$$F_\tau \equiv \langle d_0 d_\tau \rangle = \langle d_0^{(-)} d_\tau^{(+)} + d_0^{(+)} d_\tau^{(-)} \rangle = F_\tau^{(+)} + F_\tau^{(-)} = F_{-\tau}^*. \tag{7.380}$$

In order to find the inverse transformation (express $F_\tau^{(\pm)}$ in terms of F_τ), it is necessary to use the spectral expansion.

Often, it is only the symmetric combination $F_\tau^{(s)}$ that is used, and it is called the CF. Below, it will be shown that the antisymmetric combination $F_\tau^{(a)}$ is closely connected with the response of the system to a coherent perturbation, i.e., with its susceptibility or the Green's function. Note that in experimental optics, one usually deals with non-coherent systems, so that, according to (7.375), CF $F_\tau^{(\pm)}$ provide a more direct description of the observed effects.

From the definitions (7.379) and (7.376), we find the microscopic formulas,

$$G_\tau^{(s)} = \sum_k c_k^2(2\langle N_k\rangle + 1)\cos(\omega_k\tau), \tag{7.381}$$

$$G_\tau^{(a)} = \sum_k c_k^2 \sin(\omega_k\tau) = \mathrm{Im}G_\tau^{vac}, \tag{7.382}$$

$$F_\tau^{(s)} = \sum_{m>n} |d_{mn}|^2(\rho_{mm} + \rho_{nn})\cos(\omega_{mn}\tau), \tag{7.383}$$

$$F_\tau^{(a)} = \sum_{m>n} |d_{mn}|^2 \Delta_{nm} \sin(\omega_{mn}\tau). \tag{7.384}$$

Below, it will be shown (see (7.390), (7.394)) that in the case of a linear or equilibrium system, the only difference between various types of CFs is the different contributions of zero-point fluctuations into these functions.

7.7.4 *Spectral representation*

The Fourier transforms of stationary CFs are called *the spectral densities* (of the fluctuations of the corresponding observables, such as, for instance, the electric field or the dipole moment of an atom). It is clear from definitions (7.373) that the Fourier transform $F_\omega^{(+)}$ of the normally ordered CF $F_\tau^{(+)}$ is nonzero only at $\omega > 0$, while the Fourier transform of the anti-normally ordered CF, only at $\omega < 0$. Therefore, they can be combined into a single function F_ω, the Fourier transform of F_τ,

$$\begin{gathered} F_\omega \equiv \int d\tau e^{i\omega\tau} F_\tau/2\pi, \;\; F_\omega = F_\omega^* = F_\omega^{(s)} + iF_\omega^{(a)} = F_\omega^{(+)} + F_\omega^{(-)} \geqslant 0, \\ F_\omega^{(\pm)} = F_\omega\theta(\pm\omega), \;\; F_\omega^{(s)} = (F_\omega + F_{-\omega})/2, \;\; F_\omega^{(a)} = (F_\omega - F_{-\omega})/2i, \end{gathered} \tag{7.385}$$

and similarly for the field CFs $G_\omega^{(\pm)}$. Here, we have also introduced the Fourier transforms of symmetric and anti-symmetric real CFs, which have the following properties:

$$F^{(s)}_{\omega} = F^{(s)*}_{\omega} = F^{(s)}_{-\omega}, \quad F^{(a)}_{\omega} = -F^{(a)*}_{\omega} = -F^{(a)}_{-\omega}. \tag{7.386}$$

Substituting into (7.375) the inverse transformation,

$$F^{(\pm)}_{\tau} = \int_0^{\infty} d\omega e^{\mp i\omega\tau} F_{\pm\omega}, \tag{7.387}$$

with the factor $\hbar\omega$, we obtain the spectral expansion of the power,

$$\mathcal{P} = \frac{2\pi}{\hbar}\int_0^{\infty} d\omega\omega(F_{\omega}G_{-\omega} - F_{-\omega}G_{\omega}) = \frac{2\pi}{\hbar}\int_{-\infty}^{\infty} d\omega\omega F_{\omega}G_{\omega}, \tag{7.388}$$

where all spectral CFs are positive. Thus, we have separated positive and negative terms, corresponding to emission and absorption. If we pass to the Fourier transforms of symmetric and anti-symmetric CFs, (7.388) takes the form

$$\mathcal{P} = \frac{4\pi}{i\hbar}\int_0^{\infty} d\omega\omega(F^{(s)}_{\omega}G^{(a)}_{\omega} - F^{(a)}_{\omega}G^{(s)}_{\omega}), \tag{7.389}$$

where such separation is absent.

The explicit expressions for the spectral CFs in terms of the microscopic parameters can be easily found through the Fourier transformation of (7.376) and (7.381)–(7.384),

$$\begin{aligned}
G^{(+)}_{\omega} &= \sum_k c_k^2 \langle N_k\rangle \delta(\omega - \omega_k), \\
G^{(-)}_{\omega} &= \sum_k c_k^2 (\langle N_k\rangle + 1)\delta(\omega + \omega_k), \\
G^{(s)}_{\omega} &= \sum_k c_k^2 \left(\langle N_k\rangle + \frac{1}{2}\right)[\delta(\omega - \omega_k) + \delta(\omega + \omega_k)], \\
G^{(a)}_{\omega} &= \frac{i}{2}\sum_k c_k^2 [\delta(\omega - \omega_k) - \delta(\omega + \omega_k)], \\
F^{(+)}_{\omega} &= \sum_{m>n} |d_{mn}|^2 \rho_{mm}\delta(\omega - \omega_{mn}), \\
F^{(-)}_{\omega} &= \sum_{m>n} |d_{mn}|^2 \rho_{nn}\delta(\omega + \omega_{mn}), \\
F^{(s)}_{\omega} &= \frac{1}{2}\sum_{mn} |d_{mn}|^2 (\rho_{mm} + \rho_{nn})\delta(\omega + \omega_{mn}), \\
F^{(a)}_{\omega} &= \frac{1}{2i}\sum_{mn} |d_{mn}|^2 \Delta_{nm}\delta(\omega + \omega_{mn}).
\end{aligned} \tag{7.390}$$

Comparing the last expression with Eq. (4.57) for the linear susceptibility (polarisability) α_ω of a single atom at $\gamma = 0$, one can see that the anti-symmetric spectral CF of a stationary system coincides (up to the factor $2\pi/i\hbar$) with the imaginary part of the susceptibility of the system with respect to a coherent excitation,

$$\alpha''_\omega = (2\pi/i\hbar)F^{(\alpha)}_\omega. \tag{7.391}$$

The response function α_τ and its Fourier transform are called the *retarded part of the Green's function* of the system. Note that in order to determine the susceptibility of matter, experimentally or theoretically, one requires the field to be in a coherent quantum state (Sec. 7.5). And vice versa, the definition of the susceptibility of the vacuum (Sec. 4.1) is based on the coherent state of the matter. From the comparison between (7.390) and (7.365), it follows that $G^{(a)}_\omega = (\mathrm{Im}G^{vac}_\tau)_\omega$, i.e., the role of α in the case of the field is played by the anti-normally ordered CF of the vacuum.

Recall that α''_ω, in its turn, unambiguously determines α'_ω (Sec. 4.1). *Thus, the kinetic parameters of a stationary system are in one-to-one correspondence with its unperturbed fluctuation characteristics.*[aa] *In the case of equilibrium systems, this relation can be inverted, so that each one of the four CFs of the system determines the other ones.*

7.7.5 *Equilibrium systems. FDT*

At thermodynamic equilibrium, we have Planck's distribution for the photons and Boltzmann's distribution for the populations,

$$\langle N_k\rangle = \mathcal{N}(\omega_k), \quad \rho_{mm} = \Delta_{nm}\mathcal{N}(\omega_{mn}), \tag{7.392}$$

where

$$\mathcal{N}(\omega) = \left[\exp\frac{\hbar\omega}{\kappa T} - 1\right]^{-1} \equiv \mathcal{N}_\omega. \tag{7.393}$$

Due to the δ functions in (7.390), the temperature factors of $\mathcal{N}$ can be removed from the sum; as a result, all CFs can be expressed in terms of each other or in terms of the susceptibility imaginary part. For instance, in the case of an atom,

$$\begin{aligned} F^{(+)}_\omega &= \frac{\hbar}{\pi}\mathcal{N}_\omega\alpha''_\omega\theta_\omega, \\ F^{(-)}_{-\omega} &= \frac{\hbar}{\pi}(\mathcal{N}_\omega + 1)\alpha''_\omega\theta_\omega = F^{(+)}_\omega \exp\frac{\hbar\omega}{\kappa T}, \\ F^{(s)}_\omega &= \frac{\hbar}{\pi}\left(\mathcal{N}_\omega + \frac{1}{2}\right)\alpha''_\omega, \end{aligned} \tag{7.394}$$

[aa]Such relations are called the Kubo formulas, see, for instance, Ref. [Zubarev (1971)].

where θ_ω equals unity at $\omega > 0$ and zero at $\omega < 0$. The last equation in (7.394) is called the *fluctuation-dissipation theorem (FDT)* or the *Nyquist-Callen-Welton theorem.*

Let the atom and the field be in equilibrium states with the temperatures T_1 and T_2, respectively. Then, according to (7.394) and similar equations for field CFs, (7.388), (7.389) take the form

$$\mathcal{P} = -4i \int_0^\infty d\omega\omega\alpha''_\omega G^{(a)}_\omega (\mathcal{N}_1 - \mathcal{N}_2), \tag{7.395}$$

where $\mathcal{N}_n \equiv [\exp(\hbar\omega/\kappa T_n) - 1]^{-1}$. If the temperatures are close, the last factor in (7.395) scales as $T_1 - T_2$, then the ratio $\mathcal{P}/|T_1 - T_2|$ determines, through α''_ω, the 'heat conductivity' of the atom-field link, which provides the heat exchange between the two thermostats.

Here, we only considered single-quantum transitions and linear susceptibility, which are described in the first orders of the perturbation theory. A similar consideration can be performed for multi-quantum transitions, nonlinear susceptibilities, and higher-order CFs. The corresponding generalizations of the FDT, obtained by Efremov and Stratonovich, are described in short in Ref. [Klyshko (1980)].

In conclusion, let us mention that the interaction between an atom and a field leads not only to the energy exchange but also to a certain shift of the energy levels. In the case of the vacuum state of the field, this shift leads to a variation of the eigenfrequencies by a value about 10^9 Hz (for the hydrogen atom). This effect is called *the Lamb shift* (see, for instance, Ref. [Allen (1975)]). If the field is in an excited state, the shift is called the *Stark effect in alternating field* (5.48). This phenomenon has to be taken into account in many problems of laser spectroscopy.

Bibliography

Akhmanov, S. A. and Khokhlov, R. V. (1964). *Problems of nonlinear optics* (VINITI, Moscow) [in Russian].

Akhmanov, S. A. and Chirkin, A. S. (1971). *Statistical Phenomena in Nonlinear Optics* (MGU, Moscow) [in Russian].

Akhmanov, S. A., D'yakov, Yu. E., and Chirkin, A. S. (1981). *Introduction to statistical radiophysics and optics* (Nauka, Moscow) [in Russian].

Akhmanov, S. A. and Koroteev, N. I. (1981). *Methods of nonlinear optics in the spectroscopy of scattered light* (Nauka, Moscow) [in Russian].

Aleksandrov, E. B. (1972). Optical manifestation of interference of non-degenerate atomic states, *Sov. Phys. Usp.* **15**, pp. 436–451.

Allen, L., and Eberly, J. H. (1975). *Optical Resonance and Two-level Atoms* (John Wiley, New York).

Andreev, A. V., Emel'yanov V. I., Il'inskii, Yu. A. (1980). Collective spontaneous emission (Dicke superradiance), *Sov. Phys. Usp.* **23**, pp. 493-514.

Apanasevich, P. A. (1977). *Foundations of the theory of interaction of light with matter* (Nauka i Tekhnika, Minsk) [in Russian].

Arecchi, F. T., Scully, M. O., Haken, H., and Weidlich, W. (1974). *Quantum Fluctuations of Laser Radiation* (Mir, Moscow) [Russian translation].

Barachevskii, V. A., Lashkov, G. I., and Tsekhomskii, V. A. (1977). *Photochromism and its applications* (Khimiya, Moscow) [In Russian].

Bertin, F. (1971). *Principles of Quantum Electronics (Russ. transl.)* (Mir, Moscow) [in Russian].

Bespalov, V. I. (ed.) (1979). *Optical Phase Conjugation in Nonlinear Media* (Institute of Applied Physics, Gor'ky) [In Russian].

Bloembergen, N. (1965). *Nonlinear Optics* (W. A. Benjamin, Inc., New York).

Bloembergen, N. (ed.) (1977). *Nonlinear Spectroscopy* (North-Holland, Amsterdam).

Braginsky, V. B. and Vorontsov, Yu. I. (1975). Quantum-mechanical limitations in macroscopic experiments and modern experimental technique, *Sov. Phys. Usp.* **18**, pp. 644–650.

Braginsky, V. B. and Vorontsov, Yu. I., Thorne, K. S. (1980). Quantum Nondemolition Measurements, *Science* **209**, pp. 547–557.

Butylkin, V. S., Kaplan, A. E., Yakubovich, E. I., Khronopulo, Yu. G. (1977). *Resonant*

Interactions of Light with Matter (Nauka, Moscow) [In Russian].
Chang, R. K. and Furtak, T. (eds.) (1981). *Surface Enhanced Raman Scattering* (Plenum, New York).
Cummins, H. Z. and Pike, E. R. (eds.) (1974). *Photon-correlation and light-beating spectroscopy (Proc. NATO Advanced Study Group)* (Plenum Press, New York).
Delone, N. B., Krainov, V. P., Khodovoi, V. A. (1975). The two-level system in a strong light field, *Sov. Phys. Usp.* **18**, pp. 750-755.
Delone, N. B. and Krainov, V. P. (1978). *Atom in a strong light field* (Atomizdat, Moscow) [in Russian]; (1985). *Atoms in Strong Light Fields* (Springer-Verlag, Heidelberg).
Dmitriev, V. G. and Tarasov, L. V. (1982). *Applied nonlinear optics* (Radio i Svyaz', Moscow) [in Russian].
Dunskaya, I. M. (1981). *Rise of quantum electronics* (Nauka, Moscow) [in Russian].
Elyutin, P. V. (1982). *Theoretical foundations of quantum radiophysics* (MGU, Moscow) [in Russian].
Fabelinsky, I. L. (1965). *Molecular Scattering of Light* (Nauka, Moscow) [In Russian]; (1968). (Plenum Press, New York).
Fain, V.M. (1972). *Photons and non-linear media* (Sov. Radio, Moscow) [in Russian].
Ginzburg, V. L. (1983). The nature of spontaneous radiation, *Sov. Phys. Usp.* **26**, pp. 713-719.
Glauber, R. J. (1965). *in Quantum Optics and Electronics, edited by C. D. A. Blandin and C. Cohen-Tannoudji* (Gordon & Breach, New York).
Gorelik, G. S. (1948). On the demodulation analysis of light, *Usp. Phys. Nauk* **34**, p. 321 [In Russian].
Grishanin, B. A. (1981). *Quantum electronics for radio physicists* (MGU, Moscow) [in Russian].
Haken, H. (1977). *Synergetics* (Springer-Verlag, Berlin).
Hanbury Brown, R. (1971). Measuring the angular diameters of stars, *Contemp. Phys.* **12**, pp. 357–377.
Kaczmarek, F. (1981). *Introduction to laser physics (Russ. transl.)* (Mir, Moscow) [in Russian]; (1978). (PWN, Warsaw) [in Polish].
Karlov, N. V. (1992). *Lectures on quantum electronics* (Translated from Russian by E. Yankovsky) (Mir, Moscow).
Klimontovich, Yu. L. (1966). *Quantum light generators* (Prosveshchenie, Moscow) [in Russian].
Khanin, Y. (1975). *Dynamics of quantum oscillators* (Sov. Radio, Moscow) [in Russian]; Khanin, Y. (1995) *Principles of Laser Dynamics* (Elsevier, Amsterdam).
Kielich, S. (1980). *Nonlinear molecular optics (Russ. transl.)* (Nauka, Moscow) [in Russian]; (1977). (PWN, Warsaw) [in Polish].
Klauder, J. R. and Sudarshan, E. C. G. (1968). *Fundamentals of Quantum Optics* (W. A. Benjamin, Inc., New York).
Klimontovich, Yu. L. (1980). *Kinetic Theory of Electromagnetic Processes* (Nauka, Moscow) [In Russian]; (1982). (Springer, Heidelberg).
Klimontovich, Yu. L. (1982). *Statistical Physics* (Nauka, Moscow) [In Russian]; (1986). (Harwood, New York).
Klyshko, D. N. (1980). *Photons and Nonlinear Optics* (Nauka, Moscow) [In Russian]; (1988). (Gordon & Breach, New York).

Landau, L. D. and Lifshitz, E. M. (1964). *Statistical Physics* (Nauka, Moscow) [in Russian]; (1980). 3rd ed. (Pergamon Press, Oxford).
Landau, L. D. and Lifshitz, E. M. (1973). *Field Theory* (Nauka, Moscow) [in Russian]; (1975). (Pergamon Press, Oxford).
Landau, L. D. and Lifshitz, E. M. (1982). *Electrodynamics of Continuous Media* (Nauka, Moscow) [in Russian]; (1984). 2nd ed. (Pergamon Press, Oxford).
Lax, M. (1968). *Fluctuation and Coherence Phenomena in Classical and Quantum Physics* (Gordon & Breach, New York).
Letokhov, V. S. and Chebotaev, V. P. (1975). *Principles of nonlinear laser spectroscopy* (Nauka, Moscow) [In Russian].
Letokhov, V. S. (1983). *Nonlinear Selective Photoprocesses in Atoms and Molecules* (Nauka, Moscow) [in Russian].
Loudon, R. (2000). *The Quantum Theory of Light*, 3rd edn. (Oxford University Press, Oxford).
Louisell, W. H. (1964). *Radiation and Noise in Quantum Electronics* (McGraw-Hill, New York).
Macomber, J. D. (1976). *The Dynamics of Spectroscopic Transitions* (John Wiley, New York).
Maitland, B. and Dunn, M. (1969). *Laser Physics* (North Holland, Amsterdam).
Manykin, E. A., and Samartsev, V. V. (1984). *Optical Echo Spectroscopy* (Nauka, Moscow) [In Russian].
Migulin, V. V., Medvedev, V. I., Mustel', E. R., and Parygin, V. N. (1978). *Fundamentals of the oscillation theory* (Nauka, Moscow) [in Russian].
Minogin, V. G. and Letokhov, V. S. (1986). *Laser radiation pressure on atoms* (Nauka, Moscow) [In Russian].
Nye, J. F. (1957). *Physical Properties of Crystals* (Clarendon Press; Oxford University Press, Oxford).
Pantell, R. and Puthoff, H. (1969). *Fundamentals of Quantum Electronics* (John Wiley, New York).
Paul, H. (1980). Photon counting as a tool for basic research, *Soviet Journal of Quantum Electronics* **7**, pp. 1437.
Perina, J. (1972). *Coherence of Light* (Van Nostrand Reinhold, London).
Piekara, A. (1973). *A new face of optics: Introduction to quantum electronics and nonlinear optics (Russ. transl.)* (Sov. Radio, Moscow) [in Russian]; (1968). (PWN, Warsaw) [in Polish].
Rabinovich, M. I., Trubetskov, D. I. (1989). *Oscillations and waves in linear and nonlinear systems* (Kluwer-Academic Publ., Amsterdam).
Rytov, S. M. (1976). *Introduction to statistical radiophysics. Part 1.* (Nauka, Moscow) [in Russian].
Schubert, M. and Wilhelmi, B. (1973). *Introduction to Nonlinear Optics* (Mir, Moscow) [Russian translation]; *Einfuhrung in die nichtlineare Optik* (Teubner, Leipzig) [in German]; (1986). *Nonlinear Optics and Quantum Electronics* (Wiley, New York).
Silin, V. P., Rukhadze, A. A. (1961). *Electromagnetic properties of plasma and plasma-like media* (Gosatomizdat, Moscow) [in Russian].
Steinfeld, J. I. (ed.) (1978). *Laser and Coherence Spectroscopy* (Plenum Press, New York).

Strakhovskii, G. M., Uspenskii, A. V. (1979). *Foundations of quantum electronics* (Vysshaya Shkola, Moscow) [in Russian].
Svelto, O. (2010). *Principles of Lasers* (Springer, New York).
Tarasov, L. V. (1976). *Physical foundations of quantum electronics* (Sov. Radio, Moscow) [in Russian].
Tarasov, L. V. (1981). *Physics of processes in coherent optical radiation sources* (Radio i Svyaz', Moscow) [in Russian].
Vinogradova, M. B., Rudenko, O. V., and Sukhorukov, A. P. (1979). *Wave theory* (Nauka, Moscow) [in Russian].
Walls, D. F. (1983). Squeezed states of light, *Nature* **306**, pp. 141–146.
Walther, H. (ed.) (1976). *Topics in Applied Physics, Volume 2: Laser Spectroscopy of Atoms and Molecules* (Springer-Verlag, Berlin-Heidelberg-New York).
Yariv, A. (1976). *Introduction to Optical Electronics*, 2nd ed. (Holt, Rinehart & Winston, New York).
Yariv, A. (1989). *Quantum Electronics*, 3rd ed. (John Wiley, New York).
Zel'dovich, Ya. B., Pilipetsky, N. F., and Shkunov, V. V. (1985). *Principles of Phase Conjugation* (Nauka, Moscow) [In Russian]; (1985). (Springer Verlag, Berlin).
Zernike, F., Midwinter, J. E. (1973). *Applied Nonlinear Optics* (Wiley, New York).
Zhabotinskii, M. E. (ed.) (1969). *Kvantovaya elektronika. Malen'kaya entsiklopediya (Small Encyclopedia of Quantum Electronics)* (Sov. Entsiklopediya, Moscow) [in Russian].
Zubarev, D. N. (1971). *Nonequilibrium Statistical Thermodynamics* (Nauka, Moscow) [In Russian]; (1974). (Consultant Bureau, New York).

Bibliography expanded by the Editors:

Agrawal, G. (2007) *Nonlinear Fiber Optics*, 4th ed. (Academic Press, San Diego).
Bachor, H.-A. and Ralf, T. (2004). *A Guide to Experiments in Quantum Optics* (Wiley-VCH Verlag GmbH & Co. KGaA, Weinheim).
Bouwmeester, D., Ekert, A., Zeilinger, A. (Eds.) (2000). *The Physics of Quantum Information* (Springer, Berlin).
Boyd, R. (2008) *Nonlinear optics*, 3rd ed. (Academic Press, San Diego).
Bruss, D. (2002) Characterizing entanglement, *Journ. of Math. Phys.*, **43**, p. 4237.
Goodman, J. (1985). *Statistical Optics* (McGraw-Hill, New York).
Grynberg, G., Aspect, A., Fabre, C. (2010). *Introduction to quantum optics: From the semiclassical approach to quantized fields* (Cambridge University Press, New York).
Klyshko, D. N. (1994). Quantum optics: quantum, classical, and metaphysical aspects, *Phys. Usp.* **37**, pp. 1097–1122.
Klyshko, D. N. (1998). Basic quantum mechanical concepts from the operational viewpoint, *Phys. Usp.* **41**, pp. 885–922.
Leonhardt, U. (1997). *Measuring the quantum state of light* (Cambridge University Press, Cambridge).
Mandel, L. and Wolf, E. (1995). *Optical Coherence and Quantum Optics* (Cambridge University Press, Cambridge).

Milloni, P, Eberly, J. (2009) *Laser Physics* (Wiley, New Jersey).

Nielsen, M. A. and Chuang I. L. (2000). *Quantum Computation and Quantum Information* (Cambridge University Press, Cambridge).

Peres, A. (1993). *Quantum Theory: Concepts and Methods* (Kluwer, Amsterdam).

Schleich, W. P. (2001). *Quantum Optics in Phase Space* (Wiley, New York).

Scully, M. O. and Zubairy, M. (1997). *Quantum Optics* (Cambridge University Press, Cambridge).

Schrödinger, E. (1935). Die gegenwartige Situation in der Quantenmechanik, *Naturwissenschaften* **23**, p. 807–812.

Walls, D.F. and Milburn, G. J. (1994). *Quantum Optics* (Springer)

Yamamoto, Y. and Imamoglu A. (1999). *Mesoscopic quantum optics*. (Wiley, New York).

Index